MACROMOLECULAR PHYSICS

Macromolecular Physics

Editor:

Maria Emilova Velinova, Ph.D.

ArclerPress

www.arclerpress.com

Macromolecular Physics

Editor: Maria Emilova Velinova, Ph.D.

708 3rd Avenue, 6th Floor
New York
NY 10017
United States of America
www.arclerpress.com

ISBN: 978-1-68094-439-6

Library of Congress Control Number: 2016951838

About the Editor

Maria Emilova Velinova, Ph.D.

Maria Velinova is Ph.D. holder in Quantum chemistry at the University of Sofia since April 2012. Her major research experience is in the field of Computational Chemistry, especially in statistical mechanics methods applied to different sorts of biomolecules. Member of the Laboratory of Quantum and Computational Chemistry at the University of Sofia.

Contents

List of Contributors

T. E. Gorelik
Institute of Physical Chemistry, University of Mainz, Jakob Welder Weg 11, Mainz, 55128, Germany.

C. Czech
Institute of Inorganic and Analytical Chemistry, Goethe-University Frankfurt am Main, Max-von-Laue-Str. 7, Frankfurt am Main, 60438, Germany

S. M. Hammer
Institute of Inorganic and Analytical Chemistry, Goethe-University Frankfurt am Main, Max-von-Laue-Str. 7, Frankfurt am Main, 60438, Germany

M. U. Schmidt
Institute of Inorganic and Analytical Chemistry, Goethe-University Frankfurt am Main, Max-von-Laue-Str. 7, Frankfurt am Main, 60438, Germany

Wenxiao He
Department of Chemical and Biological Engineering, Chalmers University of Technology, Göteborg SE-41296, Sweden

Yu Fu
Department of Chemical and Biological Engineering, Chalmers University of Technology, Göteborg SE-41296, Sweden

Martin Andersson
Department of Chemical and Biological Engineering, Chalmers University of Technology, Göteborg SE-41296, Sweden

S. A. Jang
Department of Advanced Materials Science and Engineering, Hanbat National University, 125, Dongseo-daero, Yuseong-gu, Daejeon 305-719, South Korea

H. J. Lee
Department of Advanced Materials Science and Engineering, Hanbat National University, 125, Dongseo-daero, Yuseong-gu, Daejeon 305-719, South Korea

C. V. Thompson
Department of Materials Science and Engineering, Massachusetts Institute of Technology, Cambridge, Massachusetts 02139, USA

C. A. Ross
Department of Materials Science and Engineering, Massachusetts Institute of Technology, Cambridge, Massachusetts 02139, USA

Y. J. Oh
Department of Advanced Materials Science and Engineering, Hanbat National University, 125, Dongseo-daero, Yuseong-gu, Daejeon 305-719, South Korea

Takeaki Araki
Department of Physics, Kyoto University, Sakyo-ku, Kyoto 606-8502, Japan.

Francesca Serra
Dipartimento di Chimica, Biochimica, Biotecnologie per la Medicina, Universià di Milano, Milano, Italy

Hajime Tanaka
Institute of Industrial Science, University of Tokyo, Komaba 4-6-1, Meguro-ku, Tokyo 153-8505, Japan

Xinchen Kang
Beijing National Laboratory for Molecular Sciences, Key Laboratory of Colloid and Interface and Thermodynamics, Institute of Chemistry, Chinese Academy of Sciences, Beijing 100190, China.

Wenting Shang
Beijing National Laboratory for Molecular Sciences, Key Laboratory of Colloid and Interface and Thermodynamics, Institute of Chemistry, Chinese Academy of Sciences, Beijing 100190, China.

Qinggong Zhu
Beijing National Laboratory for Molecular Sciences, Key Laboratory of Colloid and Interface and Thermodynamics, Institute of Chemistry, Chinese Academy of Sciences, Beijing 100190, China.

Jianling Zhang
Beijing National Laboratory for Molecular Sciences, Key Laboratory of Colloid and Interface and Thermodynamics, Institute of Chemistry, Chinese Academy of Sciences, Beijing 100190, China.

Tao Jiang
Beijing National Laboratory for Molecular Sciences, Key Laboratory of Colloid and Interface and Thermodynamics, Institute of Chemistry, Chinese Academy of Sciences, Beijing 100190, China.

Buxing Han
Beijing National Laboratory for Molecular Sciences, Key Laboratory of Colloid and Interface and Thermodynamics, Institute of Chemistry, Chinese Academy of Sciences, Beijing 100190, China.

Zhonghua Wu
Beijing Synchrotron Radiation Facility, Institute of High Energy Physics, Chinese Academy of Sciences, Beijing 100049, China

Zhihong Li
Beijing Synchrotron Radiation Facility, Institute of High Energy Physics, Chinese Academy of Sciences, Beijing 100049, China

Xueqing Xing
Beijing Synchrotron Radiation Facility, Institute of High Energy Physics, Chinese Academy of Sciences, Beijing 100049, China

Leyla-Cann Sögütoglu
Radboud University, Institute for Molecules and Materials, Heyendaalseweg 135, 6525 AJ Nijmegen, The Netherlands.

René R. E. Steendam
Radboud University, Institute for Molecules and Materials, Heyendaalseweg 135, 6525 AJ Nijmegen, The Netherlands.

Hugo Meekes
Radboud University, Institute for Molecules and Materials, Heyendaalseweg 135, 6525 AJ Nijmegen, The Netherlands.

Elias Vlieg
Radboud University, Institute for Molecules and Materials, Heyendaalseweg 135, 6525 AJ Nijmegen, The Netherlands.

Floris P. J. T. Rutjes
Radboud University, Institute for Molecules and Materials, Heyendaalseweg 135, 6525 AJ Nijmegen, The Netherlands.

Craig J. Callahan
EPSRC Centre for Innovative Manufacturing in Continuous Manufacturing and Crystallisation, Centre for Oscillatory Baffled Reactor Applications (COBRA), School of Engineering and Physical Science, Heriot-Watt University, Edinburgh, EH14 4AS, UK

Xiong-Wei Ni
EPSRC Centre for Innovative Manufacturing in Continuous Manufacturing and Crystallisation, Centre for Oscillatory Baffled Reactor Applications (COBRA), School of Engineering and Physical Science, Heriot-Watt University, Edinburgh, EH14 4AS, UK

Zhong Liu
Key Laboratory of Salt Lake Resources and Chemistry, Qinghai Institute of Salt Lakes, Chinese Academy of Sciences, Xining 810008, China
EaStChem, School of Chemistry, University of St Andrews, St Andrews, KY16 9ST, UK.

Chang-Yang Chiang
EaStChem, School of Chemistry, University of St Andrews, St Andrews, KY16 9ST, UK.

Wu Li
Key Laboratory of Salt Lake Resources and Chemistry, Qinghai Institute of Salt Lakes, Chinese Academy of Sciences, Xining 810008, China

Wuzong Zhou
EaStChem, School of Chemistry, University of St Andrews, St Andrews, KY16 9ST, UK.

Aleksandr Perevedentsev
Department of Physics and Centre for Plastic Electronics, Imperial College London, London, United Kingdom

Paul N. Stavrinou
Department of Physics and Centre for Plastic Electronics, Imperial College London, London, United Kingdom

Donal D. C. Bradley
Paul Smith
Department of Physics and Centre for Plastic Electronics, Imperial College London, London, United Kingdom

Y. Yamasaki
Department of Applied Chemistry, Waseda University, 3-4-1 Okubo, Shinjuku-ku, Tokyo 169-8555, Japan

K. Hasegawa
Department of Applied Chemistry, Waseda University, 3-4-1 Okubo, Shinjuku-ku, Tokyo 169-8555, Japan

T. Osawa
Department of Applied Chemistry, Waseda University, 3-4-1 Okubo, Shinjuku-ku, Tokyo 169-8555, Japan

S. Noda
Department of Applied Chemistry, Waseda University, 3-4-1 Okubo, Shinjuku-ku, Tokyo 169-8555, Japan

Bin Zhang
Institute of Physics, University of Freiburg, 79104 Freiburg, Germany.
Hermann Staudinger Graduate School, University of Freiburg, 79104 Freiburg, Germany
School of Materials Science & Engineering, Zhengzhou University, Zhengzhou 450002, People's Republic of China

Jingbo Chen
School of Materials Science & Engineering, Zhengzhou University, Zhengzhou 450002, People's Republic of China

Hui Zhang
Institute of Physics, University of Freiburg, 79104 Freiburg, Germany.

Moritz C. Baier
Chair of Chemical Materials Science, Department of Chemistry, University of Konstanz, 78464 Konstanz, Germany

Stefan Mecking
Chair of Chemical Materials Science, Department of Chemistry, University of Konstanz, 78464 Konstanz, Germany

Renate Reiter
Institute of Physics, University of Freiburg, 79104 Freiburg, Germany.
Freiburg Materials Research Centre, University of Freiburg, 79104 Freiburg, Germany

Rolf Mülhaupt
Institute of Macromolecular Chemistry, University of Freiburg, 79104 Freiburg, Germany
Freiburg Materials Research Centre, University of Freiburg, 79104 Freiburg, Germany

Günter Reiter
Institute of Physics, University of Freiburg, 79104 Freiburg, Germany.
Freiburg Materials Research Centre, University of Freiburg, 79104 Freiburg, Germany

Zhenyu Wan
ARC Photovoltaics Centre of Excellence, University of New South Wales (UNSW)

Shujuan Huang
ARC Photovoltaics Centre of Excellence, University of New South Wales (UNSW)

Martin A Green
ARC Photovoltaics Centre of Excellence, University of New South Wales (UNSW)

Gavin Conibeer
ARC Photovoltaics Centre of Excellence, University of New South Wales (UNSW)

Laila M. Al-Harbi
Chemistry Department, Faculty of Science, King Abdulaziz University, Jeddah, Saudi Arabia

Daniel Hermida-Merino
BM26/DUBBLE, CRG, ESRF-The European Synchrotron, CS40220, 38043 Grenoble Cedex 9, France
Dutch Polymer Institute DPI, P.O. Box 902, Eindhoven 5600 AX, The Netherlands

Gemma E. Newby
ID09b, ESRF-The European Synchrotron, CS40220, 38043 Grenoble Cedex 9, France

Ian W. Hamley
Department of Chemistry, University of Reading, Whiteknights, Reading, RG6 6AD, UK.

Wayne Hayes
Department of Chemistry, University of Reading, Whiteknights, Reading, RG6 6AD, UK.

Andrew Slark
Henkel UK Limited, Wexham Road, Slough, SL2 5DS, UK

C. G. Schäfer
Ernst-Berl Institut für Technische und Makromolekulare Chemie, Technische Universität Darmstadt, Alarich-Weiss-Str. 4, D-64287 Darmstadt, Germany.

T. Winter
Ernst-Berl Institut für Technische und Makromolekulare Chemie, Technische Universität Darmstadt, Alarich-Weiss-Str. 4, D-64287 Darmstadt, Germany.

S. Heidt
[a]Ernst-Berl Institut für Technische und Makromolekulare Chemie, Technische Universität Darmstadt, Alarich-Weiss-Str. 4, D-64287 Darmstadt, Germany.

C. Dietz
Center of Smart Interfaces and Department of Materials Sciences, Technische Universität Darmstadt, Alarich-Weiss-Str. 10, 64287 Darmstadt, Germany

T. Ding
NanoPhotonics Centre, Cavendish Laboratory, University of Cambridge, Cambridge CB3 0HE, UK

J. J. Baumberg
NanoPhotonics Centre, Cavendish Laboratory, University of Cambridge, Cambridge CB3 0HE, UK

M. Gallei
Ernst-Berl Institut für Technische und Makromolekulare Chemie, Technische Universität Darmstadt, Alarich-Weiss-Str. 4, D-64287 Darmstadt, Germany.

Preface

Today's macromolecular physics is an interdisciplinary science that has been extended beyond its traditional essentials to fields as energy materials, biological systems, optics, electronics and nanotechnology. The key for all this possible applications lies on the macromolecular crystals and their particular properties.

The book has been divided into three parts. The first one is dedicated to the microscopic structure of crystals, their morphology and defects from which depend on the physical and mechanical properties of macromolecular crystals. The defects of structure induce interesting application of the crystals and open up novel possibilities for defect engineering as unusual catalysts, self-organization of soft matter by defects, memory effects of topological origin and flow control of liquid crystals and colloid motion by defects. As methods to obtain the structure and morphology of the sample, are discussed the electron diffraction spectroscopy and the transition electron microscopy.

The second part of the book concerns the discussion of the topics as the primary and secondary nucleation of crystals, their growth, annealing of crystals and solution, and melts crystallization of macromolecules. All these processes are examined in several types of sample presented with more details in the selected articles.

Lastly, the third part of the book summarizes the knowledge of the melting of different macromolecular crystals. It is also presented an interesting application of melt-shear organization technique in order to prepare large-area inorganic/organic opal photonic crystal (OPC) and polymer inverse opal photonic crystal (IOPC), which show wide range of potential applications within the fields of optical (bio)sensing platforms, catalysis, display technologies, actuation systems and security materials.

Editor
Maria Emilova Velinova, Ph.D.

CHAPTER 1

CRYSTAL STRUCTURE OF DISORDERED NANOCRYSTALLINE αII-QUINACRIDONE DETERMINED BY ELECTRON DIFFRACTION

T. E. Gorelik [a], C. Czech [b], S. M. Hammer [b] and M. U. Schmidt [b]

[a]Institute of Physical Chemistry, University of Mainz, Jakob Welder Weg 11, Mainz, 55128, Germany.

[b]Institute of Inorganic and Analytical Chemistry, Goethe-University Frankfurt am Main, Max-von-Laue-Str. 7, Frankfurt am Main, 60438, Germany

ABSTRACT

The nanocrystalline α^{II}-phase of the industrially produced organic pigment quinacridone was studied by 3D electron diffraction of its crystals with a thickness of only 10 nm. The diffraction data showed strong diffuse scattering along one direction indicating severe stacking disorder. The average crystal structure was obtained from electron diffraction data using direct methods. In α^{II}-quinacridone, the molecules are connected by a pair of hydrogen bonds thereby forming molecular chains, which are stacked, resulting in the formation of layers. The layers exhibit a stacking disorder with a mixture of herringbone and parallel arrangements, which explains the diffuse scattering. The crystal structure was confirmed by dispersion-corrected DFT calculations.

INTRODUCTION

Quinacridone (Pigment Violet 19, **1**, Fig. 1) is an industrial organic pigment with an annual production of several thousands of tons and a sales volume of more than 100 million euros per year.

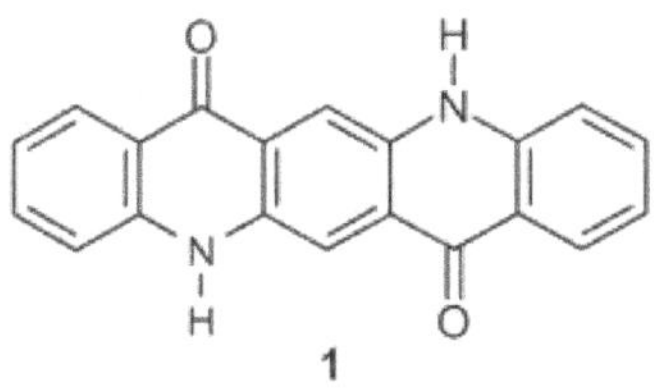

1

Figure 1: Chemical structure of quinacridone, 1.

Citation: T. E. Gorelik, C. Czech, S. M. Hammer and M. U. Schmidt. Crystal structure of disordered nanocrystalline αII-quinacridone determined by electron diffraction. CrystEngComm, 2016, 18, 529-535. DOI: 10.1039/C5CE01855B

This compound is used for colouration of automobile coatings, paints, plastics and high-grade printing inks.[1] Quinacridone is insoluble in water and all solvents. In its application medium, as a coating, the quinacridone powder is not dissolved, but finely dispersed, whereby the crystal structure is maintained.

Quinacridone exists in four polymorphic forms[2] which differ in colour and stability. The synthesis, depending on the conditions, results in the dull dark reddish-violet α^{I}-phase, or in the carmine-red α^{II}-phase. Treatment of the α^{I}- or α^{II}-phase with NaOH in suspension leads to the reddish violet β-phase. Heating a suspension in organic solvents leads to the red γ-phase. A large series of other phases has been described (B_I, γ′, γ^{I}, γ^{II}, γ^{III}, γ^{IV}, δ, Δ, two different ε-phases, and ζ),[3] but a close inspection of their X-ray powder patterns revealed that all of them are either impure β- or γ-phases, or phase mixtures.[2] The β- and γ-phases are more stable than the α phases and show high photostability, i.e. they are not bleached by light and weathering.[4]Both the β- and the γ-phases are commercially sold as pigments.[5]

The quinacridone molecule itself is yellow;[6] the reddish to violet shades of the crystal phases are apparently caused by hydrogen bonding and exciton coupling in the solid state.

The crystal structures of the β- and γ-phases were determined from single-crystal X-ray analyses.[7] In 1994, the structure of the α^{I}-phase was, to our knowledge, the first example of an organic structure solved by a combination of crystal structure prediction and X-ray powder diffraction.[2,8] In the α^{I}- and β-phases, the molecules are connected by double hydrogen bonds, thereby forming molecular chains. In the γ-phase, the molecules form a criss-cross pattern (Fig. 5).

The α^{II}-phase is a nanocrystalline powder. Its X-ray powder pattern consists of a few peaks and humps (Fig. 2). All attempts to improve the crystallinity, by means of recrystallization or solvent treatment, failed: either the crystallinity did not change, or the material transformed into the more stable β- or γ-phases.

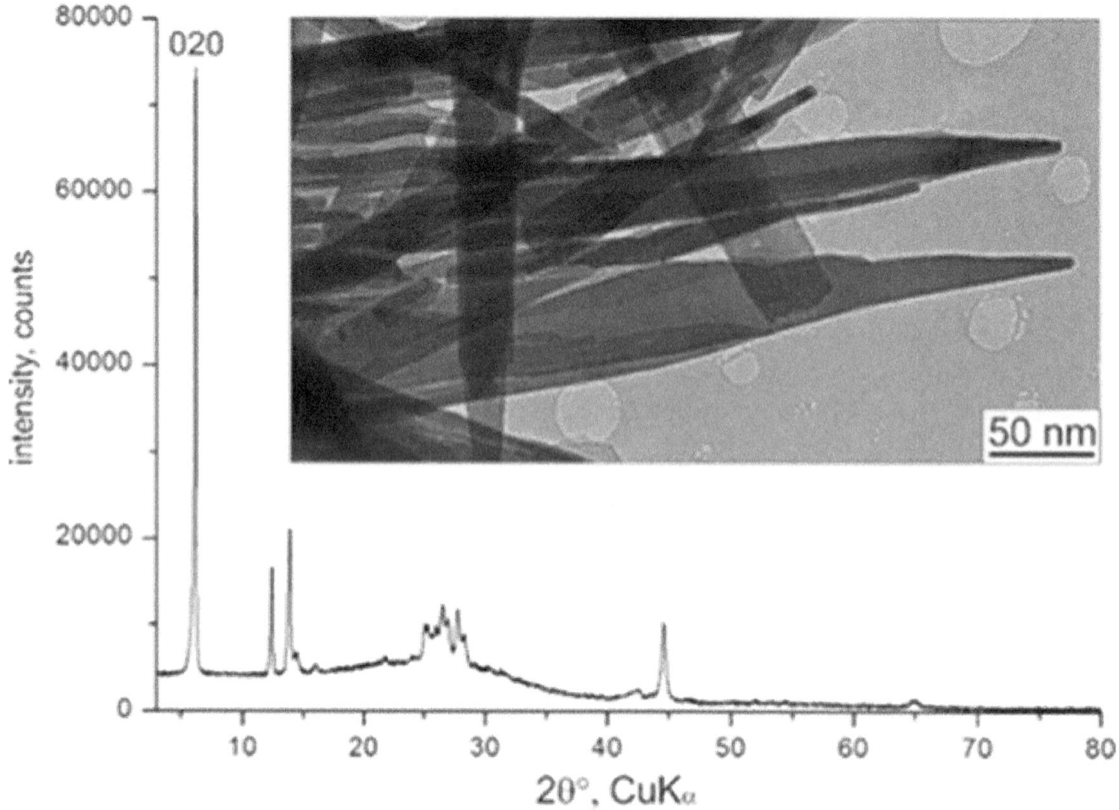

Figure 2 X-ray powder diagram and TEM image of α^{II}-quinacridone crystals. The 020 reflection is marked

In 1996, Lincke suggested a criss-cross structure for the α^{II}-phase. He constructed a structural model with two independent molecules in a triclinic unit cell, and performed a manual fit to the X-ray powder data.[10]However, the fit was not of satisfactory quality. Later, lattice-energy minimisation by dispersion-corrected DFT methods revealed that this structure was not correct.[2]

Infrared spectroscopy indicates that the α^{II}-phase possesses a chain structure like that of the α^{I}- and β-phases, and not a criss-cross-structure like that of the γ-phase. Pair-distribution function analysis[11] shows that the local structure of the α^{II}-phase is similar to that of the α^{I}-phase, which again points to a chain-like structure.

The X-ray powder pattern could not be indexed reliably, and all attempts to solve the structure using the X-ray powder data were unsuccessful. The use of synchrotron diffraction data did not help either, because the peak widths are caused by the small domain size of the crystals, and not by the instrumental parameters.

In recent years, the employment of electron diffraction for crystal structure analysis has experienced a revival with the development of automated techniques for data collection and processing.[12] Since the interaction of electrons with matter is much stronger than that of X-rays, and due to the flexibility in focussing the electron beam, sensible single-crystal electron diffraction data can be collected from crystals of size down to 50 nm. An electron diffraction experiment is typically done using a transmission electron microscope (TEM). The crystal is tilted around a goniometer axis and the electron diffraction patterns are collected sequentially, thereby mimicking a single-crystal X-ray experiment. From a set of collected 2D diffraction patterns, a 3D reciprocal volume can be reconstructed, giving information on the lattice parameters, crystal symmetry, and reflection intensities for structure determination. Despite the problems associated with dynamic and multiple scattering in electron diffraction, a series of inorganic crystal structures have been determined by electron diffraction within the last few years.[13]

The study of organic compounds represents a particular challenge to electron crystallography, due to the fast deterioration of soft matter under the electron beam. Electron diffraction is usually used in combination with other methods (powder X-ray diffraction, NMR, simulations) as additional information to support the structure analysis.[14] Nevertheless, reports on successful structure analyses of molecular crystals from electron diffraction data alone have recently started to appear.[15]

Therefore, here we used electron diffraction for the structure determination of nanocrystalline α^{II}-quinacridone.

EXPERIMENTAL

Materials

A sample of α^{II}-quinacridone was obtained from Clariant (Frankfurt am Main), and used as received.

TEM sample preparation

A small quantity of α^{II}-quinacridone was suspended in hexane using an ultrasonic bath. A drop of the suspension was then placed onto a holey carbon-coated copper grid and dried in air.

Electron diffraction experiments

Electron diffraction experiments were done using a TECNAI F30 transmission electron microscope operating at 300 kV equipped with an STEM unit. Electron diffraction data were collected at room temperature in nanodiffraction mode using the dedicated automated diffraction tomography (ADT) module in STEM mode. The beam diameter for nanodiffraction collection was 100 nm. The data were recorded onto a 1k GATAN 794 MSC CCD camera and a US4000 GATAN CCD camera. Seven diffraction tilt series were collected in total, each of them within the 120° total tilt range with a tilt step of 1°. Since the crystals were larger than the spot size used (100 nm), we could minimize the sample beam damage by repeatedly shifting the electron beam along the crystal during the measurements, thereby always recording data from a "fresh" part of the crystal. Electron diffraction data processing was done using the ADT3D software (NANOMEGAS, Belgium) reinforced by in-house written MatLab scripts.

Structure Solution

The structure was solved using direct methods implemented in the SIR software. 1031 independent input reflections within the resolution limit of 0.8 Å were used. For the space group$P2_1/c$, the $R_{int}(F)$ of the data set was 28.79%, completeness was 68%, and B(iso)/u(iso) = 3.501/0.0443. The final residual of the structure solution R(F) was 35.01%.

Structure Refinement

The structure refinement on the basis of electron diffraction data was performed in SHELX using distances and planarity restraints/constraints on the molecular geometry. In total, 37 parameters were refined using 1031 reflections. Two thermal factors became negative (the oxygen atom and a carbon atom of the middle phenyl ring), others were in a reasonable range. The final structure had an R(I) factor of 64%. In this case, the refinement could not serve as a proof of the structure correctness. The high figures of merit for electron diffraction compared to X-ray diffraction are due to significant intensity perturbations caused by the diffuse scattering, multiple scattering and other uncertainties associated with electron diffraction data.

X-ray Powder Data

X-ray powder data were recorded at room temperature in transmission geometry on a STOE-STADI-P diffractometer equipped with a curved Ge (111) primary monochromator and a linear position-sensitive detector, using Cu-$K_{\alpha 1}$ radiation (λ = 1.5406 Å).

DFT calculations

DFT calculations were performed with CASTEP[16] with the PBE functional[17] combined with the semi-empirical dispersion correction by Grimme.[18] The convergence criteria for energies, forces, cell stress and Cartesian displacements were 0.001 kJ per mol per atom, 3 kJ mol^{-1} $Å^{-1}$, 0.5 kbar and 0.001 Å.

RESULTS AND DISCUSSION

The investigated sample of α^{II}-quinacridone consisted of long lath-habit crystals with a length of up to a few microns, a width of less than 100 nm and a thickness of only 5 to 10 nm (Fig. 2).

Seven electron diffraction data sets from different crystals were collected. For each data set, individual diffraction patterns were combined to obtain 3D reciprocal volumes. The 3D volumes of all studied crystals were similar and consisted mostly of parallel diffuse streaks (Fig. 3b). The streaks were arranged in a regular pattern (Fig. 3a). Most of the diffuse streaks were continuous and did not show any pronounced intensity maxima that could have been interpreted as Bragg reflections (Fig. 3c). Only the central column consisted of sharp reflections without any diffuse scattering. This type of diffraction pattern is typical for a sheet structure with severe stacking disorder.[19]

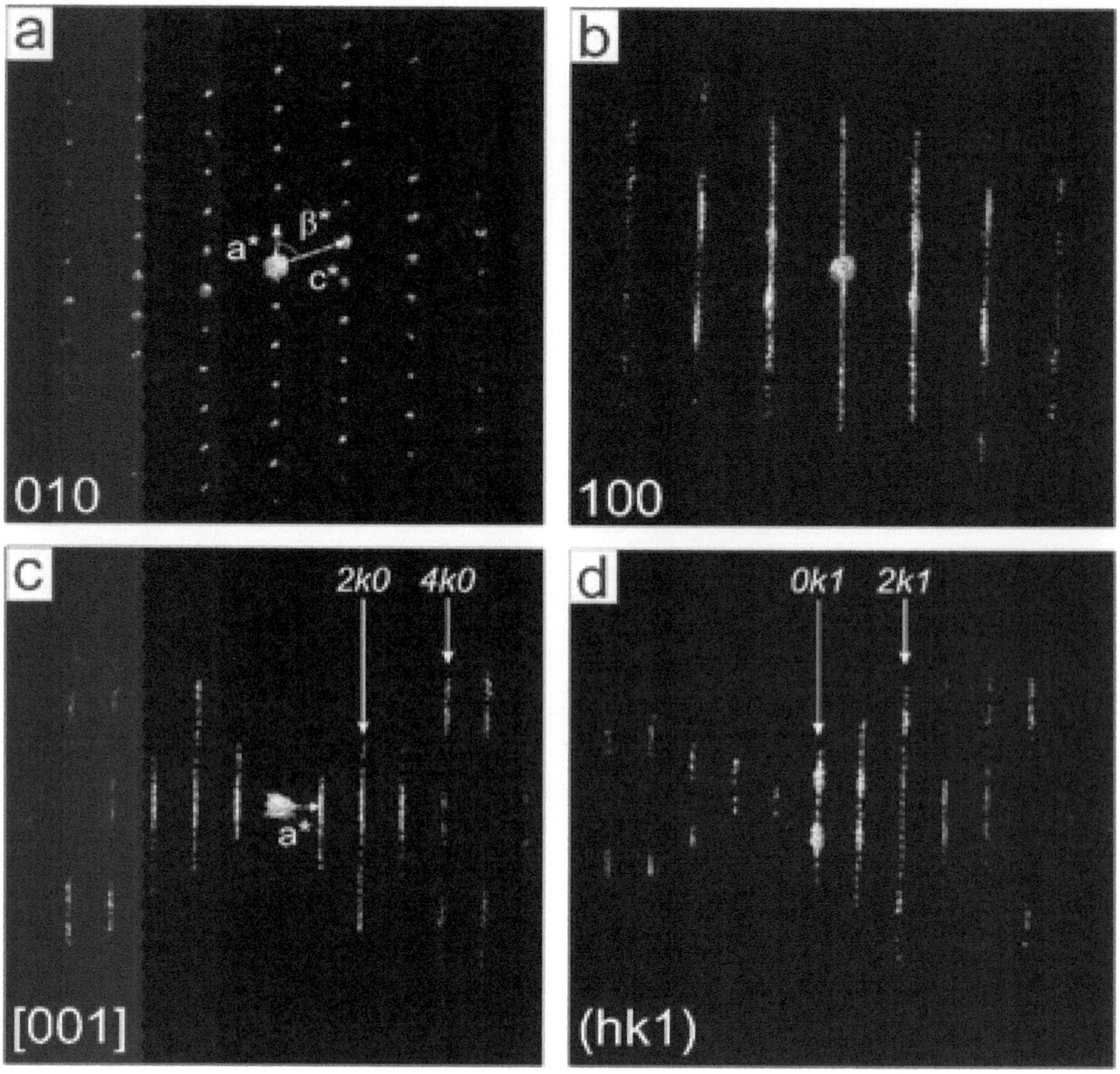

Figure 3 3D reciprocal volume of an α^{II}-quinacridone crystal: a) projection of the volume along the b* direction showing resolved periodicity along the a* and c* axes; b) projection of the volume along the a* direction (the b* axis comprising diffuse scattering lines is vertical); c) central cut of the reciprocal volume including the a* and b* (vertical) axes, the periodicity along the vertical lines is not evident; d) l = 1 layer cut through the reciprocal space, reflections rows showing discrete periodic spots are marked.

Due to the absence of sharp reflections in most parts of the reciprocal volume, the determination of the unit cell parameters was challenging and had to be done manually in several steps.

- The direction of the diffuse streaks was assigned to the ***b**** axis.
- a*, c* and β* could be obtained from the projection shown in Fig. 3a.
- The values of b*, α* and γ* were difficult to determine, due to strong diffuse scattering. Furthermore, most crystals were oriented with their ***b**** axes parallel to the electron beam, so that the 0k0 reflections could not be observed directly. Accidentally, a crystal with a different orientation was found, which allowed the recording of an electron diffraction pattern including the ***b**** axis, revealing systematic extinctions of 0k0 reflections with the reflection rule k = 2n. Since triclinic space groups do not have systematic extinctions, the crystal system must be at least monoclinic. As the angle β* is oblique, ***b*** must be the unique axis, consequently, α = γ = 90°.
- The monoclinic crystal system is confirmed by the overall distribution of the intensities of the 3-dimensional diffraction patterns, which follows a 2/m symmetry (Laue class 2/m = C_{2h}), corresponding to a monoclinic crystal system.
- The accurate length of the ***b*** vector was finally determined from the (020) reflection clearly resolved in the X-ray powder diagram (Fig. 1).

The resulting unit-cell parameters were a = 7.1 Å, b = 28.4 Å, c = 3.9 Å, β = 110°, α = γ = 90°, see Table 1. The unit cell volume of 734 $Å^3$ corresponds to two molecules per unit cell. The systematic extinctions of the 0k0 reflections show the presence of a 2_1 screw axis.

Table 1 Crystal data for quinacridone polymorphs (room-temperature data)[9]

Crystal phase	α^I	α^{II}	β	γ
Ref.	2	Present work	2	2
Space group, Z	$P\bar{1}$, 1	$P2_1/c$, 2	$P2_1/c$, 2	$P2_1/c$, 2
a, Å	3.802(2)	7.1	5.692(1)	13.697(9)
b, Å	6.612(3)	28.4	3.975(1)	3.881(3)
c, Å	14.485(6)	3.9	30.02(4)	13.4020(10)
α, °	100.68(8)	90	90	90
β, °	94.40(6)	110	96.76(6)	100.44(1)
γ, °	102.11(5)	90	90	90
V, Å^3	346.7(1)	734	674.5(9)	700.6(7)

The quinacridone molecule itself possesses C_{2h} symmetry. In crystal structures, molecules with C_{2h} symmetry are located on crystallographic inversion centres with a frequency of 95%.[20] Correspondingly, the space group was likely to be $P2_1/c$.[21] For ordered crystals without diffuse scattering, the three settings of the space group – $P2_1/c$, $P2_1/a$ and $P2_1/n$ – could easily be distinguished through the systematic extinctions of the h0l reflections (l = 2n, or h = 2n, or h + l = 2n). However, in α^{II}-quinacridone the diffuse streaks running through all h0lreflections made it impossible to detect any systematic extinction. Hence, the structure solution was performed in parallel in all three space groups.

Diffuse lines were cut at the calculated reflections positions, and the intensities of the reflections were extracted. This very rough treatment of the diffuse scattering gave, however, good results. The structure was solved by direct methods as implemented in SIR.[22] The resulting electron density maps in $P2_1/c$ are shown in Fig. 4a and b. The position and orientation of the molecules can be clearly seen (Fig. 4a). Although the individual atomic positions within the molecules were not well resolved (Fig. 4b), the maps allowed building of a structure model which was chemically sensible and described the overall electron density (Fig. 4c and d).

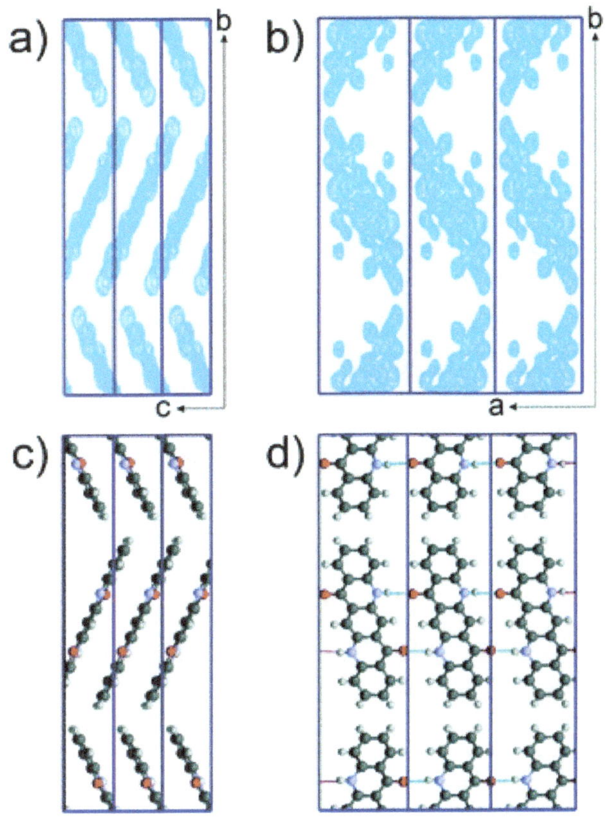

Figure 4 Structure of α^{II}-quinacridone solved from electron diffraction data by direct methods. Scattering potential maps obtained as a structure solution by SIR viewed along the a (a) and c(b) directions; corresponding structural model viewed along the a (c) and c (d) directions

The validation of structures solved from electron diffraction data remains a debatable issue. Structure refinement against electron diffraction data (SHELX) can hardly serve as a proof of the correctness of a structure. The refinement routines are not adopted to electron diffraction with its intrinsic multiple scattering. In the case of α^{II}-quinacridone, diffuse scattering passing through the reflections brought additional uncertainty in the intensities. As a result, the values of the refinement residuals were exceptionally high.

Recently, an alternative approach to structure validation was proposed – structure minimization using the dispersion-corrected density functional theory (DFT-D).[23] A test on 241 single-crystal structures showed that upon a full optimization with DFT-D, including the optimization of the lattice parameters, all structures, which changed by less than 0.25 Å (Cartesian root mean square deviation of all atoms except hydrogen) should be regarded as reliable.[24] For the $P2_1/c$ structure of α^{II}-quinacridone, the deviation was only 0.11 Å. Hence the structure was confirmed.

From the electron diffraction data, the structure could also be solved in $P2_1/a$ and $P2_1/n$. The resulting structures showed identical hydrogen-bonding patterns and differed from the $P2_1/c$structure only by a mutual shift of neighbouring molecules in the b-direction by (1/2,0,1/2) and (1/2,0,0), respectively. However, the DFT calculations revealed that the energy of the $P2_1/c$structure is 5.0 kJ mol^{-1} lower than that of the $P2_1/n$ structure and 6.8 kJ mol^{-1} lower than that of the $P2_1/a$ structure. Hence, it is safe to assume that the $P2_1/c$ structure represents the correct packing of the molecules.

In the α^{II}-phase of quinacridone, the molecules are linked through a pair of hydrogen bonds into almost planar chains (Fig. 5). The chains are packed into a layer. A similar arrangement of molecules is also observed in the α^{I} and β-phases. However, these three polymorphs differ in the mutual arrangements of their layers: in the α^{I}-phase, all molecules are parallel, in the α^{II}-phase, the molecules form a herringbone pattern, and in the β-phase, neighbouring layers are mutually rotated by 69.4°, see Fig. 5. A herringbone packing of planar chains, as is now found for the α^{II}-phase, has not been observed in any quinacridone derivative before, but is known to exist in other pigment classes, e.g. in diketopyrrolopyrrole (Pigment Red 254)[25] and thiazine-indigo (Pigment Orange 80).[26]

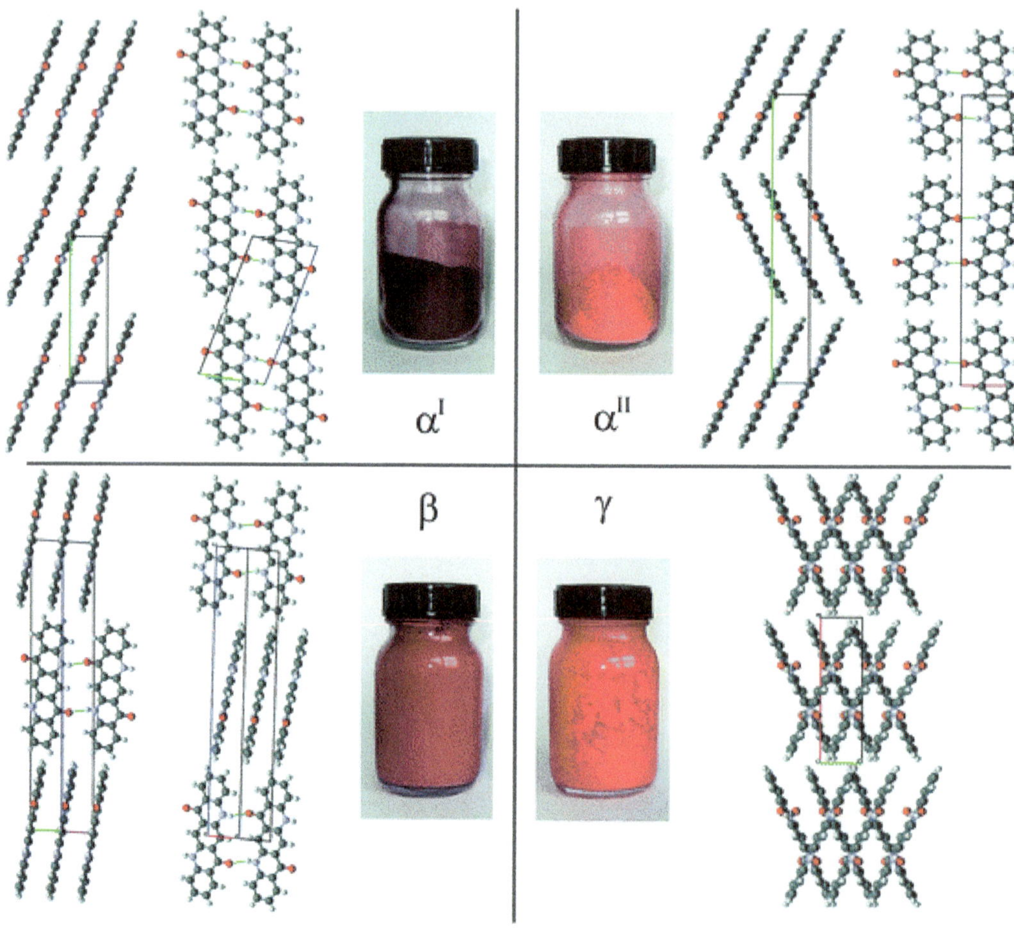

Figure 5 Colours and crystal structures of quinacridone polymorphs.

A closer inspection of the crystal structures of the α^{I} and α^{II}-phases of quinacridone reveals that not only the chains, but also the mutual arrangement of the chains within the layers, are almost identical in both phases. The main difference between the polymorphs is the stacking of the layers: either parallel (α^{I}) or herringbone (α^{II}). Both motifs can be randomly combined, resulting in a disordered structure with a random mixed sequence, see Fig. 6. The simulated electron diffraction pattern of this structural model (Fig. 6a) reproduces the experimental diffraction intensity distribution quite well (Fig. 6b), explaining the observed diffuse scattering. It is therefore likely that the real structure of the α^{II}-phase consists of a herringbone arrangement with a considerable number of stacking faults exhibiting a parallel arrangement of neighbouring layers.

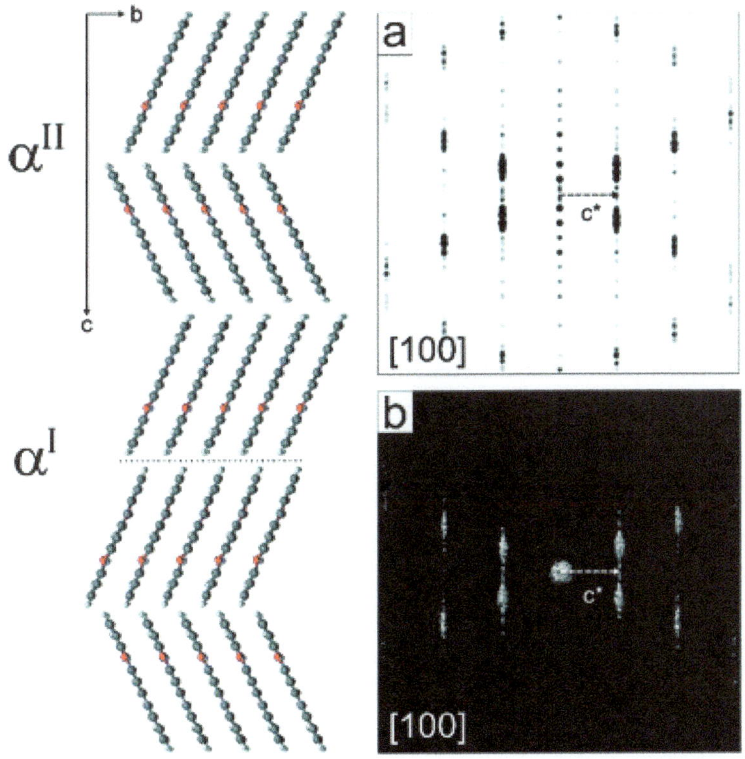

Figure 6 Structural model of α^{II}-quinacridone including disorder: random layer stacking; at the dotted line there is a parallel stacking (as in the α^{I}-phase) instead of the herringbone pattern (as in the idealized α^{II}-structure); a) simulated electron diffraction pattern from the model; b) experimental electron diffraction pattern of α^{II}-quinacridone

This stacking disorder is also supported by the lattice-energy minimizations with DFT-D calculations: the lattice-energy of a mixed-stacked structure consisting of 2/3 herringbone and 1/3 parallel arrangements is only 0.1 kJ mol^{-1} higher than the energy of pure herringbone (ideal α^{II}) and 1.0 kJ mol^{-1} lower than the energy of the pure parallel packing (α^{I}). The β- and γ-phases are energetically much more favourable (β: −4.3 kJ mol^{-1}, γ: −5.5 kJ mol^{-1}), which reflects the experimental stability order $\gamma > \beta \gg \alpha^{I}, \alpha^{II}$.

The complexity of the real structure of α^{II}-quinacridone explains the problems with the structure analysis from powder X-ray diffraction data. The low crystallinity reduces the quality of the data in general, and the diffuse scattering leads to intensity at positions not matching the Bragg positions.

Different samples of α^{II}-quinacridone showed slightly different X-ray powder diffraction patterns depending on the synthetic conditions, indicating different fractions of parallel and herringbone stacking. This indicates that, the ordered structures of the α^{I} and α^{II} phases are the end members of a continuous series of disordered structures consisting of different amounts of each packing type.

Finally, we would like to emphasise the role of electron diffraction in the structure analysis of nanocrystalline materials. In the structural investigation of α^{II}-quinacridone, all traditional methods – X-ray single crystal and powder diffraction – failed and only the employment of electron diffraction ultimately allowed us to determine the crystal structure.

CONCLUSIONS

Electron diffraction data of α^{II}-quinacridone showed severe diffuse scattering. The average crystal structure solved by direct methods showed a herringbone molecular packing not observed previously for any of the quinacridone derivatives. The crystal structure was confirmed by DFT-D energy minimization. The diffuse scattering observed in the data could be explained by a model with stacking disorder containing a mixed sequence of herringbone and parallel arrangements of neighbouring layers. The correctness of the model was proved by DFT-D energy minimizations.

ACKNOWLEDGEMENTS

We would like to thank our colleagues from Clariant, especially Dr. Dieter Schnaitmann, Dr. Wolfgang Schwab and Thorsten Schmiermund, for 20 years of collaboration on quinacridone. X-ray powder patterns and IR spectra were recorded by Edith Alig (Goethe University, Frankfurt). Synchrotron data of α^{II}-quinacridone, used for PDF analyses, were measured by Dr. Nadine Schrodt (née Rademacher) and Dr. Ann-Christin Dippel at beamline P02.1 (beamline scientist in charge: Dr. Hanns-Peter Liermann) of the light source PETRA III at DESY. For the pair-distribution function analyses, we thank Dr. Lothar Fink (Goethe-University). We thank Dr. Ute Kolb (University of Mainz) for providing access to the TEM facility in Mainz. The research was funded by the German Research Foundation (DFG) priority program 1415.

REFERENCES

1. W. Herbst and K. Hunger, Industrial Organic Pigments, Wiley-VCH, Weinheim, 3rd edn, 2004 .
2. E. F. Paulus, F. J. J. Leusen and M. U. Schmidt, CrystEngComm, 2007, **9**, 131–143
3. (a) E. E. Jaffe, European Pat., 305328, 1989 (b) K. Hiromitsu, S. Masashi and N. Yoichi, Jap. Pat., 52-010324, 1977 ; (c) F. Bäbler and E. E. Jaffe, Eur.

Pat., 530143, 1993; (d) E. E. Jaffe, Eur. Pat., 267877, 1988 ; (e) W. Deuschel, F. Gundel, H. Wuest and E. Daubach, US Pat., 3074950, 1963 (f) K. Hashizume, M. Miyatake, M. Shigemitsu, I. Kumano, H. Katsura and M. Oshima, Deutsches Pat., 1805266, 1970; (g) F. Bäbler, Eur. Pat., 1074587, 2001 (h) G. Lincke, J. Mater. Sci., 1997, **32**, 6447–6451; (i) H. R. Schweizer, US Pat., 3272821, 1962 ; (j) R. S. Tyson and L. Schapiro, Deutsches Pat., 2435219, 1975 .

4. The bleaching of red coatings, which is frequently observed on aged cars, is not caused by a bleaching of the pigment, but by a decomposition of the coating (so-called chalking), resulting in a white, opaque layer which hides the red colour of the pigments.

5. Quinacridone is internationally registered in the Colour Index as C.I. Pigment Violet 19. Trade names are e.g. ®Hostaperm Red Violet ER02 for the β-phase and ®Hostaperm Red E5B02 for the γ-phase (manufacturer: Clariant).

6. The yellow colour of individual quinacridone molecules is visible in a very dilute solution of quinacridone in boiling dimethylsulfoxide at 189 °C, showing a pale yellow shade. Also high-level quantum-mechanical calculations confirm the yellow colour of an individual molecule. Below about 150 °C quinacridone is insoluble in water and in all solvents.

7. (a) E. F. Paulus, E. Dietz, A. Kroh and F. Prokschy, 12th European Crystallographic Meeting, Moscow, 20–29 August, 1989, Collected Abstracts, vol. 2, pp. 23–24 (β- and γ-phases); (b) N. Nishimura, T. Senju and J. Mizuguchi, Acta Crystallogr., Sect. E: Struct. Rep. Online, 2006, **62**, o4683–o4685 (β-phase); (c) G. D. Potts, W. Jones, J. F. Bullock, S. J. Andrews and S. J. Maginn, J. Chem. Soc., Chem. Commun., 1994, 2565–2566 (γ-phase); (d) J. Mizuguchi, T. Sasaki and K. Tojo, Z. Kristallogr. - New Cryst. Struct., 2002, **217**, 249–250 (γ phase).

8. F. J. J. Leusen, Presentation at the 15th European Crystallographic Meeting ECM15, Dresden, 18 August–2 September, 1994, abstract in Z. Kristallogr., 1994, 8, 161

9. For the α^{II}-phase, the standard deviation of the lattice parameters are not included for the unit cell determined by electron diffraction, as specific sources of errors apply for electron diffraction experiments. The accuracy of distance measurements in electron diffraction is determined by a systematic error due to camera length calibration and a statistical error due to lens hysteresis while switching between imaging and diffraction mode; both are below 2%.

10. G. Lincke and H.-U. Finzel, Cryst. Res. Technol., 1996, **31**, 441–452.

11. M. U. Schmidt, presentation on the European Powder Diffraction Conference (EPDIC), Darmstadt, Germany, 2010.

12. (a) U. Kolb, T. E. Gorelik, C. Kübel, M. T. Otten and D. Hubert, Ultramicroscopy, 2007, **107**, 507–513; (b) U. Kolb, T. E. Gorelik and M. T. Otten, Ultramicroscopy, 2008, **108**, 763–772 ; (c) D. Zhang, P. Oleynikov, S. Hovmöller and X. Zou, Z. Kristallogr., 2010, **225**, 94–102.

13. (a) I. Rozhdestvenskaya, E. Mugnaioli, M. Czank, W. Depmeier, U. Kolb, A. Reinholdt and T. Weirich, Mineral. Mag., 2010, **74**, 159–177; (b) C. S. Birkel, E. Mugnaioli, T. Gorelik, U. Kolb, M. Panthöfer and W. Tremel, J. Am. Chem. Soc., 2010, **132**, 9881–9889 ; (c) J. Jiang, J. L. Jorda, J. Yu, L. A. Baumes, E. Mugnaioli, M. J. Diaz-Cabanas, U. Kolb and A. Corma, Science, 2011, **333**, 1131–1134; (d) M. Gemmi, I. Campostrini, F. Demartin, T. E. Gorelik and C. M. Gramaccioli, Acta Crystallogr., Sect. B: Struct. Sci., 2012, **68**, 15; (e) T. Willhammar, J. L. Sun, W. Wan, P. Oleynikov, D. L. Zhang, X. D. Zou, M. Moliner, J. Gonzalez, C. Martínez, F. Rey and A. Corma, Nat. Chem., 2012, **4**, 188–194; (f) J. Plášil, L. Palatinus, J. Rohlíček, L. Houdková, M. Klementová, V. Goliáš and P. Škácha, Am. Mineral., 2014, **99**, 276–282.

14. (a) A. V. Yakimanski, U. Kolb, G. N. Matveeva, I. G. Voigt-Martin and A. V. Tenkovtsev, Acta Crystallogr., Sect. A: Found. Crystallogr., 1997, **53**, 603–614; (b) I. G. Voigt-Martin, Z. X. Zhang, U. Kolb and C. Gilmore, Ultramicroscopy, 1997, **68**, 43–59; (c) T. Gorelik, M. U. Schmidt, J. Brüning, S. Bekö and U. Kolb, Cryst. Growth Des., 2009, **9**, 3898–3903;

15. (d) T. Gorelik, G. Matveeva, T. Schleuß, A. F. M. Kilbinger, J. van de Streek, A. Bohle, G. Brunklaus and U. Kolb, CrystEngComm, 2010, **12**, 1824–1836.

16. (a) U. Kolb, T. E. Gorelik, E. Mugnaioli and A. Stewart, Polym. Rev., 2010, **50**, 385–409; (b) T. E. Gorelik, J. van de Streek, A. F. M. Kilbinger, G. Brunklaus and U. Kolb, Acta Crystallogr., Sect. B: Struct. Sci., 2012, **68**, 171–181; (c) C. H. Förster, T. E. Gorelik, U. Kolb, V. Ksenofontov and K. Heinze, Eur. J. Inorg. Chem., 2015, 920–924.

17. S. J. Clark, M. D. Segall, C. J. Pickard, P. J. Hasnip, M. J. Probert, K. Refson and M. C. Payne, Z. Kristallogr., 2005, **220**, 567–570.

18. J. P. Perdew, K. Burke and M. Ernzerhof, Phys. Rev. Lett., 1996, **77**, 3865.

19. S. J. Grimme, Comput. Chem., 2006, **27**, 1787–1799.

20. (a) T. R. Welberry, Diffuse X-ray Scattering and Models of Disorder, Oxford University Press, Oxford, 2004; (b) M. Gardon, C. B. Pinheiro and G. Chapuis, Acta Crystallogr., Sect. B: Struct. Sci., 2003, **59**, 527–536; (c) H.-B. Bürgi, M. Hostettler, H. Birkedal and D. Schwarzenbach, Z. Kristallogr., 2005, **220**, 1066–1075 ; (d) T. R. Welberry and D. J. Goossens, Acta Crystallogr., Sect. A: Found. Crystallogr., 2008, **64**, 23–32; (e) H. Krüger, S. Stöber, T. Welberry, R. Withers and J. Fitz Gerald, Acta Crystallogr., Sect. B: Struct. Sci., 2011, **67**, 476–485; (f) J. L. Teteruk, J. Glinnemann, T. E. Gorelik, A. Linden and M. U. Schmidt, Acta Crystallogr., Sect. B: Struct. Sci., Cryst. Eng. Mater., 2014, **70**, 296–305.

21. E. Pidcock, W. D. S. Motherwell and J. C. Cole, Acta Crystallogr., Sect. B: Struct. Sci., 2003, **59**, 634–640.

22. The space group $P2_1/m$ was not considered: the occupation of inversion centres by molecules in $P2_1/m$ is very rare (N. U. Chernikova, V. K. Belskiy and P. M. Zorkiy, Zh. Strukt. Khim., 1990, **31**, 148–153, in Rus.),

and the positioning on the mirror planes was sterically impossible with the obtained lattice parameters.

23. M. C. Burla, R. Caliandro, M. Camalli, B. Carrozzini, G. L. Cascarano, C. Giacovazzo, M. Mallamo, A. Mazzone, G. Polidori and R. Spagna, J. Appl. Crystallogr., 2012, **45**, 357–361.

24. M. A. Neumann, F. J. J. Leusen and J. Kendrick, Angew. Chem., 2008, **120**, 2461–2464 (Angew. Chem., Int. Ed. , 2008 , **47** , 2427–2430).

25. J. van de Streek and M. A. Neumann, Acta Crystallogr., Sect. B: Struct. Sci., 2010, **66**, 544–558.

26. J. Mizuguchi, A. Grubenmann and G. Rihs, Acta Crystallogr., Sect. B: Struct. Sci., 1993, **49**, 1056–1060.

27. C. Buchsbaum, E. F. Paulus and M. U. Schmidt, Z. Kristallogr., 2011, **226**, 822–831

CHAPTER 2

MORPHOLOGICAL CONTROL OF CALCIUM PHOSPHATE NANOSTRUCTURES USING LYOTROPIC LIQUID CRYSTALS

Wenxiao He , Yu Fu and Martin Andersson

Department of Chemical and Biological Engineering, Chalmers University of Technology, Göteborg SE-41296, Sweden.

ABSTRACT

Organisms use "soft" organic compartments to control the morphology of the embedded "hard" minerals. Here we present a simple method using liquid crystal (LC) phases as "soft" and "inert" templates to prepare nanostructured calcium phosphates (CaPs), which are inorganics of known bioefficacy. Specifically, 6 nm-thick CaP nanowires and CaP sheets that precisely replicate reverse hexagonal (H_2) and lamellar (L_α) LCs have been successfully synthesized and we attribute this to the sufficient spatial regulation offered by the negative (H_2) or flat curvature (L_α) of the aqueous domain. A normal hexagonal (H_1) phase possesses a positive curvature of the aqueous domain, therefore limited spatial restriction. For this reason, precise replication of the H_1 phase by CaP has not been possible. Interestingly, the dynamic nature of the template allowed the construction of micron-sized brushite objects with a laminated structure decorating a specific facet, possibly as a result of epitaxial overgrowth of nano-sized brushite subunits.

INTRODUCTION

Skeletal biominerals possess distinctive morphological, compositional, crystallographic features and organizational order, and thereby exhibit superior material properties. An eminent example is the mineral phase in mammalian bone. Bone apatite nanoplatelets (a highly substituted polymorph of calcium phosphate, CaP) uniaxially align within the collagen matrix, which provide mammals optimal protection, support and other specialized functions.[1] The synthesis of these biominerals mostly proceeds within micro- to nanoscopic compartments, i.e.structural matrix assembled by insoluble biomacromolecules.[1–4] In addition to the spatial confinement, soluble components including ions and nonstructural biomacromolecules synergistically modulate the mineralization process.[1–4] To elucidate the geometric effects in tuning the morphology and polymorphism of biominerals, herein, we report the "softly" confined morphogenesis of CaPs into well-aligned nanowires, nanosheets, ordered laminated structures in "inert" liquid crystal (LC) phases, as well as a site-specific-decoration strategy to construct CaP superstructures with ordered nano-features. The precise control of CaP nanostructures by gentle spatial confinement and the possibility to add-on the structure onto pre-existing architectures offers new opportunities for rational design of nano-inorganics and hybrid materials for advanced applications including artificial bone, biocompatible sensing and nanophotonic devices.

With compositions close to the bone mineral and widely recognized bioefficacy, CaP has been the predominant mineral

Citation: Wenxiao He , Yu Fu and Martin Andersson. Morphological control of calcium phosphate nanostructures using lyotropic liquid crystals. J. Mater. Chem. B, 2014, 2, 3214-3220. DOI: 10.1039/C4TB00095A family utilized in biomedical research and industry, where researchers have endeavored to obtain CaP materials with well-defined structures.[5,6] Amorphous calcium phosphate (ACP), dicalcium phosphate (dihydrate: DCPD, brushite; anhydrous: DCPA, monetite), octacalcium phosphate (OCP), and calcium deficient hydroxyapatite (CDHA) are the CaP phases that can be formed under near-physiological conditions (in the pH range of 6–8 at room temperature and air pressure) and therefore retain biological relevance.[5,7–10] The anisotropic crystal growth of CaP polymorphs is counterproductive to morphological control.[11] Moreover, the competition between CaP phases during precipitation further introduces complexity for controlled morphogenesis. Strategies including utilizing active additive/template–CaP interaction, biological templates and hard templates have been adopted to actively and/or geometrically regulate CaP growth.[12–17] However, these approaches have several limitations: as in the case of using active additive/templates, specially designed macromolecules that could direct CaP nucleation and crystallization are needed; while in the case of biological templates, an additional precursor infiltration step as well as matrices from restricted sources like demineralized biological samples are required; whereas for hard templates, the structural rigorousity of the templates cannot truly reflect "soft" biological confinements and superstructures cannot be developed from them. On the other hand, synthetic hydrogels with structural flexibility have long been used as scaffolds for biomimetic mineralization, yet, the lack of an ordered nanostructure in most systems does not allow precise control over CaP morphologies. LC phases possessing well-organized nanometric aqueous domains are ideal matrices for biomimetic mineralization, as in natural bone the collagen matrix is itself essentially a LC. Yet there are few reports on the successful replication of CaP from LC phases, as forces holding the LC structure (hydrophobic interaction between self-assembled amphiphilic polymers) can be too weak to restrain anisotropic growth of crystalline CaPs.[18] To overcome this, our previous strategy was to biomimetically employ an amorphous intermediate phase. Specifically, reverse hexagonal (H_2) LC was used to synthesize uniform ACP nanospheres, from which apatite nanoparticles with well-defined morphology and bone-resembling features were prepared through controlled crystallization.[19]

The current study focuses on the manipulation of CaP morphogenesis and polymorphism in confined nanospace provided by H_2, lamellar (L_α) and normal hexagonal (H_1) LCs. We show that a commercially available nonionic block-copolymer can be used to construct matrices with varied degree of spatial confinement, so that it can guide CaP growth into ACP, monetite (DCPA) or brushite (DCPD) with morphologies replicating the matrix (Fig. 1). The choice of nonionic block-copolymer assemblies may on one hand purely provide spatial restriction during mineralization without active polymer–crystal interaction, thus the technique and outcome can be translated to other compartmentalized matrices; on the other hand, it may also allow re-arrangement of the template structure during particle growth. The flexibility of this system enables us to produce brushite crystals with hierarchical structures (Fig. 1, H_1).

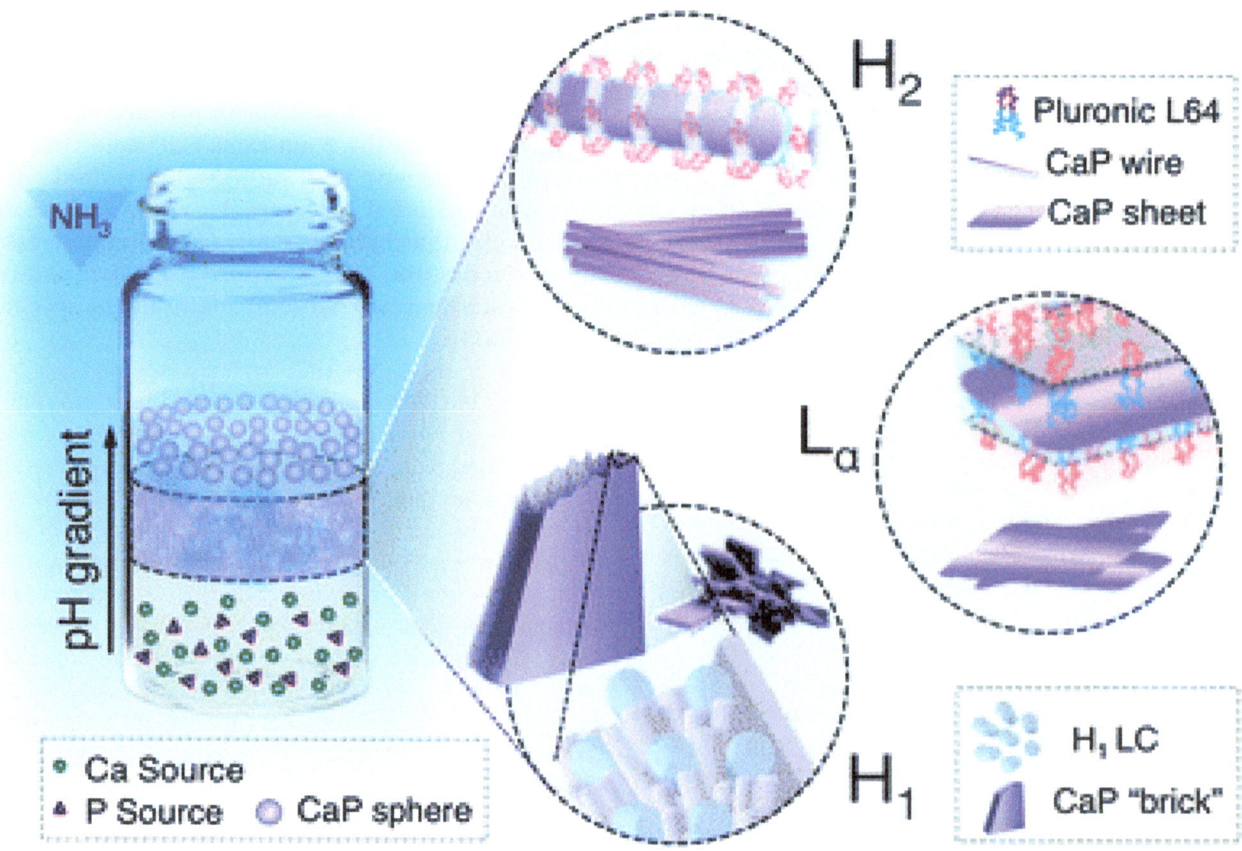

Figure. 1 Illustration of the morphological development of CaP particles in H_2, L_α and H_1 LC phase, where a kinetic window exists for CaP growth to replicate the template structure or to build up the CaP superstructure. The sizes of CaPs and templates are exaggerated for illustration.

EXPERIMENTAL

LC gel Preparation

All chemicals were of analytical grade, purchased from Sigma-Aldrich (except Pluronic F127, which was kindly donated by BASF) and used without further purification. Pluronic L64 ($EO_{13}PO_{30}EO_{13}$) and F127 ($EO_{100}PO_{70}EO_{100}$) were utilized to form different LC phases. $Ca(NO_3)_2{\cdot}4H_2O$ (0.84 M) and H_3PO_4 with a Ca/P molar ratio of 1.67 were dissolved in Milli-Q water. To form a LC gel with the desired structure, pre-mixed precursor solution, Pluronics and oil were mixed evenly according to the recipes listed in Table 1.[20,21] Gel samples with a thickness of ca. 1 mm, 2.5 mm or 35–40 mm were prepared by filling glass vials (20 mL, 57.00 × 27.50 mm, VWR) with ca. 0.4 g, 1 g or 15 g of the as-formed liquid crystalline (LC) gel having the desired structure. In one of the glass vials, pH paper (pH range: 1–11 or 1–14) was cut into small pieces and mixed with the LC gel, which was later on used as a pH indicator for the reaction (as shown in Fig. S1c, ESI†).

Table 1 Recipes of LC gel with varied phase structures prepared from L64–water–p-xylene and F127–water–butanol systems[20,21]

No.	Phase structure[a]	Composition (wt%)		
		L64	***p*-Xylene**	**Precursor solution**
***a* L_α, H_1, and H_2 denote lamellar, normal hexagonal and reverse hexagonal LC phases, respectively.**				
L-1	H_2	70	15	15
L-2	L_α	70	0	30
L-3	L_α	55	10	35
L-4	H_1	55	0	45
		F127	Butanol	Precursor solution
F-1	H_1	60	16	24

Reaction in Ammonia Atmosphere

After preparation, the glass vials were tightly sealed and stored for 24 h to reach equilibrium. An additional sonication step (sonication bath, 60 min) was applied to the Pluronic F127-based H_1 gel sample. After that, the gels were placed in the desiccator cabinet (7 × 12 × 12 inch) with an ammonia atmosphere, as illustrated in the ESI Fig. S1a,† where the beaker in the lower part contained varied concentrations and volumes of ammonium hydroxide solution. By diffusion of ammonia gas into the LC gel, CaP precipitated from the aqueous domain of the LC phase. After 24–72 h, the glass vials were taken out from the desiccator cabinet and gel samples were collected accordingly to their depth from upper gel surface (Fig. S1c, ESI†) or pH ranges (pH value of gel layers at certain depths was read from the pH indicator, as shown in Fig. S1c†). The collected gel samples were dispersed in ethanol and centrifuged (Hettich ALC) for 10 min (1585 × g, 2500 rpm). Then the supernatant was removed gently and the sedimentation was re-dispersed in ethanol. This purification process was repeated at least 3 times.

Characterization

Transmission electron microscopy (TEM) analysis was performed on a JEOL 1200EX II microscope operating at 120 kV. The TEM specimens were prepared on a lacey carbon coated copper grid by evaporating a drop of CaP suspension in ethanol. Scanning electron microscopy (SEM) was also performed on the TEM sample grid (by sticking onto SEM stub using carbon tape), using a LEO ULTRA 55 FEG at an accelerating voltage of 2 or 5 kV. X-ray powder diffraction (XRD) patterns were recorded using a Bruker D8 Advance X-ray diffractometer with Cu-Kα_1 radiation (λ = 1.54056 Å). All the crystalline phases were identified and semi-quantitatively analysed based on reference intensity ratio (RIR) values using the Search/Match module of DIFFRAC EVA (Bruker AXS) and ICDD database. Small Angle X-ray Scattering (SAXS) experiments were performed on the prepared LC gels before and after the synthesis using synchrotron radiation. The experiments were performed at the I911 beamline at MAX-lab (Lund, Sweden). The wavelength was 0.91 Å and a two-dimensional MarCCD 165 mm detector was used. Samples were sealed in a slit (1 mm thick) enclosed by two parallel transparent cellophane films within a stainless steel sample holder mounted in an evacuated sample chamber. The data were collected for 60–180 s and the used q-range ($q = 4\pi \sin \theta/\lambda$) was 0.1–4.0 nm^{-1}.

RESULTS AND DISCUSSION

TEM studies of CaP particles grown from a Pluronic L64-based H_2 LC gel presented varied morphologies depending on the sampling time point and position. CaP nanospheres with a diameter of roughly 11 nm formed exclusively in the upper 3 mm-thick gel layer (Fig. 2a) after 48 h of reaction. The corresponding selected area electron diffraction (SAED) pattern (Fig. 2b) shows that they are ACP, which agrees with our earlier observation with half the precursor concentration.[19]

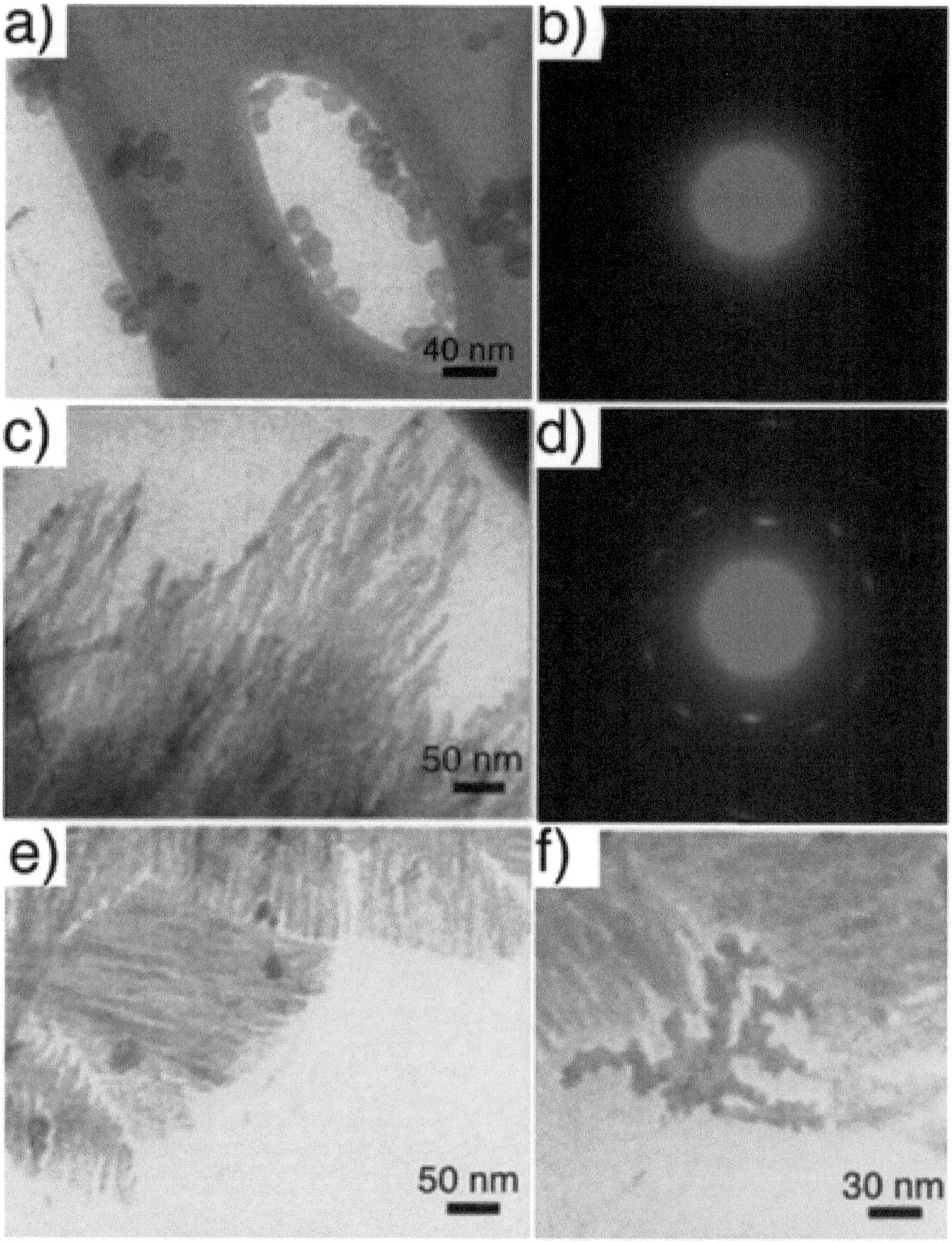

Figure 2 TEM images of CaP particles obtained from H_2 LC gel when collected from (a) upper 3 mm-thick gel after 48 h and (b) the corresponding SAED pattern, (c) gel layer 3–6 mm below the surface after 72 h and (d) the corresponding SAED pattern, (e) and (f) same condition as (c) but with a gel equilibration time of 10 h instead of 24 h. Gel recipe: L-1 inTable 1. Gel thickness: 35–40 mm. Ammonia dose: 300 mL 35 wt% NH_3OH solution.

In the gel layer 3–6 mm below the surface, CaP particles with unsettled morphologies (Fig. S2a, ESI†) were harvested after 24 h of reaction, indicating an early structural development. Whereas after 48 h of reaction, CaP nanowires with a diameter of ca. 6 nm (Fig. S2b, ESI†) were formed, together with ACP nanospheres. With a prolonged reaction time of 72 h, the nanowire shape of CaPs (Fig. 2c) retained in the same sampling position. The electron diffraction pattern of these nanowire arrays (Fig. 2d) unveils the crystallinity of the sample. The arc-like diffraction spots indicate the presence of crystallographic orientation within the nanowire arrays. However, it has been difficult to resolve the CaP phase by XRD analysis (Fig. 3a), in which the diffuse pattern implies relative low abundance and poor crystallinity of the nanowires. Yet the star-marked region in the diffractogram might suggest a monetite phase of the nanowires. These nanowire arrays seem to align in the gel as observed under optical microscopy (Fig. S3, ESI†), in which each hair-like feature represents an assembly of CaP nanowires.

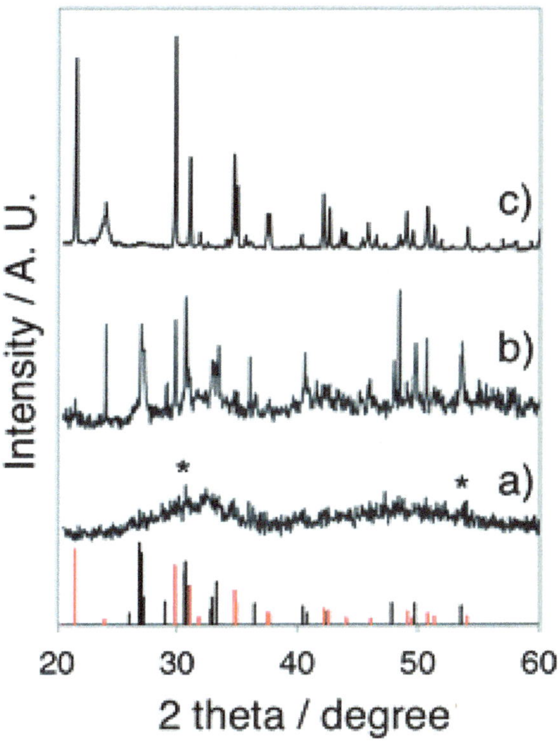

Figure 3 XRD pattern of (a) the CaP nanowires obtained from the gel layer 3–6 mm below the surface of a 35–40 mm thick H_2 LC gel after 72 h of reaction (gel recipe: Table 1 L-1, ammonia dose: 300 mL 35 wt% NH_3OH solution), (b) the CaP sheets obtained from the gel layer with pH 9 of a 35–40 mm thick L_α sample after 24 h of reaction (gel recipe: Table 1 L-3, ammonia dose: 80 mL 35 wt% NH_3OH solution), and (c) the CaP bricks obtained from the gel layer 12–15 mm below the surface (pH 9–10) of a 35–40 mm thick H_1 LC after 48 h of reaction (gel recipe: Table 1 F-1, ammonia dose: 80 mL 35 wt% NH_3OH solution.). Bottom: the standard monetite (black) and brushite (red) data. Diffractograms have been vertically shifted for clarity.

Similar experiments with a shorter gel equilibration time (instead of the 24 h used for general cases) were performed to test the templating effect of the LC gel, as a limited gel equilibration time would result in smaller ordered domains with different orientations in the original LC templates. As expected, with a gel equilibration time of 10 h, smaller domains of CaP wire stacks (Fig. 2e) were obtained from gel layer of 3–6 mm after 72 h of reaction. In addition, dendrimeric CaP particles were also found in between the ordered domains (Fig. 2f), which reflect the defects in the H_2 LC template due to incomplete equilibration.

The disparity in particle morphologies and polymorphism in different gel layers can be linked to differences in the pH profile within the aqueous domains of the LC gel. The spherical morphology and slightly larger particle diameter (than that of the original water channels which is ca. 6 nm) of ACPs formed in the upper layer implies a fast kinetic precipitation.[19] While in the intermediate layers, the morphology of the water domains were replicated directly to crystalline nanowires, suggesting a moderate pH increase that is slow enough for nucleation in an acidic environment, yet fast enough to allow a kinetic driven crystal growth with controlled shape. The moderate pH rise in the intermediate layer was due to the presence of CaP particles formed in the upper gel layer that slowed down the ammonia penetration. The particles formed in a deeper gel layer, i.e. with an even slower pH increase, presented ill-templated shapes (data not shown) due to higher tendency toward crystallization.

To further elucidate the sensitivity of CaP morphogenesis to pH-rate, we performed experiments using the L64-based L_α LC phase with a gel thickness of ca. 1 mm. An ammonia dose of 800 mL 0.196 wt% NH_3OH solution (in the pH beaker within desiccator) resulted in spherical CaP aggregates (Fig. 4a) in the L_α gel (final pH 10) after 24 h of reaction, corresponding to the upper layer observed in the thick gel samples. With a lower ammonia dose (800 mL 0.068 wt% NH_3OH solution), foiled CaP sheets (Fig. 4b) that replicated the water domain of the $L_\alpha$gel (final pH 8–9) were formed. The absence of diffraction spots and rings in the electron diffractogram (the inset in Fig. 4b) shows that these CaP sheets were amorphous. For comparison, we also performed experiments using 35–40 mm thick L_α gel samples for 24 h and with a more concentrated ammonia atmosphere (80 mL 35 wt% NH_3OH solution in pH beaker). As expected, spherical CaP aggregates and foiled CaP sheets were found in the gel layer with pH 10 and pH 8–9, respectively (data not shown). Interestingly, some of the CaP sheets presented more rigid folding lines (as indicated by black arrows in Fig. 4c) than the others. The barely visible diffuse diffraction ring in the SAED pattern (the inset in Fig. 4c) might suggest the presence of crystallinity in these sheets. However, a clear electron diffraction pattern of this area was not attainable due to the extremely small sample thickness and lability under an electron beam. The XRD pattern (Fig. 3b) reveals that these rigid sheets are dicalcium phosphate (DCP) either with (DCPD, brushite) or without lattice water (DCPA, monetite) with the DCPA/DCPD mass fraction ratio of roughly 2.4.

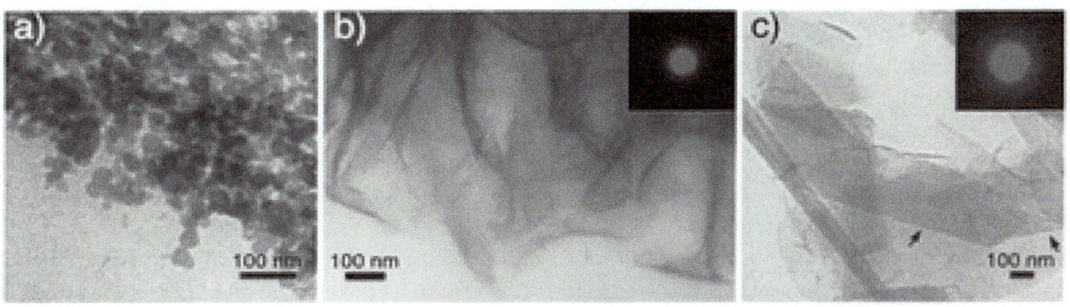

Figure 4 TEM images of CaPs obtained after 24 h from ca. 1 mm-thick L_α LC gel with ammonia doses of (a) 800 mL 0.196 wt% NH_3OH solution, (b) 800 mL 0.068 wt% NH_3OH solution (SAED pattern inserted), and (c) from the gel layer with pH 8–9 of a 35–40 mm thick sample with 80 mL 35 wt% NH_3OH solution in a pH beaker (SAED pattern inserted). Gel recipe: L-2 (a and b) and L-3 (c) in Table 1.

With a continuous aqueous domain and a high water content, controlled morphogenesis of CaP in the L64-based H_1 LC phase with a sample thickness of ca. 2.5 mm and ammonia dose of 800 mL 0.33 wt% NH_3OH solution in the pH beaker seemed to be a failure, as submillimeter-sized crust-like CaP aggregates (Fig. 5a) were formed in the upper gel layer (pH 8–9) that greatly blocked the further penetration of ammonia gas. Nevertheless, a higher magnification reveals the laminated details of the CaP aggregates (Fig. 5b), where stacks of CaP plates closely pack together. The debris from broken pieces (Fig. 5c) further discloses the thin flake-like morphology of CaP subunits. In addition, the CaP plates within a stack are packed so well that their edges seem to align into flat faces with straight ridges (Fig. 5d). A possible construction principle of this laminated structure would be the epitaxial overgrowth of highly oriented layered CaP crystal subunits on an underlying substrate composed of large CaP single crystals that expose a specific facet facing different directions.

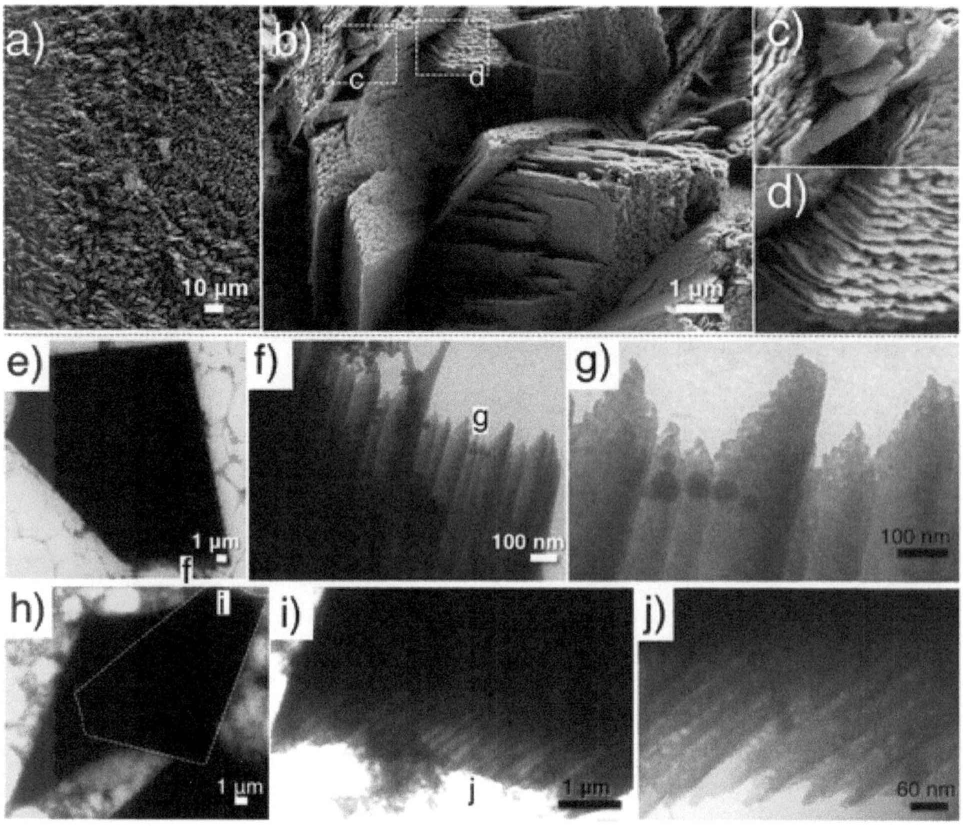

Figure 5 (a–d) SEM images of laminated CaP sheets obtained from the upper crust of a 2.5 mm thick H_1 LC after 24 h of reaction; (e–j) TEM images of CaP bricks obtained from the layer 12–15 mm below the gel surface (pH 9–10) of a 35–40 mm thick H_1 LC after 48 h of reaction. Gel recipe: L-4 (a–d), F-1 (e–j) in Table 1. Ammonia dose: (a–d) 800 mL 0.33 wt% NH_3OH solution, and (e–j) 80 mL 35 wt% NH_3OH solution. Squares and inserted letters mark the magnified region.

To demonstrate the proposed construction principle of laminated CaP sheets, experiments using a Pluronic F127-based H_1 LC phase with an additional sonication step prior to the reaction in the desiccator were conducted. The F127-based H_1 LC contains butanol as the oil phase that would assist ammonia penetration through a thick gel system, thus causing a smooth pH rise within the gel layer.[21] Furthermore, the additional sonication step introduces crystal nuclei to the system (Fig. S4a, ESI†), which then grow into larger crystal aggregates (Fig. S4b and c†). This approach allowed us to investigate the facets CaP subunit overgrowth that developed. Specifically, ca. 35–40 mm thick H_1 LC samples (recipe: F-1, Table 1) were first subjected to sonication for 60 min, and then reacted in a concentrated ammonia atmosphere (80 mL 35 wt% NH_3OH solution in pH beaker) for 48 h and from which the gel layers of 12–15 mm below the surface (pH 9–10) were collected. Fig. 5e and h show the typical TEM images of as-synthesized CaP "bricks", in which a trapezium-like pentagon shape was presented. A higher magnification discloses the inner hierarchy of the shorter parallel "edge" of the trapezium (Fig. 5f, g, i and j).

The SEM studies depict the topography of such CaP "bricks" in detail (Fig. 6a–h), where smooth facets (a- and e-facet and the corresponding twin facets marked in Fig. 6a–d, g and h), grainy facets (b- and twin facets marked in Fig. 6a, f and g), striated facets (d- and twin facet marked in Fig. 6a, d, e and h), as well as laminated sides (c- and twin facets marked in Fig. 6a, c–f and h; occasionally exposed f- and twin facets marked in Fig. 6a and f) were presented together in the same brick. The narrow peaks in the XRD pattern confirm that the CaP brick comprises of highly crystalline brushite (Fig. 3c, phase pure). The a-, b-, d- and e-facet of the brick marked in Fig. 6a could be (hkl) indexed as (010), (111), ($\bar{1}$20), (11$\bar{1}$) and the corresponding mirror facets with {010} as the twin plane, respectively, of pure brushite.[22,23] We highlight that the c-side from SEM investigation (Fig. 6a) constantly happens to be the shorter parallel "edge" of the trapezium from TEM studies (Fig. 5e and h), where the laminated feature of the c-side (Fig. 6c–f and h) reconciles with the inner hierarchy presented on the shorter parallel "edge" in TEM images (Fig. 5f, g, i and j).

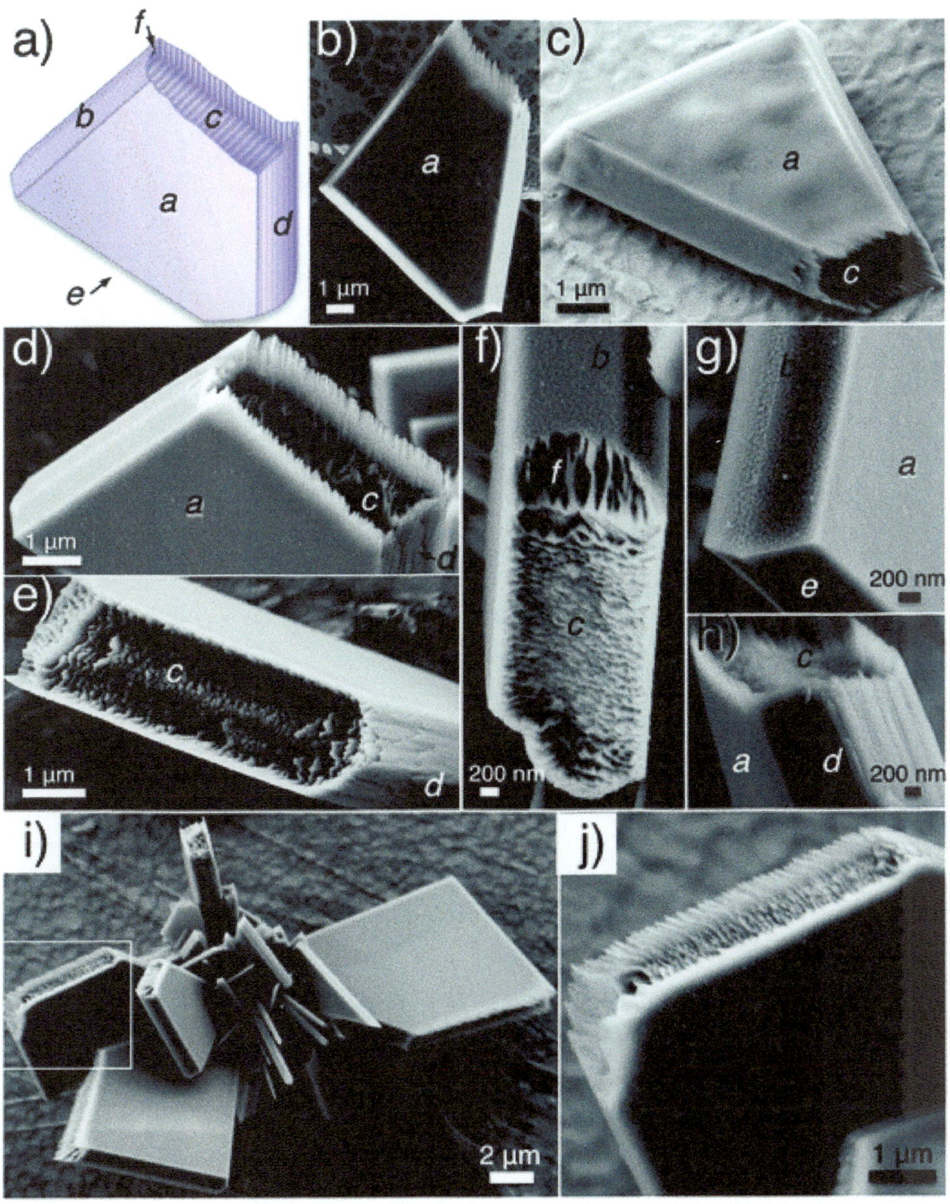

Figure 6 (a–h) SEM micrographs revealed the all-around details of the CaP bricks and (i and j) SEM images of a CaP object composed of several CaP bricks. Samples were prepared and collected with the same condition as the ones shown in Fig. 5e–j. Inserted italic letters index the different "facets" of the brick. Squares mark the magnified region.

In addition to single brushite bricks, 15–25 micron-sized objects comprised of several bricks were observed abundantly in the sample (Fig. 6i and j). The evolution of these superstructures, disclosed by SEM, starts from a crystalline cluster formed by sonication-induced spontaneous precipitation within the highly saturated acidic precursor solution (Fig. S4a, ESI†). This then grows into large polycrystalline aggregates under further sonication (Fig. S4b and c†); and finally, each petal of the polycrystalline aggregate develops crystalline facets as well as a laminated structure on the c-side under an ammonia atmosphere (Fig. S4d, ESI†).

In an attempt to distinguish the laminated structure from the bulk single crystal of a brushite brick, the as-prepared brushite objects were subjected to a gentle grinding process using an agate mortar (5 min) in the presence of several drops of ethanol. Fig. 7a and c show the typical morphology of brushite objects after grinding. SEM images at higher magnification (Fig. 7b and d) unveil the c-side of brushite bricks after grinding, where the spaghetti-like patterns in Fig. 7b are the edges of disordered brushite subunits on the c-side after grinding. While in Fig. 7d, both bumpy solid surfaces (upper-left region) and attached distorted flakes with thickness on the order of 20 nm (lower-right region) on c-side are observed. The former can be assigned to the bulk single crystal of the brushite brick while the latter are distinctly the remaining brushite subunits after grinding. All of these features favor for our assumption of an epitaxially grown layer of brushite flake subunits on top of the brushite bulk brick, as illustrated in Scheme S1.† The epitaxial growth of ordered subunits from the crystalline bulk has been previously observed for calcium carbonate.[24,25] Additionally, CaP with channel-like features resulting from the packing of brushite subunits has also been reported.[26] However, these studies all require the interplay between ionic polymers and the crystals, while in our system, the structure is dependent solely on spatial confinement.

Figure 7 SEM images of the as-prepared CaP objects after gentle grinding (agate mortar, 5 min, with drops of ethanol). Squares mark the magnified region.

Finally, SAXS analysis confirmed the presence of the LC phase (F127 based H_1 LC, L64 based H_2 and L_α LC, respectively) in the early stage of CaP structural development, as SAXS patterns (Fig. 8) with peak positions associated with the different LC structures were maintained in the collected gel layers (pH 7 or pH 8) after reaction for 48 h (L_α) or 72 h (H_1 and H_2), together with an initial slope indicating the particle formation within each LC.

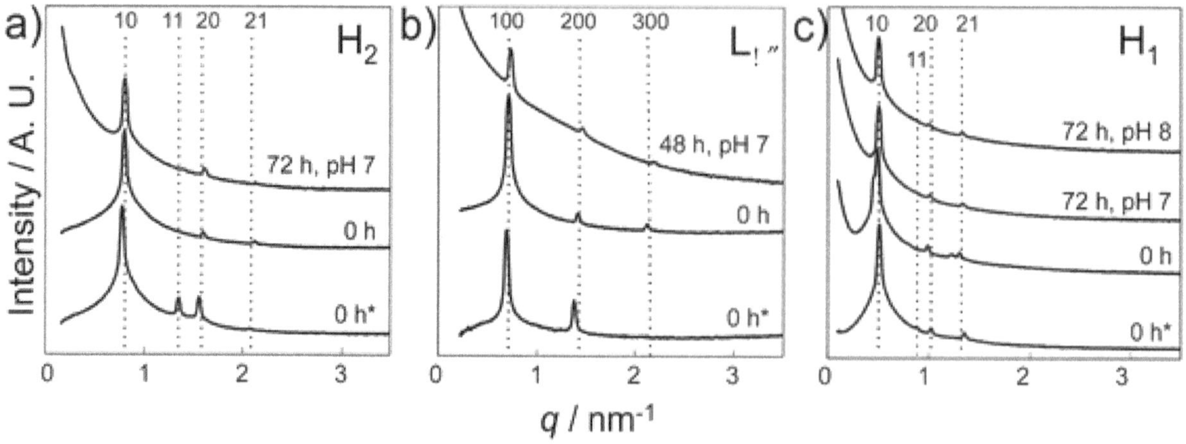

Figure 8 SAXS results of various LC gels before and after the reaction. (a) H_2 LC gel prepared using L64 (recipe: Table 1 L-1), (b) L_α LC gel prepared using L64 (recipe: Table 1 L-2), and (c) H_1 LC gel prepared using F127 (recipe: Table 1 F-1). Gel thickness: 35–40 mm. Samples were exposed to a NH_3 atmosphere (300 mL 35 wt% NH_3OH solution in pH beaker) for 0 and 48 h or 72 h and were collected according to pH (0 h* refers to the LC gel prepared without Ca and P sources, while the rest refers to LC gels prepared with a precursor concentration of 0.84 M Ca^{2+}). SAXS data have been vertically shifted for clarity.

In the current system, ACP, DCPA and DCPD are the three major phases competing in the process of precipitation. ACP, with the isotropic nature, i.e. possibility to replicate any shape, is favorable for controlled morphogenesis, while the anisotropicity of DCPs encumbers it, as the preferential growth could break down the LC template organized via hydrophobic interaction.[11] We show that it is possible to precisely template CaP in H_2 (Fig. 2c, e and f) and L_α (Fig. 4b and c) LC phases and we associate this to the negative (H_2) or flat curvature (L_α) of the aqueous domains, as they offer sufficient spatial regulation on the growth of ACP and even DCPs. However, accurate CaP replication of the H_1 LC structure was not attainable, due to the positive curvature of the aqueous domain in the H_1 phase, providing less spatial confinement. Never the less, we demonstrate that LC re-assemblability provides the template with the ability to build up crystalline CaP with organized nano-patterns (Fig. 5 and 6).

CONCLUSION

In summary, crystalline CaP nanowires and nanosheets (both ACP and DCP) that respectively replicate a H_2 and L_α LC have been synthesized by a simple gas diffusion method with commercially available reactants. In addition, the method can be used to produce brushite objects with superstructure in a H_1 LC. The generation of a laminated structure on brushite objects is facet-specific and possibly the result of brushite subunit epitaxial overgrowth. Control over the CaP morphology was achieved solely by spatial confinement, allowing the possibility of this methodology to be used with alternative components.

CONFLICT OF INTEREST

The authors declare no competing financial interest.

ACKNOWLEDGEMENTS

We acknowledge the Swedish Research Council and NanoSphere Centre/Formas for funding. We also thank MAX-lab (Lund, Sweden) for the allocation of beam time and assistance with SAXS measurements.

REFERENCES

1. A.-W. Xu, Y. Ma and H. Colfen, J. Mater. Chem., 2007, **17**, 415.
2. H. A. Lowenstam and S. Weine, On Biomineralization, Oxford University Press, New York, 1989 d.
3. B. R. Heywood and S. Mann, Adv. Mater., 1994, **6**, 9.
4. S. Weiner and P. M. Dove, An Overview of Biomineralization Processes and the Problem of the Vital Effect, in Reviews in Mineralogy and Geochemistry; Biomineralization, ed. P. M. Dove, J. J. De Yoreo and S. Weiner, Mineralogical Society of America, 2003, vol. 54, p. 1.
5. S. V. Dorozhkin and M. Epple, Angew. Chem., Int. Ed., 2002, **41**, 3130.
6. M. Vallet-Regí and E. Ruiz-Hernández, Adv. Mater., 2011, **23**, 5177.
7. R. Z. LeGeros, Z. Kardiol., 2001, **3**, 116.
8. L. J. Wang and G. H. Nancollas, Chem. Rev., 2008, **108**, 4628.
9. L. Wang and G. H. Nancollas, Dalton Trans., 2009, 2665.
10. A. Navrotsky, Proc. Natl. Acad. Sci. U. S. A., 2004, **101**, 12096.
11. H. Cölfen, Angew. Chem., Int. Ed., 2008, **47**, 2351.
12. S. R. Hall, H. Bolger and S. Mann, Chem. Commun., 2003, 2784.
13. H. Tlatlik, P. Simon, A. Kawska, D. Zahn and R. Kniep, Angew. Chem., Int. Ed., 2006, **45**, 1905.
14. X. Wang, J. Zhuang, Q. Peng and Y. D. Li, Adv. Mater., 2006, **18**, 2031.
15. F. Nudelman, K. Pieterse, A. George, P. H. H. Bomans, H. Friedrich, L. J. Brylka, P. A. J. Hilbers, G. de With and N. A. J. M. Sommerdijk, Nat. Mater., 2010, **9**, 1004.
16. F. Nudelman and N. A. J. M. Sommerdijk, Angew. Chem., Int. Ed., 2012, **51**, 6582.
17. K. Bleek and A. Taubert, Acta Biomater., 2013, **9**, 6283.
18. S. Eftekharzadeh and S. I. Stupp, Chem. Mater., 1997, **9**, 2059.
19. W. X. He, P. Kjellin, F. Currie, P. Handa, C. S. Knee, J. Bielecki, L. R. Wallenberg and M. Andersson, Chem. Mater., 2012, **24**, 892
20. P. Alexandridis, U. Olsson and B. Lindman, Macromolecules, 1995, **28**, 7700.
21. P. Holmqvist, P. Alexandridis and B. Lindman, Langmuir, 1997, **13**, 2471.
22. F. Abbona, F. Christensson, M. F. Angela and H. E. L. Madsen, J. Cryst. Growth, 1993, **131**, 331.
23. J. L. Giocondi, B. S. El-Dasher, G. H. Nancollas and C. A. Orme, Philos. Trans. R. Soc., A, 2010, **368**, 1937.
24. D. Volkmer, M. Harms, L. Gower and A. Ziegler, Angew. Chem., Int. Ed., 2005, **44**, 639.
25. H. Imai, T. Terada and S. Yamabi, Chem. Commun., 2003, 484.
26. A. Shkilnyy, J. Brandt, A. Mantion, O. Paris, H. Schlaad and A. Taubert, Chem. Mater., 2009, **21**, 1572

CHAPTER 3 CRYSTALLOGRAPHIC ANALYSIS OF THE SOLID-STATE DEWETTING OF POLYCRYSTALLINE GOLD FILM USING AUTOMATED INDEXING IN A TRANSMISSION ELECTRON MICROSCOPE

S. A. Jang[1], H. J. Lee[1], C. V. Thompson[2], C. A. Ross[2] and Y. J. Oh[1]

[1]Department of Advanced Materials Science and Engineering, Hanbat National University, 125, Dongseo-daero, Yuseong-gu, Daejeon 305-719, South Korea

[2]Department of Materials Science and Engineering, Massachusetts Institute of Technology, Cambridge, Massachusetts 02139, USA

ABSTRACT

We analyzed the effect of crystallographic anisotropy on the morphological evolution of a 12-nm-thickgoldfilm during solid-state dewetting at high temperatures using automated indexing tool in atransmission electron microscopy. Dewetting initiated at grain-boundary triple junctions adjacent to large grains resulting from abnormal grain growth driven by (111) texture development. Voids at the junctions developed shapes with faceted edges bounded by low-index crystal planes. The kinetic mobility of the edges varied with the crystal orientation normal to the edges, with a predominance of specific edges with the slowest retraction rates as the annealing time was increased.

INTRODUCTION

Thin films on solid surfaces are energetically metastable because of their high surface-area-to-volume ratio, and they will dewet or disintegrate into agglomerates at high temperatures. The dewetting can be guided or controlled at the nanoscale by using templates with topography, pre-patterning the films, or employing an irradiating interfering laser beam;[1–3] this nanoscale control may provide a path to incorporating dewetted films in applications such as patterned magnetic recording media,[4,5] plasmon resonance waveguides,[6,7] plasmon-enhanced photovoltaic devices,[8–11] and catalysts for nanowire growth.[12]

In polycrystalline metallic films, solid-state dewetting (SSD), which occurs below the melting point, starts with hole formation at a grain-boundary triple junction, proceeds with the retraction of the void edges by capillary-driven diffusion, and is completed with the formation of particles due to a Rayleigh-like instability of ligaments.[13,14] This phenomenology has been treated in various analytical models.[15–17]However, because of anisotropy in the surface energy and diffusion coefficients at different edges, the dewetting process is affected by the crystallographic orientation of the original film. This has been mostly explored using single-crystal films, where the surface-energy anisotropy typically results in faceted void edges and affects the edge retraction rate or direction and further pinch-off behavior. For example, for (001)-oriented Si films on silica annealed at 800 °C, facets that are rectangularly bounded by edges parallel to the {110} crystal planes form initially; then, the hole corners rapidly retract and branch to develop star-like-shaped holes.[18–20] Ye and Thompson[21] measured the edge retraction rate as a function of the initial in-plane orientation of the edge using pre-patterned single-crystal Ni films, and presented the results as "a kinetic Wulff plot" that exhibited sharp minima of the retraction rate at certain in-plane orientations.

Citation: Jang, S. A., H. J. Lee, C. V. Thompson, C. A. Ross, and Y. J. Oh. "Crystallographic Analysis of the Solid-State Dewetting of Polycrystalline Gold Film Using Automated Indexing in a Transmission Electron Microscope." APL Materials 3, no. 12 (December 1, 2015): 126103

In polycrystalline films, on the other hand, grain boundaries can provide an additional diffusion path, and texture development and grain growth that are concurrent with the dewetting process complicate our understanding of the effects of the crystallographic anisotropy. To our knowledge, the lack of research on polycrystalline films is due to the limitations of appropriate analytical methods. In this regard, the analysis of Muller and Spolenak using electron backscattered diffraction (EBSD) in conjunction with scanning electron microscopy (SEM) provided a general understanding of the dewetting evolution in light of crystallographic orientations.[22] However, to precisely determine the effects of crystallographic anisotropy and grain boundaries on initial hole growth and edge retraction, a more detailed analysis using higher-resolution analysis tools is needed.

The present study aims to elucidate the effect of crystallographic anisotropy on morphological evolution during dewetting in thin Au films using an automatic crystal orientation and phase mapping system integrated into a transmission electron microscope. Polycrystalline films on non-reactive substrates or templates have been used to investigate the assembly of ordered nanoparticles via the SSD of thin metal films; thus, we expect that the elucidation of a precise crystallographic phenomenology during SSD in polycrystalline films will provide some technological insight to enable control of both the particle shape and size distribution and to further the texturing of agglomerated particles.

A 12-nm-thick Au film was deposited by RF magnetron sputtering onto a 10-nm-thick silicon nitride support films for TEM (PELCO®). The films were then annealed at 400 °C for 1-120 min in a rapid thermal annealing (RTA) furnace in forming gas (Ar + 4%H_2) for *ex situ* TEM observation of solid-state dewetting. A crystal orientation map was acquired from the TEM samples by beam scanning with a step size of 2.4 nm using an automatic crystal orientation and phase mapping system in TEM (NanoMEGAS ASTAR™). For *in situ* observation of dewetting process, some as-deposited Au films were annealed in a heating holder for *in situ* TEM.

Fig. 1 shows the microstructural evolution during heating in *in situ* TEM. A 12-nm-thick Au film on a silicon nitride support film was heated to 400 °C in the *in situ* TEM stage at a rate of 50 °C/min with a 10-min hold at each 100 °C increment; an image was recorded at the end of each hold step. As shown in Fig. 1, at temperatures below 300 °C, the grains grew very slowly and no voids were observed. However, when the temperature was increased to 400 °C, the grains grew rapidly and tiny voids formed at the junctions of the grain boundaries. Newly formed voids, as shown in Fig. 1(c), appeared rectangular, but additional images indicated that they exhibited a circular or irregular shape. The void initiation begins after a substantial amount of grain growth has occurred and when the grain boundary movement has subsided. Upon further annealing in the TEM stage at 400 °C, as shown in Figs. 1(d) and 1(e), the holes grew with highly faceted edges, formed nanowires via narrowing ligaments, and finally decomposed into isolated particles.

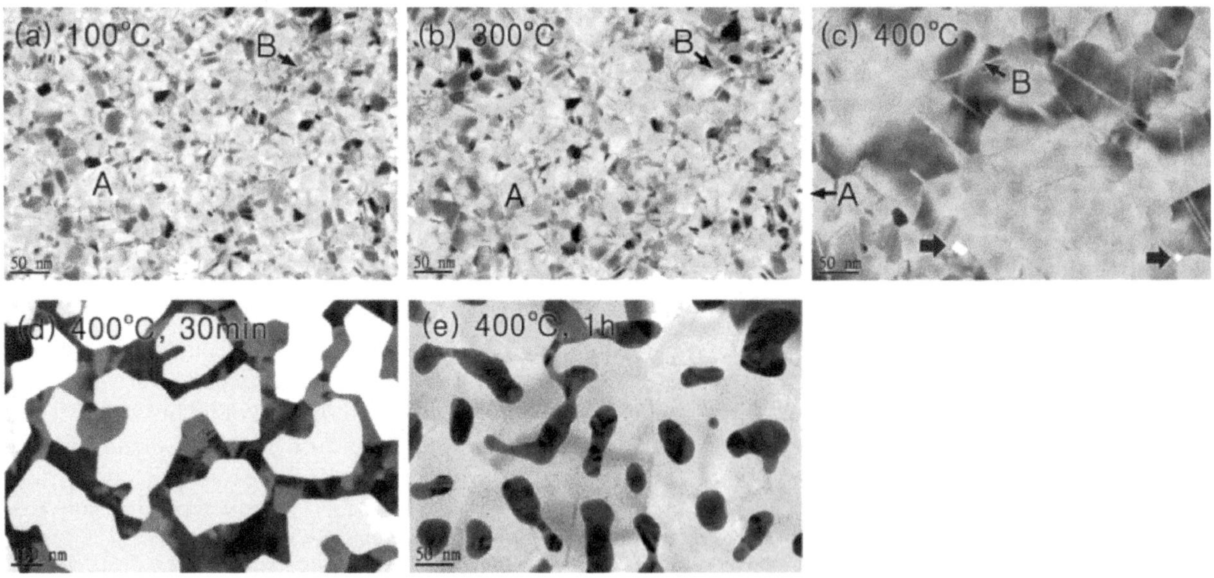

Figure 1. Grain growth in a 12-nm-thick Au film on a 10-nm-thick silicon nitride film support observed during *in situ* annealing in the TEM stage. The temperature of the sample holder was increased in 100 °C increments (with an average increase rate of 50 °C/min) and held for 10 min before a subsequent temperature increase. Images were recorded at the end of the hold time for (a)–(c), but (d) and (e) were acquired immediately after and 30 and 60 min after having reached 400 °C, respectively. A and B in the images indicate the same locations.

Figs. 2(a) and 2(b) present a crystal orientation mapping image obtained in TEM; this image shows the grains in the Au film around an early tiny hole at the grain boundary. Coarse grains with a {111} out-of-plane texture grew abnormally at the expense of fine grains without other textures, and holes formed at triple junctions adjacent to the large grains. According to previous considerations,[23,24] normal grain growth of a film with thickness h stagnates when the grain size reaches ~2.5 h (=R_{stag}), but hole formation occurs at grain boundary grooves[25,26] and requires grain growth above a critical grain size (R_c), which is given by a force balance at the grain boundary junction; thus, for hole formation and growth in the film, R_{stag} should be greater than R_c. In this respect, $R_{stag} < R_c$ is assumed for the fine-grain regions in Fig. 2(a), whereas the hole in the figure was formed during subsequent abnormal grain growth driven by the {111} texture development. Further studies of the crystallographic changes during dewetting were carried out using crystal orientation mapping at various stages of the dewetting process. Figs. 2(c)-2(e) show an orientation mapping of the same film shown in Figs. 2(a) and 2(b), but for a region containing a larger number of small holes (<100 nm). The overall film developed a {111} out-of-plane texture, as shown in the normal orientation map in (e). The hole edges are curved with straight segments, and some edges extend into the film along high-angle boundaries. There are more voids in regions with greater (111) texture, which supports the theory that abnormal grain growth triggers hole formation.

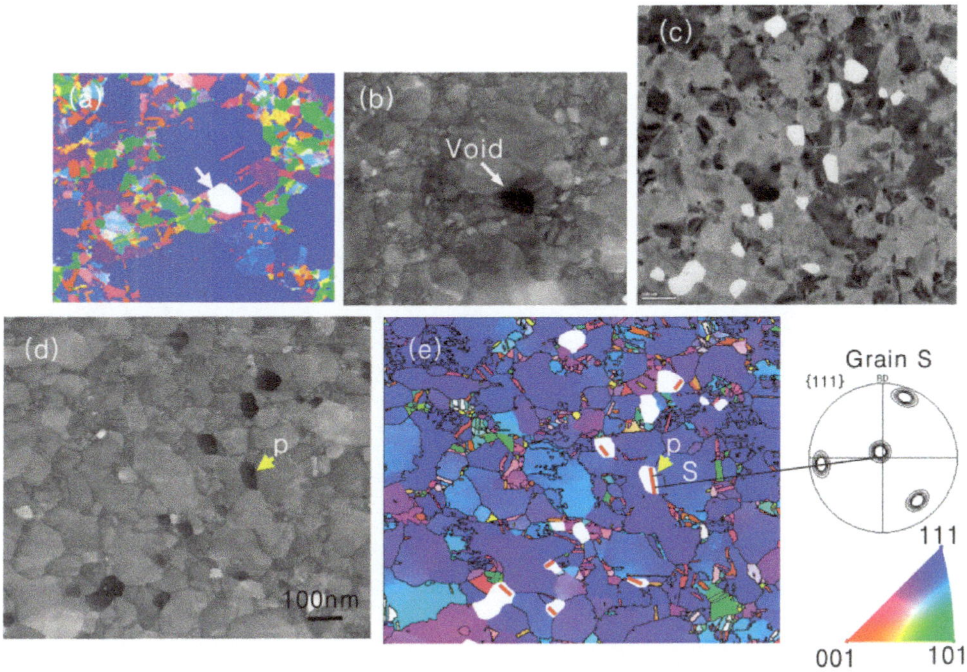

Figure 2. Crystal orientation mapping using an automatic crystal orientation and phase mapping system for TEM around holes in a 12-nm-thick Au film after furnace-annealed for 1 min at 400 °C under an Ar/4%H_2 flow: ((a) and (e)) crystal orientation map (film normal direction); ((b) and (d)) correlation index map; (c) BF-TEM image for (d) and (e). The analysis was performed at two different locations in the film ((a) and (b) and (c)–(e)). The maps ((b) and (d)) emphasize the grain boundaries, and the black lines in (e) denote the grain boundaries with a misorientation ≥10°. The image on the right in (e) is a (111) pole figure analyzed for grain S. If the faceted hole boundary plane, p, is assumed to be parallel to the {111} crystal plane, the line connecting the center to one of the {111} poles in the pole figure for grain S must be normal to the void edge. In this manner, the void edges that are assumed to be parallel to the {111} or {443} crystal planes are denoted by red lines in the holes in (e).

Changes in the in-plane texture itself may affect hole formation in the film. Assuming that the stress effect is negligible, the free energy change[27] that drives dewetting can be given by

$$\Delta E_{dew} = (\gamma_f + \gamma_i - \gamma_s)\,\Delta A_w + \gamma_e \Delta A_e, \qquad (1)$$

where γ_s and γ_f are the surface energies for the substrate and the film parallel to the substrate surface, γ_i is the interfacial energy between the substrate and film, and γ_e is the surface energy of the film edge at a void. ΔA_W and ΔA_e denote the changes in the areas of the film's top surface and edge surface at a void and have negative and positive values in the dewetting process, respectively. For dewetting to occur, the following conditions must be met: $\gamma_f + \gamma_i > \gamma_s$ and $\Delta E_{dew} < 0$. In-plane texture development would lower the values of γ_f and γ_i and thus reduce the driving energy for film dewetting. However, our observations revealed that the holes were mostly formed at large grains with (111) in-plane texture, which implies that the above-mentioned grain size requirement is more important than the effect of the in-plane texture itself. Thermal strain, which is normally compressive, may also affect void formation because it can be relaxed around voids.[24,28–30] However, the role of thermal strain is controversial[24,31] and thus will be dealt in further mechanistic studies with the variation in dewetting temperature.

Fig. 3 presents the crystal orientation map of a film with larger voids, after annealing at 400 °C for 10 min. The film exhibits narrower ligaments but highly faceted voids, and some voids are coalescing. The film was strongly (111) textured. If the faceted void boundaries are assumed to consist of low-index crystal planes,[18,22] the in-plane projections of the low-index crystal planes of the grain around each faceted hole should be perpendicular to the faceted void edge. Fig. 3(b), for example, shows the pole figures obtained for the A and B grains adjacent to the faceted voids. The void edges at A and B are exactly perpendicular to the in-plane projections of the (111) and (110) planes, respectively. Accordingly, the film boundaries at the hole edges of A and B are assumed to be associated at least with the (111) and (110) crystal planes, respectively. However, the edges perpendicular to the in-plane projection of (111), such as the edge at A, can also be perpendicular to other low-index crystal planes. See supplementary material[32] for detailed information on the crystal planes that bound the void edges. Using this procedure, the crystallographic plane of the faceted edges can be determined and represented by numbers, as illustrated in Fig. 3(b). The (111) plane was prevalent, but other low-index planes were also observed. Most of the faceted hole boundaries in Fig. 3(a) were indexed, and their frequencies are plotted as a bar chart in Fig. 3(c). Adjacent plane orientations were grouped, as denoted by different colors, and the planes that could not be grouped are presented in gray. The surface energy of each plane relative to that of the (111) plane was obtained from the literature[33,34] and is also listed in the figure. As shown in Fig. 3(c), the (111) and (443) planes, with the lowest and second-lowest energies per area, are the most prevalent and account for 44.5% of the total hole boundary length. Using the same procedure, we analyzed the portion of the two lowest energy planes of the total boundary length for the voids in Fig. 2(e) with a shorter annealing time: the hole boundaries with (111) or (443) are denoted as red lines in the holes in Fig. 2(e) and account for 15% of the total hole boundary length. This result may suggest that the hole boundaries with higher energy retracted faster and gradually disappeared, leaving the lower energy boundaries. The energy consideration in Eq. (1) also supports this conclusion: the surface energy at the film edge becomes prominent as dewetting proceeds, and hence a lower γ_e, at fixed γ_f and γ_i, increases the driving energy for dewetting.

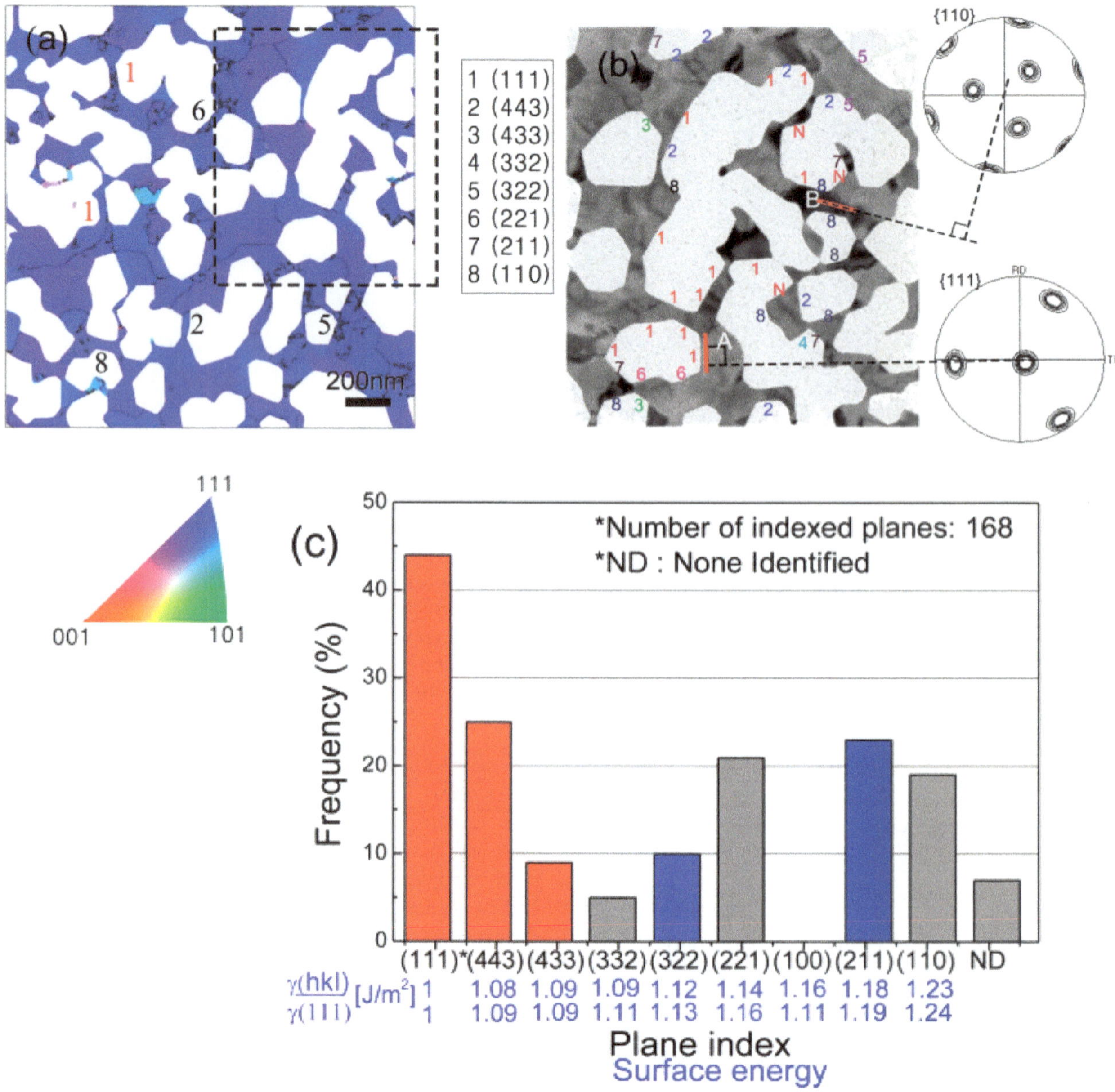

Figure 3 A crystal orientation map (film normal direction) (a) of a 12-nm-thick Au film after annealing for 5 min at 400 °C under flowing Ar/4%H_2in a furnace. The numbers in the BF image (b) represent the crystal planes that are indexed for the faceted hole boundaries using the pole figures of the grains around the holes. The result is plotted in (c). The surface energies in (c) are sourced from Refs. 33 and 34. The numbers in (a) also represent the crystal planes of the film boundary of the narrow ligaments. Image (b) corresponds to the area indicated in (a).

The plot in Fig. 3(c), however, shows that the faceted hole boundaries still develop a large portion of other crystal planes with somewhat higher surface energies, and a correlation between the surface energy and plane percentage is not evident except for (111) and (443). As shown in Fig. 3(a), some ligaments formed nanowires with a uniform width in which both edge boundaries consisted of the same low-index crystal planes. In Fig. 3(a), the edge boundaries are parallel to the planes (322), (221), or (211), in addition to (111) and (443). This trend indicates that the stability of a hole facet is not simply determined by its surface energy. In Ye and Thompson's result obtained using a pre-patterned single-crystal Ni film on MgO,[21] the edge retraction rate exhibited sharp minima at certain initial in-plane orientations of the edge. A two-dimensional edge retraction model for highly anisotropic faceted edges in the same film included other important contributions to the edge retraction rate, such as the surface self-diffusivity on the top facets of the rim, the film thickness, and the film contact angle on the substrate, in addition to the surface energies.[35] In the present work on Au films on nitride, it is uncertain which factor is the most important for determination of the retraction rate of the hole facets. However, it is clear that the relative frequency of the plane indexes normal to the edges shown in Fig. 3(c) can be regarded as a relative measure of the kinetic stability of the edges against dewetting.

Fig. 4 compares the shapes of the corners of faceted holes. As shown in the top image of the holes in Fig. 4(b), the hole corners at grain boundaries appear sharp, whereas the corners in the grains are blunt. Figs. 4(a) and 4(c) present high-resolution images in which the beam direction is parallel to the [011] crystal orientation of the film. The hole edges are parallel to low-index crystal planes. In Fig. 4(a), the in-plane directions normal to the two faceted edges in the grain are [11 $^-$111$^-$1] and [111$^-$ 111$^-$], and the corner of the edges is rounded by monoatomic steps in the {111} atomic layers. This rounding can be understood in view of the decrease in chemical potential with increasing curvature of the corner. However, the corner at the grain boundary is not rounded or intrusive along the grain boundary, as shown in Fig. 4(c), in contrast to the intrusive corner of the early tiny voids shown in Figs. 2(c)-2(e). Hence, unlike the early stage of void growth, after the voids form edges with a low-index in-plane normal direction, the grain-boundary diffusion appears to no longer play a significant role in void growth.

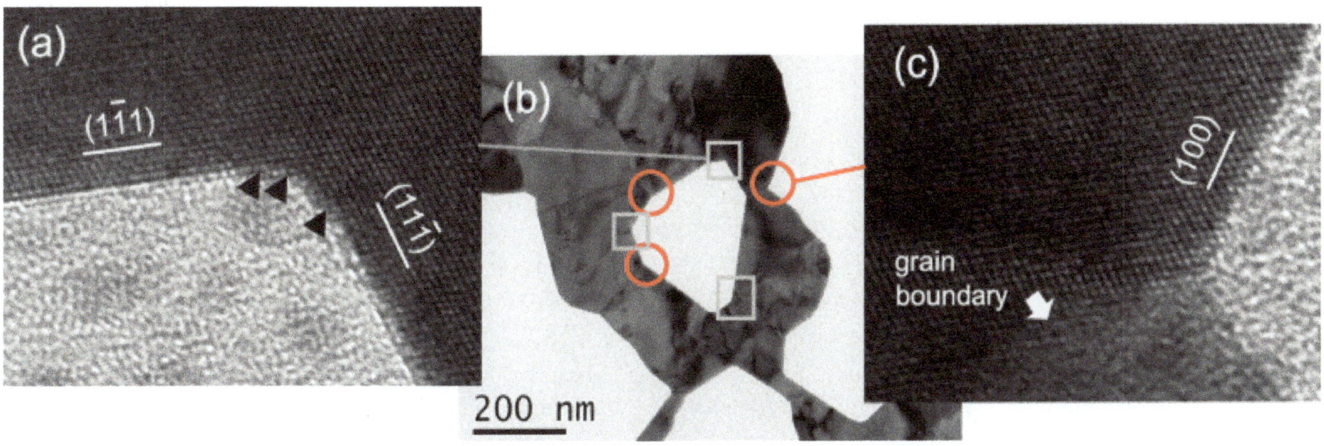

Figure 4 TEM images of hole corners with and without a grain boundary. The arrows in (a) indicate atomic (111) steps at the corner. The circles and rectangles indicate corners with and without a grain boundary, respectively.

In summary, we have reported the effects of crystal orientation on the dewetting process of polycrystalline Au films, as determined using an automated crystallographic orientation and indexing tool incorporated into a TEM. Void initiation at the grain boundaries generally requires grain growth above a critical size. In this study, hole formation initiated after abnormal grain growth driven by (111) texture development. The holes initially grew with faceted edges lying along different low-index crystal planes. The retraction rate of the edges varied with the crystal orientation normal to the edges, which resulted in the predominance of edges with specific crystal planes as the annealing time was increased. This observed phenomenology can provide insight into strategies for nanoscale control in the solid-state dewetting processes of metallic thin films with strong texture development and surface energy anisotropy for applications in patterned magnetic recording media, catalysts, and photonic devices.

ACKNOWLEDGMENTS

This work was partly supported by the National Science Foundation under Contract No. DMR1104610 and a National Research Foundation of Korea (NRF) grant funded by the Korean government (MEST) (No. 2010-0010001). We thank Dr. S. H. Kang in KAERI for supporting a crystallographic analysis. We are also grateful for the automated crystallographic orientation and indexing facility in TEM provided by Korea Basic Science Institute (KBSI) and *in situ* TEM by National Nanofab Center (NNFC).

REFERENCES

1. Y. J. Oh, J. H. Kim, C. V. Thompson, and C. A. Ross, *Nanoscale* **5**, 401 (2013). http://dx.doi.org/10.1039/C2NR32932H
2. J. D. Fowlkes, L. Kondic, J. Diez, Y. Wu, and P. D. Rack, *Nano Lett.* **11**, 2478 (2011).http://dx.doi.org/10.1021/nl200921c
3. J. Trice, C. Favazza, D. Thomas, H. Garcia, R. Kalyanaraman, and R. Sureshkumar, *Phys. Rev. Lett.* **101**, 017802 (2008).http://dx.doi.org/10.1103/PhysRevLett.101.017802
4. Y. J. Oh, C. A. Ross, Y. S. Jung, Y. Wang, and C. V. Thompson, *Small* **5**, 860 (2009).http://dx.doi.org/10.1002/smll.200801433
5. C. A. Ross, *Annu. Rev. Mater. Res.* **31**, 203 (2001). http://dx.doi.org/10.1146/annurev.matsci.31.1.203
6. S. A. Maier, P. G. Kik, and H. A. Atwater, *Phys. Rev. B* **67**, 205402 (2003).http://dx.doi.org/10.1103/PhysRevB.67.205402

7. A. Femius koenderink, *Nano Lett.* **9**, 4228 (2009).http://dx.doi.org/10.1021/nl902439n
8. E. Moulin, J. Sukmanowski, P. Luo, R. Carius, F. X. Royer, and H. Stiebig, *J. Non-Cryst. Solids* **354**, 2488 (2008). http://dx.doi.org/10.1016/j.jnoncrysol.2007.09.031
9. K. R. Catchpole and S. Pillai, *J. Lumin.* **121**, 315 (2006). http://dx.doi.org/10.1016/j.jlumin.2006.08.053
10. Yu. A. Akimov, K. Ostrikov, and E. P. Li, *Plasmonics* **4**, 107 (2009).http://dx.doi.org/10.1007/s11468-009-9080-8
11. T. L. Temple, G. D. K. Mahanama, H. S. Reehal, and D. M. Bagnall, *Sol. Energy Mater. Sol. Cells* **93**, 1978 (2009).http://dx.doi.org/10.1016/j.solmat.2009.07.014
12. Y. F. Guan, R. C. Pearce, A. V. Melechko, D. K. Hensley, M. L. Simpson, and P. D. Rack, *Nanotechnology* **19**, 235604 (2008).http://dx.doi.org/10.1088/0957-4484/19/23/235604
13. D. Kim, A. L. Giermann, and C. V. Thompson, *Appl. Phys. Lett.* **95**, 251903 (2009).http://dx.doi.org/10.1063/1.3268477
14. R. Saxena, M. J. Frederick, G. Ramanath, W. N. Gill, and J. L. Plawsky, *Phys. Rev. B* **72**, 115425 (2005).http://dx.doi.org/10.1103/PhysRevB.72.115425
15. D. J. Srolovitz and S. A. Safran, *J. Appl. Phys.* **60**, 255 (1986).http://dx.doi.org/10.1063/1.337691
16. R. H. Brandon and F. J. Bradshaw, *Technical Report*, Royal Aircraft Establishment, Farnborough,1966.
17. H. Wong, P. W. Voorhees, M. J. Miksis, and S. H. Davis, *Acta Mater.* **48**, 1719 (2000).http://dx.doi.org/10.1016/S1359-6454(00)00016-1
18. P. Sutter, W. Ernst, Y. S. Choi, and E. Sutter, *Appl. Phys. Lett.* **88**, 141924 (2006).http://dx.doi.org/10.1063/1.2186741
19. R. Nuryadi, Y. Ishikawa, and M. Tabe, *Appl. Surf. Sci.* **159-60**, 121 (2000).http://dx.doi.org/10.1016/S0169-4332(00)00051-9
20. J. Ye and C. V. Thompson, *Appl. Phys. Lett.* **97**, 071904 (2010).http://dx.doi.org/10.1063/1.3480419
21. J. Ye and C. V. Thompson, *Acta Mater.* **59**, 582 (2011). http://dx.doi.org/10.1016/j.actamat.2010.09.062
22. C. M. Muller and R. Spolenak, *Acta Mater.* **58**, 6035 (2010).http://dx.doi.org/10.1016/j.actamat.2010.07.021
23. H. J. Frost, C. V. Thompson, and D. T. Walton, *Acta Metall. Mater.* **38**, 1455 (1990).http://dx.doi.org/10.1016/0956-7151(90)90114-V
24. C. V. Thompson, *Annu. Rev. Mater. Res.* **42**(1), 9 (2012).http://dx.doi.org/10.1146/annurev-matsci-070511-155048
25. E. Jiran and C. V. Thompson, *J. Electron. Mater.* **19**, 1153 (1990).http://dx.doi.org/10.1007/BF02673327
26. F. Y. Génin, W. W. Mullins, and P. Wynblatt, *Acta Metall. Mater.* **42**, 1489 (1994).http://dx.doi.org/10.1016/0956-7151(94)90167-8
27. K. Thürmer, E. D. Willaims, and J. E. Reutt-Robey, *Phys. Rev. B* **68**, 155423 (2003).http://dx.doi.org/10.1103/PhysRevB.68.155423
28. R. Abermann, *Vacuum* **41**, 1278 (1990).http://dx.doi.org/10.1016/0042-207X(90)93933-A
29. R. P. Vinci, E. M. Zielinski, J. C. Bravman, and W. D. Nix, *J. Appl. Phys.* **88**, 1389 (2000).http://dx.doi.org/10.1063/1.373828
30. D. J. Srolovitz and M. G. Goldiner, *JOM* **47**, 31 (1990). http://dx.doi.org/10.1007/BF03221433
31. P. R. Gadkari, A. P. Warrant, R. M. Todi, R. V. Petrova, and K. R. Coffey, *J. Vac. Sci. Technol., A* **23**, 1152 (2005).http://dx.doi.org/10.1116/1.1861943
32. See supplementary material at http://dx.doi.org/10.1063/1.4937432 for detailed information on the crystal planes that bound the void edges.
33. G. D. Barmparis and I. N. Remediakis, *Phys. Rev. B* **86**, 085457 (2012).http://dx.doi.org/10.1103/PhysRevB.86.085457
34. Y. N. Wen and J. M. Zhang, *Solid State Commun.* **144**, 163 (2007).http://dx.doi.org/10.1016/j.ssc.2007.07.012
35. R. V. Zucker, G. H. Kim, W. C. Carter, and C. V. Thompson, *C. R. Phys.* **14**, 564 (2013).http://dx.doi.org/10.1016/j.crhy.2013.06.005

CHAPTER 4

DEFECT SCIENCE AND ENGINEERING OF LIQUID CRYSTALS UNDER GEOMETRICAL FRUSTRATION

Takeaki Araki [a], Francesca Serra [b] and Hajime Tanaka [c]

[a]Department of Physics, Kyoto University, Sakyo-ku, Kyoto 606-8502, Japan.

[b]Dipartimento di Chimica, Biochimica, Biotecnologie per la Medicina, Università di Milano, Milano, Italy

[c]Institute of Industrial Science, University of Tokyo, Komaba 4-6-1, Meguro-ku, Tokyo 153-8505, Japan.

ABSTRACT

Spontaneous symmetry breaking while preserving flow ability is a remarkable feature of nematic liquid crystals. When a nematic liquid crystal coexists with a solid, the surface field of the solid tends to anchor the director direction on the surface: anchoring effects. If geometrical frustration between nematic ordering and anchoring is strong enough, stable topological defects are formed. Defects in an ordered state are usually regarded as undesirable features. However, recent studies reveal that defects stabilized by a topological constraint from the solid surfaces are actually quite useful and open up novel possibilities for defect engineering of liquid crystals: self-organization of soft matter by defects, memory effects of topological origin, and control of flow of nematic liquid crystals and colloid motion by defects. For example, defect reconfiguration accompanying the change in the topology costs a very high energy far beyond the thermal energy, which overwhelms a typical energy scale in soft matter. This provides extreme stability for structures assembled by defects and information memorized in defect topology. Furthermore, effects of topological defects can easily be removed perfectly by a nematic-to-isotropic transition, which provides switchable functions. Defects also affect the motion of colloids immersed in a liquid crystal and flow behaviour of a liquid crystal. Here we review recent developments in science and engineering of topological defects in nematic liquid crystals, mainly based on our numerical simulation studies.

INTRODUCTION

Liquid crystals have spontaneous orientational order, which gives rise to elasticity, but the lack of positional order provides an ability to flow.[1] For example, a nematic liquid crystal (NLC) is a state of spontaneously broken "continuous orientational symmetry" of the isotropic liquid. Orientational order in NLCs is described by a tensorial ordering field.[1] Hydrodynamic Goldstone modes appear as a consequence of the broken continuous symmetry, restoring the lost continuous symmetry of the isotropic disordered liquid.[2–4] These modes in NLCs are thermal fluctuations of the director about its equilibrium value. The modes are "massless" and their excitation does not cost any energy for vanishing wavenumber. This results in large fluctuations of the local optical axis, which scatter light very strongly. This is why NLCs look turbid and milky if there is no external field selecting the direction of the director. Another important consequence of the continuous symmetry breaking is the appearance of topological defects. Topological defects are formed when the orientational field is strongly distorted and at defect cores molecules are oriented randomly as in the isotropic disordered liquid. Topological defects in liquid crystals have

Citation: Takeaki Araki, Francesca Serra and Hajime Tanaka. Defect science and engineering of liquid crystals under geometrical frustration. Soft Matter, 2013,9, 8107-8120, DOI: 10.1039/C3SM50468A attracted considerable attention in view of analogies to other systems of spontaneous broken symmetry, such as magnetism, superfluidity, and cosmology.[2,5,6] The universality ensures that NLCs are model experimental systems for physics of topological defects, since we can easily visualize defects in NLCs.[7]

Besides the fundamental importance, the coexistence of the elasticity associated with orientational order and fluidity provides liquid crystals with fantastic dynamical characteristics, which not only have many technological applications but also may play important roles in biological functions. In conventional applications of liquid crystals, defects are considered to be unfavourable as their name suggests. For instance, defects in a liquid crystal display deteriorate its optical quality. Recently, however, topological defects have attracted much attention since they provide interesting applications. For example, a cholesteric blue phase, in which disclination lines (DLs) are aligned regularly with a cubic symmetry, is utilized in displays because of its fast response.[8] Also, effective interactions among colloidal particles mediated by defects fabricate self-assembled braids.[9,10]

Topological defects are in general formed spontaneously upon a symmetry breaking transition from a disordered state, but the amount of defects tends to decrease with time so as to reduce the total free energy of the system, since the local free energy density in the vicinity of defect cores is quite high.[11] However, topological defects can be stabilized under frustration. For example, when molecules have strong chirality, the director field tends to twist spontaneously. Since the twisted director field cannot fill a three dimensional space, exotic frustrated mesophases stabilized by defects, such as cholesteric blue phase and twist grain boundary phase, are formed.[2,3] In these es, frustration is of intrinsic origin.

Confinement of liquid crystals in geometries incompatible with the orientational symmetry also imposes frustration to the director field. Interactions between solid surfaces and molecules tend to align the molecules adjacent to the surfaces. When this anchoring interaction is large enough and the surfaces are curved incommensurately to the symmetry preferred by nematic order, stable topological defects are formed. Besides surface anchoring, a non-uniform electric field around charged particles can also induce a sort of anchoring effects and lead to the formation of stable defects.[12]

The stability of these topological defects formed by extrinsic geometrical frustration has many interesting features, reflecting its topological origin. For example, the stability is extremely high due to the constraint coming from the neutrality of topological charge; in other words, the change in the topology of a defect structure involves a very high energy barrier, which is far beyond the thermal energy k_BT. Since structures of soft matter are usually stabilized by an energy scale of k_BT, interactions stemming from topological defects are exceptionally strong. This allows us to construct a very stable robust structure. Interestingly, however, we can erase defects easily by phase transition from an ordered state to a disordered state, which provides the switchability of defect-mediated interactions. Another unique feature originates from the ability of liquid crystals to flow. Interestingly, the flow is significantly affected by the presence of topological defects, which elastically interact with solid surfaces. This provides us with novel possibilities of controlling the colloid motion in terms of the surface anchoring property of colloidal particles and controlling flow behaviour by using a solid surface topology. Thus the active use of topological defects in liquid crystals offers a promising route to "defect engineering" or "defect-based technology". In this article, we will mainly review our numerical studies on NLC systems, in which DLs are stabilized and entangled to objects imposing anchoring conditions.

This article is organized as follows. In Section 2, we review the basic knowledge on the stabilization mechanism of topological defects. In Section 3, we describe interactions mediated by entangled DLs in colloidal particles immersed in a NLC. These numerical studies are based on a fluid particle dynamics (FPD) method.[13] In Section 4, we show our recent Monte Carlo studies on the slow glassy behaviour and the memory effect, which are observed in NLCs confined in porous media. Although these two systems have so far been discussed separately, we discuss them together from the viewpoint of 'defect topology' in a unified manner. In Section 5, we discuss a dynamical coupling between flow and defects in NLCs, based on nematohydrodynamics. In Section 6, we will summarize our paper.

STABILIZATION OF TOPOLOGICAL DEFECTS BY GEOMETRICAL FRUSTRATION

Here we describe the basics of the stabilization of defects in liquid crystals by geometrical frustration. We consider only the es where the director field of NLCs prefers to be aligned normal to the surface: the so-called homeotropic anchoring. Since the stability of defects comes from geometrical frustration between the topology of solid surfaces and the symmetry of liquid crystals, we first discuss the topology of solid surfaces.

Topology of solid surfaces

The simplest topology of the surface is that of a sphere. A rod[14,15] or a building block[16] are interesting modifications of a sphere since it has orientational axes, which can be coupled with that of a NLC. However, these are topologically equivalent to a sphere. There are a number of isolated objects which have nontrivial topological characteristics such as a doughnut structure. Such interesting es of topological colloids have recently been reported by Senyuk et al.,[17] which open up further possibility of defect engineering. Besides these isolated closed shapes, there are many topological manifolds, which can be realized as porous solid materials. Some of these structures are shown in Fig. 1.

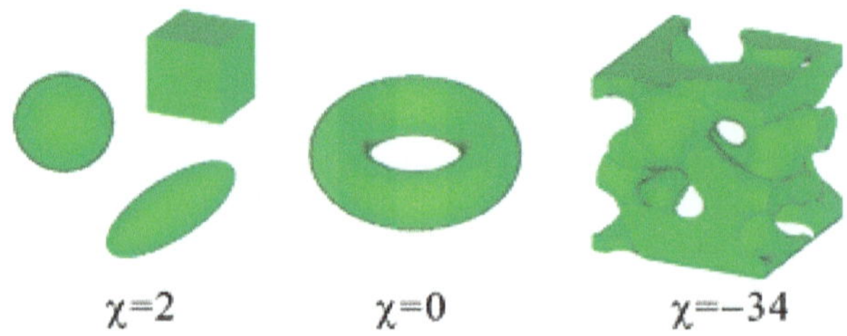

Figure 1 Examples of solid objects opening up possibilities of defect engineering. A sphere, ellipsoid, and block have Euler characteristics $\chi = 2$, whereas a torus has $\chi = 0$. A porous material usually has a large negative value of χ.

The characteristic length scale of such a structure is also important. For ordinary anchoring conditions, it should be in the order of submicrons to microns. Colloidal particles are usually used for isolated particles. A recent development allows us to make colloids with various shapes and even patchy particles[18,19] with inhomogeneous anchoring conditions.[20] On the other hand, bicontinuous porous structures may be formed by phase separation, self-organization of surfactants, or bottom-up optical polymerization techniques. Random porous structures are formed by using ordinary spinodal decomposition. We can add volume asymmetry and the resulting surface curvature asymmetry and/or anisotropy by using viscoelastic phase separation.[21]Controlled formation of ordered structures is more difficult. Templating mesoscopic self-organized ordered structures of amphiphilic systems[22] may be one promising way. Applications of two-photon polymerization are also now under investigation.[23]

Stabilization of defect structures of NLCs by a constraint from solid surfaces

Now we discuss how topological defects are stabilized by geometrical frustration imposed by the above mentioned structured solids. First, let us consider a spherical particle immersed in a NLC. If the particle surface imposes the homeotropic anchoring condition, the sphere behaves as a radial point defect having the topological charge s = +1 as shown in Fig. 2(a). To compensate this topological charge, a defect with s = −1 should be formed.[24–26] Two types of defects can satisfy this condition: a hyperbolic point defect (Fig. 2(a)) and a DL loop with s = −1/2 (Fig. 2(b)).[27] Since defects with positive s interact attractively with those with negative s, these interactions lead to the formation of defect-particle pairs with dipolar or quadrupolar symmetry, as shown in Fig. 2(a) and (b). The latter is called a Saturn-ring defect. Depending on the symmetry of the particle-defect complex, it further interacts with others; the former dipolar complexes in Fig. 2(a) spontaneously form chain-like aggregates which are parallel along the background orientation.[9,26,28] A particle accompanying the Saturn-ring defect, on the other hand, interacts with others through quadrupolar interactions.[29–31] Hence, zigzag aggregates or chain-like ones, the long-axis of which is tilted from the background director field, are formed. These effective interactions are well explained by a theory with far-field approximations, in which the particle-defect complexes are treated as separated objects.[26,32] Recently, new interactions among particles in NLCs have been found.[33–41] These interactions are mediated by entangled DLs, so that they cannot be described simply by the far-field approximation. We also numerically discovered a new stable 'figure of eight' configuration.[42] In Section 3, we will provide a brief review on this numerical study.[13]

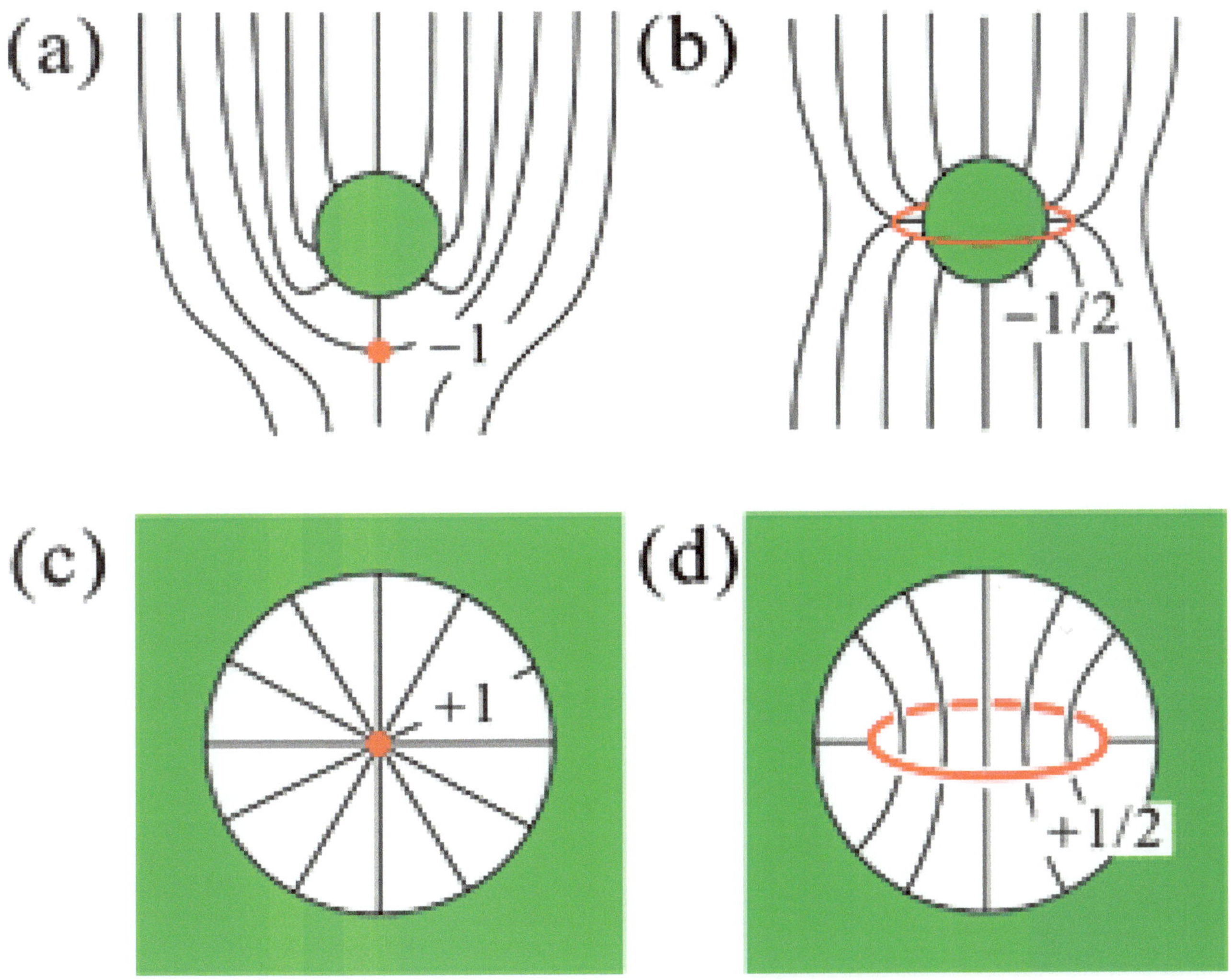

Figure 2 Possible defect structures of nematic liquid crystals around a spherical particle ((a) and (b)) and those in a spherical cavity ((c) and (d)).

Incorporating liquid crystals into cavities, porous materials, and microfluidic devices also leads to the formation of stable topological defects. In a spherical cavity with strong homeotropic anchoring, for instance, a radial point defect with $s = +1$ or a loop of DL with $s = +1/2$ are formed as shown in Fig. 2(c) and (d).[43,44] They are the counterparts of the particle-defect complexes in Fig. 2(a) and (b). In a cylindrical channel with strong homeotropic anchoring, one or two DLs run along the channel.[45–47] At a Y-shaped bifurcation, the trajectory is not uniquely determined, so that the DLs should choose one of the possible paths. We can easily imagine that defects are arranged in more entangled complex configurations for more interwoven networks.[43,44,48,49]

Provided the complete homeotropic condition in cavities, the argument developed by Stein and based on the Gauss–Bonnet theorem tells us that the sum of the topological charges of DLs remaining in a complex geometry has to coincide with a half of the Euler characteristics of the confining geometry.[50] This condition is expressed by the following equation,

$$\sum_i s_i = \chi/2,$$

where s_i is the topological charge of the i-th defect. Here χ is the Euler characteristic of the surface structure. Although the theorem was originally proved for simple, close surfaces, it was later extended to complex and open networks,[49] with the study of a system where only point defects were present. However, in the e of systems of DLs it is not easy to determine the topological charges of DLs in three dimensional systems; therefore there are difficulties in exploiting this theorem to characterize DLs remaining in complex geometries.

It has been reported that a NLC confined in a porous medium exhibits glassy slow dynamics and memory effect.[51–58] The NLCs in porous media have been studied numerically with a spin model, i.e., Lebwohl–Lasher[59] model.[54,56,60–64] Since the effects of the confinements were often introduced by quenched disorders, i.e., by fixing the spins at randomly chosen isolated lattice sites, the roles of the topology of the porous network have not been considered explicitly. Our previous study on the colloid dispersions in NLCs[42] invoked the importance of the connectivity of DLs in porous media. This motivated us to study the physical mechanism of the slow glassy dynamics and the memory effect and the roles of topology in these behaviours. This problem will be discussed in Section 4.

It is also interesting to consider a situation where the characteristic length scale of particles or pore size is very small and not so different from a defect core size. In such a situation, defects may be difficult to form. Liquid crystal microemulsions may correspond to such a e.[65] In relation to this, liquid crystals including nanoparticles and included in nanoporous materials are also an interesting topic. However, we do not discuss them in this article.

COLLOIDS IN NEMATIC LIQUID CRYSTALS

Theoretical model and simulation method

In this section, we consider colloidal particles suspended in NLC solvents. There are many interesting topological shapes of colloids such as doughnuts or knots, but here we consider the simplest shape, i.e., a sphere. Our model for this part is based on a coarse-grained description on NLCs.[1] The nematic phase is described with a tensorial order parameter $Q_{\alpha\beta}$,[1] where α and βrepresent the Cartesian coordinates, x, y and z. When the nematic phase is uniaxial, $Q_{\alpha\beta}$ is expressed by using a scalar nematic order parameter Q and the director vector n_α as

$$Q_{\alpha\beta} = \frac{3}{2}Q\left(n_\alpha n_\beta - \frac{1}{3}\delta_{\alpha\beta}\right).$$

Then the bulk state of the nematic phase is well described by the Landau–de Gennes free energy density:

$$f_{\rm LdG}(Q_{\alpha\beta}) = \frac{A}{2}Q_{\alpha\beta}Q_{\beta\alpha} - \frac{B}{3}Q_{\alpha\beta}Q_{\beta\mu}Q_{\mu\alpha} + \frac{C}{4}(Q_{\alpha\beta}Q_{\beta\alpha})^2 + \frac{L_1}{2}(\nabla_\mu Q_{\alpha\beta})^2 + \frac{L_2}{2}(\nabla_\alpha Q_{\alpha\beta})^2 - \Delta\varepsilon E_\alpha E_\beta Q_{\alpha\beta},$$

where the repeated suffixes indicate the summation over the Cartesian coordinates. B, C, L_1 and L_2 are positive material constants. A depends on the temperature T and is approximated as $A = A_0(T - T^*)$ with $A_0 > 0$. The bulk isotropic–nematic transition temperature is given by $T_{IN} = T^* + B^2/(27A_0C)$. The cubic term of $Q_{\alpha\beta}$ represents the first order nature of the transition. The fourth and fifth terms of the right-hand side of eqn (3) are related to the Frank elastic energy. The last term represents the effect of an external (electric or magnetic) field E_α and $\Delta\varepsilon$ is the anisotropy of the susceptibility.

If a NLC is in contact with solid surfaces, the free energy of the system is written as

$$\mathscr{F}_{\rm NLC} = \int_{\mathscr{V}} dr f_{\rm LdG}(Q_{\alpha\beta}) - W\int_{\mathscr{S}} ds Q_{\alpha\beta} d_\alpha d_\beta,$$

where $\int_{\mathscr{V}} dr$ and $\int_{\mathscr{S}} ds$ represent the integrals over the volume filled with the NLC and the surfaces adjacent to the nematic phase, respectively. The second term in the right-hand side of eqn (4) expresses the anchoring effect and W is the energetic cost for the anchoring per unit area. d_α is the α component of a unit vector normal to the surface. For $W > 0$, the director field prefers to align perpendicular to the surface, that is, the homeotropic anchoring we consider in this article. If the particle positions are fixed, one can obtain an equilibrium director pattern by minimizing eqn (4) with respect to $Q_{\alpha\beta}$.[26,66]

Since interactions between particles mediated by the elasticity of the nematic solvent intrinsically have many-body nature, it is very difficult to describe the forces acting on the particles analytically. Therefore, numerical minimization of the free energy that incorporates the development of the particle position has been employed to study the dynamics of colloidal dispersionin NLC solvents.[41,67] We also developed a numerical method for simulating the dynamics of colloid–NLC suspensions[42,68] including hydrodynamic effects by introducing $Q_{\alpha\beta}$ to a 'fluid particle dynamics (FPD)' method.[13,69] Our method enables us to simulate full dynamics under a coupling among particles, nematic orientational order and flow field, which are the three relevant physical variables to describe this system.[42,68] The detailed simulation method is described in ref. 42.

Binding of Colloidal Particles by Topological Defects

Now we consider nontrivial defect-mediated interactions between colloid particles immersed in a NLC. Following an argument based on the Frank elastic energy, the configuration of the accompanied defect transforms from the Saturn-ring (Fig. 2(b)) to the dipole (Fig. 2(a)) as the size of the particle is increased.[26] This prediction was confirmed experimentally.[70] In this article, we consider only the situation where each particle has a Saturn-ring defect if it is isolated.

Binding of a Particle Pair

First we consider two-body interaction. We initially place two particles of radius a = 12 (see ref. 42 for the definition of the unit length) on the x axis with an interparticle separation of $\Delta x = 26$. We use a pre-nematic state $Q_{zz} = -2Q_{xx} = -2Q_{yy} = 0.1$ as an initial condition for $Q_{\alpha\beta}$ in order to orient the background nematic order along the zaxis. Then, we quench the system to a nematic phase ($A < 0$) at $t = 0$. The equilibrium scalar order parameter has a value of $Q_B = 0.485$ in the absence of the particles. We also introduce thermal orientational fluctuations $\delta Q_{\alpha\beta}$, whose amplitude is $|\delta Q_{\alpha\beta}| = 10^{-2}$.

After equilibration ($t = 500t_Q$), where t_Q is the characteristic rotational time, we found that two topologically distinct defect configurations can be formed. Fig. 3(a) shows one of these configurations. In the figure the spheres and thick lines represent the particles and DLs, respectively. The DLs are detected by using the isosurface of the elastic energy density $f_{el} = L_1(\nabla_\mu Q_{\alpha\beta})^2/2 + L_2(\nabla_\alpha Q_{\alpha\beta})^2/2$. Therefore, the thickness of the lines roughly corresponds to the core size of defects. Here each particle accompanies a DL loop like an isolated particle (Fig. 2(b)). The two particles tend to get close to one another, accompanying bending deformations of the DLs to avoid their contact. In our previous study, we reported that the angle between the direction of the background director orientation and the line connecting the particle centers is about 50°.[42] In our recent simulation without an external field, we found the angle to be about 90°, consistent with numerical observations of Ravnik and Žumer.[37] The difference in the angle may, thus, be due to the effects of an external field.

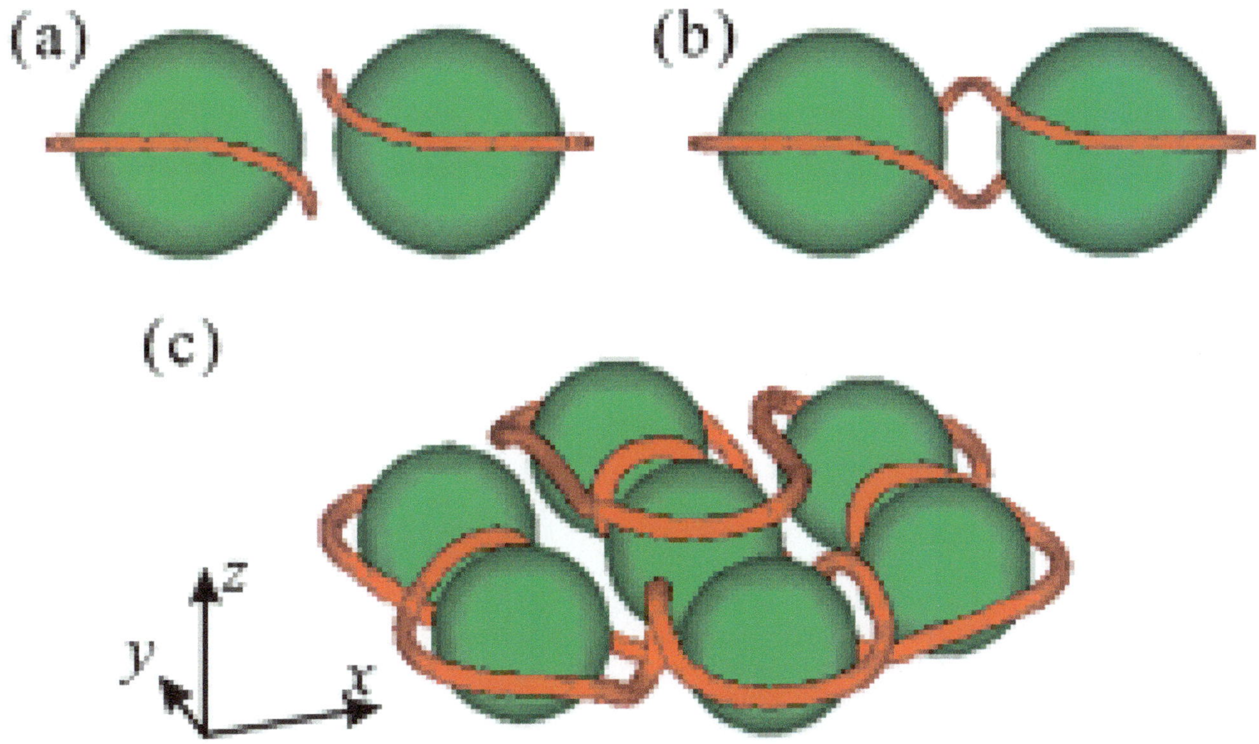

Figure 3 Simulated defect patterns around a pair of spherical colloid particles. The spheres represent the particles and the thick lines are DLs, whose topological charges are $s = -1/2$: energetically, the two Saturn-ring structure in (a) is preferred with respect to the 'figure of eight' structure in (b). Owing to an energy barrier between the two configurations, however, once the 'figure of eight' structure is formed, it can be kinetically stabilized. (c) A kinetically stabilized aggregate of seven particles bound by a single-stroke DL. The particles are likely placed in a plane perpendicular to the background director field.

In Fig. 3(b), we show the other stable defect configuration. Here, only a single DL loop having a 'figure of eight' structure remains.[42] Since the topology of the DL is different from that of the other configuration in Fig. 3(a), this 'figure of eight' structure cannot transform continuously into the other one. The DL loop tends to shrink to reduce the elastic energy cost, but cannot cross itself because of a high energy barrier for such a topological change. We can say that the particles are topologically arrested by the single closed DL loop. The particle centers sit on a plane perpendicular to the background director field. Interestingly, this defect structure has a chirality, although all the components do not have their own chiralities.[39,42] The two types of chiral structures are energetically degenerated and can thus be formed with an equal probability.

We confirmed that the stored elastic energy of the configuration in Fig. 3(a) is lower than the sum of the stored energies for two isolated particles. Since there is no energy barrier associated with this transformation, isolated particles can smoothly form the configuration in Fig. 3(a). On the other hand, the stored elastic energy of the 'figure of eight' defect structure inFig. 3(b) is much higher than that of the configuration in Fig. 3(a).[42] In this sense, the 'figure of eight' structure is a metastable configuration. Once this structure is formed, however, it remains as it is due to a large barrier associated with the discontinuous topological change in the defect structure, which is inevitably required for the transformation. So, we may say that this structure is not selected energetically, but kinetically. In other words, we need to form this structure from an isotropic state of NLC. Indeed, this structure can be created by melting transiently the nematic phase around the particles with laser heating in experiments.[36] We estimated the energy barriers to be of the order of 10^2 to $10^3 k_B T$,[42] indicating that this particle binding by a disclination line is extremely strong.

Shortly after our finding but independently, Žumer, Muševič and their coworkers reported similar topologically arrested structures.[35] They also found other stable configurations such as 'figure of omega' and 'entangled point defect'.[36–40] As described below, since this effective interaction works for more than two particles, well-fabricated colloid patterns can be created.[39]This method attracts considerable attention as a novel way to form extremely stable colloid assemblies. Before these findings by using a coarse-grained model, Guzmán et al. studied a pairwise interaction between particles by molecular dynamics simulations.[33,34,71] They found that the two Saturn-ring defects coagulate to form three ring DLs with two nodes. The particles tend to align perpendicularly to the background director and this configuration is called 'entangled hyperbolic'.[15] In our previous study,[42] we mentioned that this 'entangled hyperbolic' structure was observed only transiently in the early stage of the phase ordering and transformed to the other metastable configurations shown in Fig. 3. Similar arguments have been made by Ravnik and Žumer.[37] About this discrepancy, Hung pointed out that the defect configuration may depend on the particle size and the anchoring strength. The 'entangled hyperbolic' configuration is stable for a pair of small (nm-sized) particles.[15] This suggests an interesting possibility that the particle size has a strong influence on defect organization.

The Saturn-ring associated with a spherical particle has a topological charge $s = -1$, as it can continuously transform into a single hyperbolic point defect. Moreover, it is intuitive to think that the spherical particle acts in a uniform medium as a point defect of charge $s = +1$ and therefore it must be associated with a defect that may annihilate its extra charge. Analogously, two spherical particles have a total charge $s = +2$, therefore the 'figure of eight' loop must have $s = -2$ although it is not knotted. This was puzzling. Recently, Čopar and Žumer pointed out that this can be naturally explained by considering the three-fold symmetry of a disclination line.[72,73] We often use a line to express a disclination line, but the important information on the tensorial nature of the order parameter field is missing in this representation. Actually, the cross-section of a DL of $s = -1/2$ has the threefold symmetry due to the arrangement of the LC molecules around it. In order to calculate the topological charge, thus, it is convenient to consider a DL as a ribbon, characterized by the self-linking number, which includes both the writhe of the ribbon in space and also the twist around its own axis.[72,74] By considering the twist of the DL, it was shown that the entangled defect configurations such as 'figure of eight' have a topological charge $s = -2$ ($s = -N_p$ for many particle systems).

Simultaneous binding of many particles by a single stroke disclination line

The effective interaction mediated by entangled DLs can work also for many particle systems.[37–40,42] We showed that we can form planar aggregates of particles bound by a single-stroke DL as shown in Fig. 3(c).[42] This interaction depends on the topology of the DL. This means that it is intrinsically non-additive and cannot be expressed by a potential. Fig. 4 shows the aggregation process of many particles immersed in a NLC after being quenched from an isotropic disordered state at $t = 0$ without any external field. The system contains $N_p = 400$ particles in a cubic lattice space (128^3). The particle diameter is $2a = 7.2$, so that the volume fraction of the particles is $\bar{\phi} = 3.72\%$. Just after the quench, nematic ordering takes place. Local selection of orientation leads to the formation of DLs and their length quickly decreases with time. Since the particle motion (conserved dynamics) is slow compared to the defect motion (non-conserved dynamics), some defects are trapped by particles in the early stage. At this stage, most defects are strongly elongated and entangled in a complicated manner. DLs, which are shared by particles far apart, shrink to lower the elastic energy, inducing hydrodynamic motion of particles. Particles bound by a defect can interact even when other particles are located between them. Shrinking of defects while keeping their connectivity results in the formation of clusters of the particles bound by DLs[42] and chains.[37] In this way, topologically arrested metastable structures are kinetically selected as a consequence of orientational ordering from an isotropic state, despite them being in a high energy state.

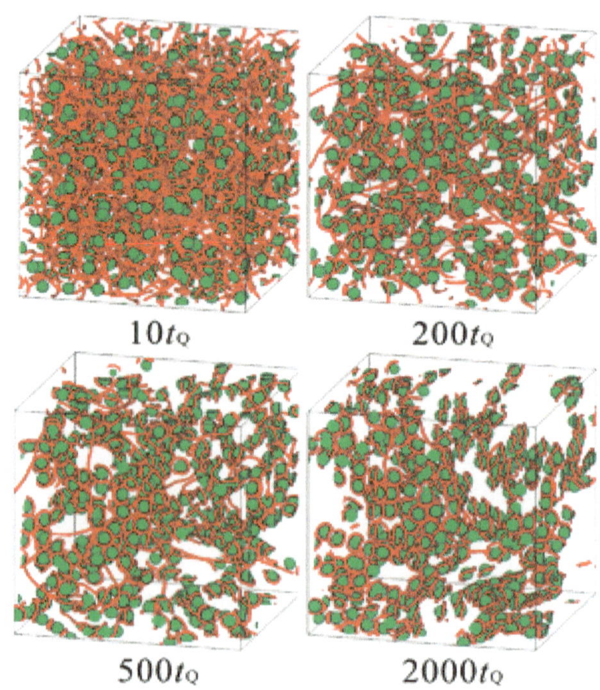

Fig. 4 The aggregation process of particles immersed in a NLC. The volume fraction of a particle is $\bar{\phi} = 3.72\%$. At $t = 0$, we quenched the system from an isotropic phase to a nematic phase.

Now we consider effects of an external field on the behaviour of defects. Fig. 5 shows the colloid distributions under (a) $E_z = 0.0$ and (b) $E_z = 0.1$ at $t = 2000t_Q$. The particle volume fractions are $\bar{\phi} = 0.93\%$, 1.86%, and 6.79%. In the absence of an external field, DLs are likely to be disconnected at the lower volume fraction. In the kinetic process, the elastic force (or tension) acting on a DL arises not only from the line itself, but also from a viscous drag force to particles trapped. When an elongated defect line cannot support the tension, it is disconnected in the process of shortening. Then, isolated Saturn-ring defects are formed around individual particles. Such particles then interact with each other via connectivity-free interactions as shown in Fig. 3(a). We note that the population of such isolated Saturn-ring defects is higher for a more dilute dispersion. This is because DLs are more strongly stretched because of larger average interparticle distances, which leads to more frequent disconnections of DLs.

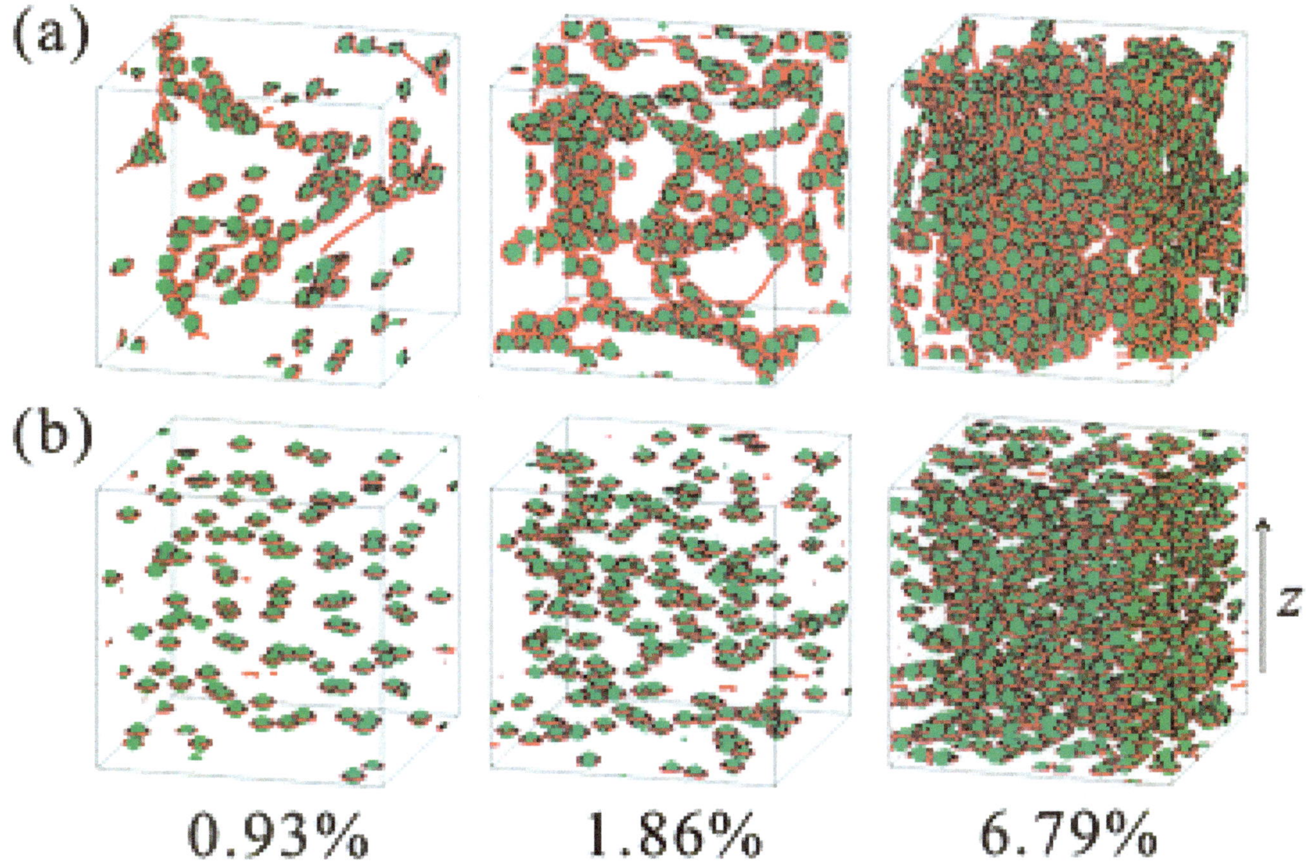

Figure 5 Snapshots of colloid distributions in NLCs at $t = 2000t_Q$ under (a) $E_z = 0$ and (b) $E_z = 0.1$. The particle volume fraction is 0.93%, 1.86% and 6.79% from the left to right panel.

The defect pattern depends not only on the particle concentration, but also the way of preparation. When a sample is quenched into a nematic phase under an electric field as shown inFig. 5(b), the director field is quickly aligned along the field in the early stage. This avoids defect entanglements. Then, no percolated DLs remain; DLs are localized around each particle, forming Saturn-ring defects for each particle. Fig. 6 plots the time development of the average particle displacement ΔR, which is calculated as

$$\Delta R^2(t) = \frac{1}{N_p} \sum_i \sum_{\alpha=x,y,z} \left\{ R^i_\alpha(t) - R^i_\alpha(0) \right\}^2,$$

where R^i_α is the position of the i-th particle. The particle displacement is larger under a weaker field. It is evident that the connectivity of the DL pattern is crucial in accelerating the aggregation dynamics. The interaction potential between isolated particles accompanying Saturn-ring defects is approximated by a quadrupole potential decaying as $|R^i - R^j|^{-5}$,[32] so that the average displacement becomes larger for a higher volume fraction for $E_z = 0.1$. On the other hand, ΔR decreases with an increase in the volume fraction for $E_z = 0$. This is because the large average displacement caused by entangled DLs is more suppressed for a higher volume fraction.

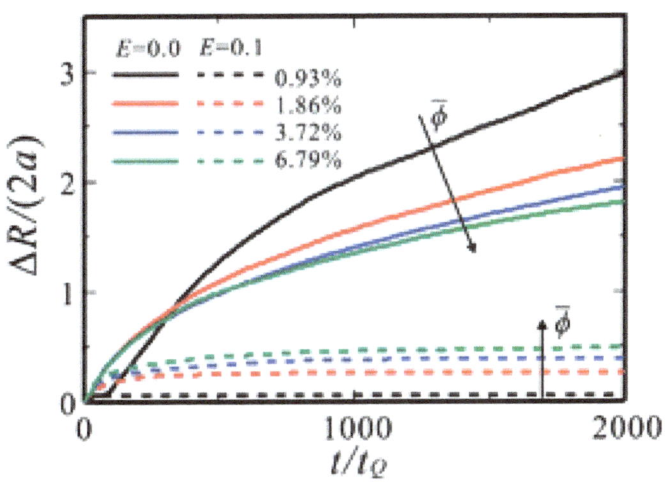

Figure 6 The time development of the particle displacement $\Delta R(t)$. We can see that particles tend to move over larger distances for a weaker external field.

Finally, we mention a very interesting study by Wood et al., which shows colloidal gel formation by defect binding. They reported experimental and numerical observations on a colloid–NLC composite.[41] They measured its mechanical strength and found that storage modulus is significantly increased when remaining DLs are percolated over the sample. A temperature quench from an isotropic phase to a nematic phase is not enough to synthesize composites of large elasticity. In order to obtain a percolated sample, or reconfigure defect structures, one has to apply strong shear to the sample.[41] They also showed that heating above the clearing point suddenly transforms a gel to a liquid, which allows an instantaneous transformation from a gel to a liquid and the resulting disappearance of the elasticity. It may be worth noting that there is a difference in the processes of phase ordering and aggregation and in the percolation threshold between the numerical observation of Wood et al. and ours, which might be of technical origin.

NEMATIC LIQUID CRYSTALS IN POROUS MEDIA

Next we consider a structure of more complicated topological characteristics. In porous media such as random nanostructures of silica gels, non-ergodic memory effects have been observed.[51–58] Long-ranged elasticity and anchoring to solid surfaces make effects of a spatial confinement of NLCs by solids very significant. For example, for a NLC sandwiched by two parallel plates imposing homeotropic anchoring, the director field can be aligned uniformly without defects. Hence, the system memorizes the unique configuration, which is selected to minimize the elastic energy. If this stable pattern is perturbed by an external field, the director field quickly recovers the original state after the removal of the field. In porous networks, on the other hand, the global nematic order cannot go back to a uniquely determined state unlike the above e and long-lived remnant order remains even after switching off the external field.

Such memory effects and slow glassy dynamics have been studied theoretically and numerically.[54,56,60–64] In these studies, point-like defects which anchor the director field are introduced to mimic a random surface field. However, this approach cannot deal with the topological characteristics of a solid matrix. We have recently succeeded in studying effects of solid matrixtopology on memory effects in collaboration with Bellini and Buscaglia.[48] By means of Monte Carlo simulations considering 'geometrical frustration', we found that DLs topologically concatenated to the solid networks play significant roles in the memory effect and thus the memory effect crucially depends on the relationship between the topological characteristics of solid surfaces and the symmetry of an external field.

We employed lattice-based Monte Carlo simulations with the following Heisenberg model like Hamiltonian for spin n^k_α:

$$\mathscr{H} = -J \sum_{(k,l)\in \mathscr{N}\mathscr{T}} \left(n^k_\alpha \cdot n^l_\alpha\right)^2 - W \sum_{k \in \mathscr{S}} \left(n_\alpha \cdot d_\alpha\right)^2 - \sum_{k \in \mathscr{N}} \left(n^k_\alpha \cdot E_\alpha\right)^2$$

The lattice space is partitioned into three portions in terms of pore structures. $\mathscr{N}$, $\mathscr{S}$ and $\mathscr{T}$ represent the space filled with NLCs, solid objects and surfaces between them. The first term of eqn (6) describes the isotropic–nematic transition.[59] In a bulk, the transition temperature is $T_{IN} \cong 0.75J$. The second and third terms represent the effects of the anchoring and external field E_α. A positive W imposes the homeotropic alignments at the surfaces. We carried out simulations with heat bath sampling at temperature T.[48] Here, we set W = J = 1.

Memory Effects due to a Topological Constraint from Solid Surfaces

Behaviour of a Defect Structure

To see the essence of memory effects of topological origin, here we mainly consider a random porous matrix (RPM) structure and a regular bicontinuous one with a cubic symmetry (BC) as the two fundamental topologies. RPM is thought to mimic a system that has been used for experiments, i.e., NLCs in Millipore filtering membranes.[53,56] To assess the importance of the bicontinuous nature of structures, we also studied a simple cubic array of isolated spheres that are not in contact (SC_n). The radius of the spheres is set to three lattice mesh, a = 3.

We quench these systems at $T = 0.1 (< T_{IN})$ and equilibrate them without a field ('zero-field cooling'). For the porous matrices we run simulations up to the time $t = 10^5 t_{MC}$. The physical meaning of the temporal evolution of Monte Carlo simulations is sometimes a matter of debate. However, the method has been known to be very powerful and useful for studying glassy systems with slow relaxation, the dynamics of which is dominated by activation processes overcoming a free energy barrier.[75,76] The zero-field-cooled states are locally well ordered because their local nematic order parameter Q is approximately the same as that of the bulk sample Q_B. At the same time, the systems are globally isotropic, and hence the tensorial order parameter averaged over the whole sample is $\langle Q_{\alpha\beta} \rangle \approx 0$. To these disordered states, we apply an electric field pulse along the z axis for a certain duration ($10^5 < t/t_{MC} \leq 2 \times 10^5$) and then switch it off at $t = 2 \times 10^5 t_{MC}$ (or $1.5 \times 10^5 t_{MC}$). Fig. 7(a) and (b) plot the temporal change in the nematic order parameter along the field of NLC, $Q(t) = \langle Q_{zz}(t) \rangle$, for RPM and BC, respectively. When the electric field is applied, the director is forced to be aligned along the field and the global order develops along that direction. The degree of the global ordering depends on the field strength. Upon removal of the field, the system relaxes to a metastable state with an intermediate degree of order, which we call a remnant order Q_M (see below for its precise definition). This is a clear indication that the system has a capability to memorize the field-induced order. Fig. 7(c) and (d) show the field-strength dependences of the order parameter in the presence of the field, Q_E, and Q_M, respectively, for RPM and BC. We can clearly see that BC possesses a higher memory capability than RPM; the difference between Q_E and Q_M for BC, which is a measure of the loss of memory, is much smaller than that for RPM.

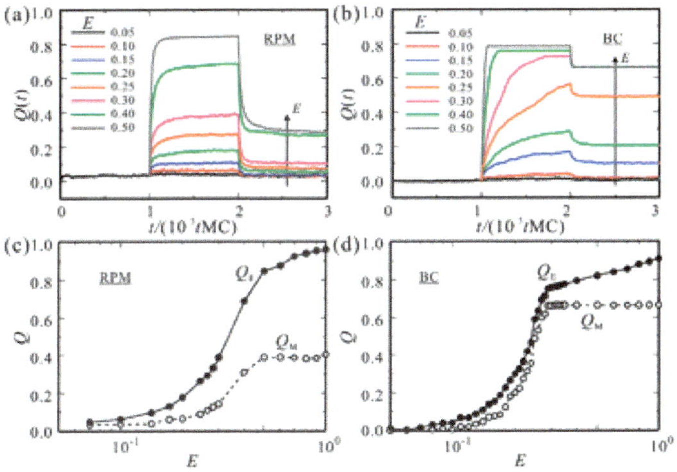

Figure. 7 Time evolution of the global nematic order along the z axis. NLCs filled in (a) RPM and (b) BC are cooled at T = 0.1 without an external field up to t = $10^5 t_{MC}$. During a period ($10^5 < t/t_{MC} < 2 \times 10^5$), the external fields E with several strengths are applied along the zaxis. The resulting temporal changes of Q_E and Q_M are plotted as functions of the field strength E for (c) RPM and (d) BC.

In Figure. 8, we show patterns of DLs after the zero-field cooling at t = $10^5 t_{MC}$ in the three employed porous matrices. We visualize DLs inside the nematic domain, following the method described in ref. 77. After zero-field cooling, many DLs wander through the pores without contacting the surfaces even after a long annealing time for all the matrix structures. Here, the defect trajectories seem to be randomly entangled. The DLs form closed loops without edges and some of them are concatenated to the porous network. In Fig. 8, some DLs appear to terminate, but actually they are connected with DLs at the opposite side because of the periodic boundary conditions.[48] After an application of a strong field E = 0.5 along the z axis, the defect pattern in BC transforms into a regular pattern consisting of loops laying perpendicularly to the z axis, whereas that in RPM changes to a different, but still randomly entangled configuration.

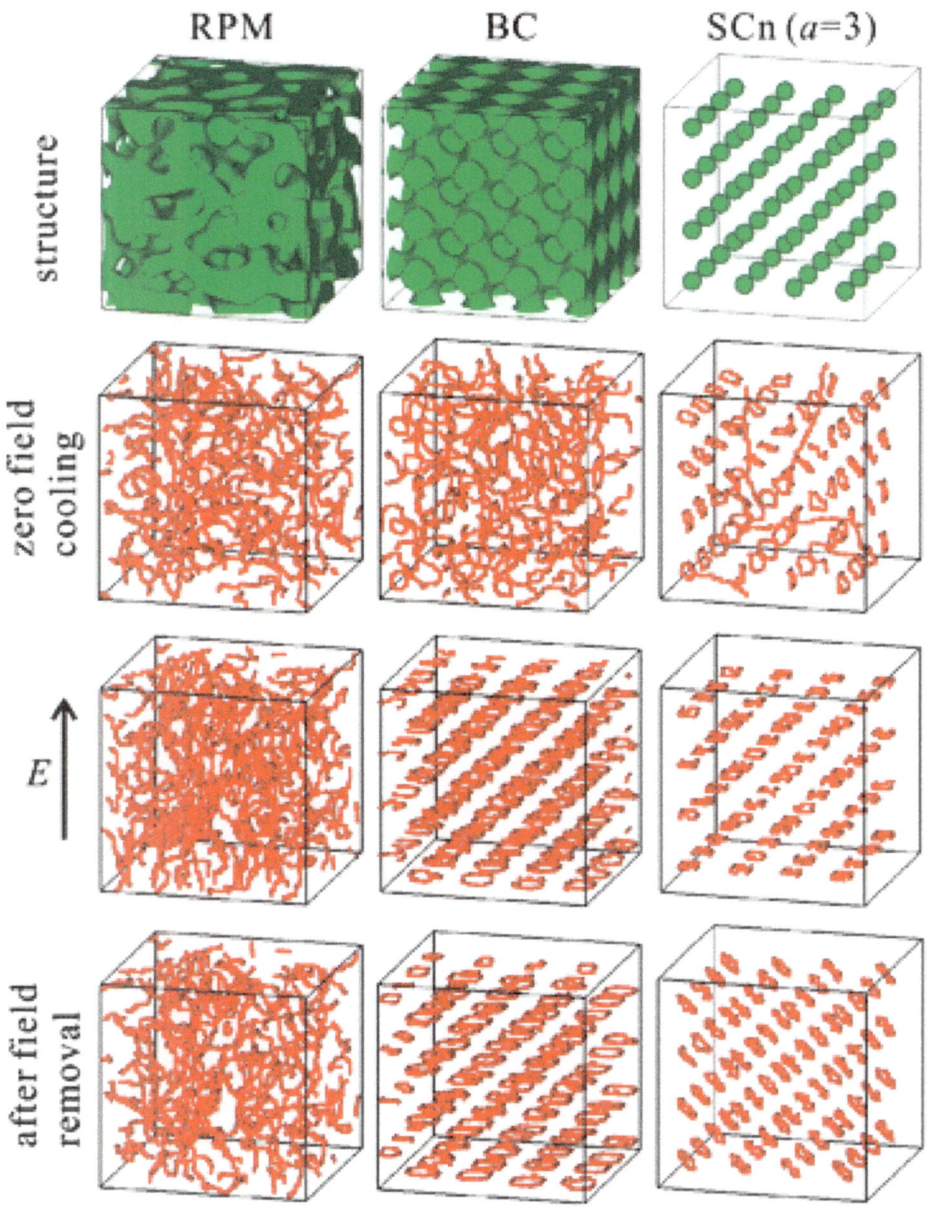

Figure 8 Patterns of disclination lines (DLs) in three types of porous matrices: A random matrix(left), a bicontinuous pattern with a cubic symmetry (center) and a simple cubic array of spheres without contact (right). Top row: the structures of the porous

structures employed. The second row: configurations of DLs after zero-field cooling, observed at $t = 10^5 t_{MC}$. The third row: configurations of DLs under an external field. Bottom row: configurations of DLs observed at $t = 3 \times 10^5 t_{MC}$. We apply a strong field ($|E| = 0.5$) along the z axis from $10^5 t_{MC}$ to $2 \times 10^5 t_{MC}$.

Obviously, frustration imposed by surface anchoring plays a crucial role in the memory effects described above. An enormously large number of possible topological states of a confined NLC suggests an apparent analogy to well-known frustrated systems such as spin and structural glasses, in which an exponentially large number of metastable states are separated by large energy barriers. The origin of frustration of our system is, however, quite different from that of these glass systems. For example, typical spin glasses are intrinsically frustrated because of the impossibility to simultaneously minimize coupling energies. In the e of confined NLCs, a topological constraint due to confinement provides frustration to an otherwise well-behaved system. The frustration is intrinsic to the system's Hamiltonian in the former, whereas extrinsic in the latter. This leads to a marked difference in the degree of order in a mesoscopic length scale: while the nematic order is well defined locally, at the length scale of the pores it is severely affected by geometrical frustration because of the impossibility to simultaneously match all the anchoring constraints without forming topological defects. Each metastable state can be characterized by its trajectory and twist of DLs wandering through the maze of a porous system. This geometrical frustration offers a possibility to design confining geometries that optimize the macroscopic function of the metastable states: 'geometrical functionalization' of materials.

Slow Glassy Dynamics

Inspection of the defect trajectories indicates that in complex geometries the onset of remnant order is associated with a change in the way DLs are concatenated to the solid matrix. In the e of a weak field, DLs respond elastically to the field without changing their topology. After the field removal, DLs relax back to their original trajectories. Hence, the remnant orders Q_M are vanishingly small as shown in Fig. 7(c) and (d). On the other hand, an application of a stronger external field supplies enough energy to reconfigure the DL trajectories, producing different closed loops through the connective paths of $\mathcal{A}$. In the limit of a very strong field, DLs are forced against the surface regions for which the normal lies in the x–y plane. The application of a strong field may even break the anchoring. However, as the field is removed, the normal anchoring is recovered and DLs relax to the local elastic energy minimum while preserving in large part of the concatenation with the solid network forced by the field, thus keeping some memory of the field effect. This leaves the system in a new aligned stable state different from the zero-field-cooled disordered stable state. It is the structure of a porous matrix that determines the degree of the remnant order in the new aligned state, i.e., what fraction of DLs is preserved after the removal of a field. In regular porous media like BC, we can realize a situation where the DLs are well aligned after a strong field is applied along a compatible direction. Since such a loop configuration (Fig. 8) is energetically favoured and thus uniquely formed in BC, the remnant order reaches a well-defined plateau for strong external fields as shown in Fig. 7(d).

Interestingly, the relaxation of the orientational order of NLCs in porous media crucially depends on the geometry and topology. Fig. 9(a) and (b) compare the pore-size ℓ dependences of the long-time relaxation behaviours of NLCs at constant $T = 0.1$ for RPM and BC, respectively. The relaxation after the removal of a strong field $E = 1$ is markedly different between RPM and BC. The relaxation of the global nematic order in BC is characterized by a single decay process and very well fitted by a stretched exponential,[48]

$$Q(t) = Q_M + \Delta Q_S \exp\{-(t/\tau_S)^{\alpha}\},$$

where Q_M is the remnant order determined as the long-time asymptote of $Q(t)$, ΔQ_S is the amplitude of the stretched exponential decay, and α is the stretching exponent ($\alpha \leq 1$). In the e of RPM, on the other hand, the decay does not saturate as clearly as in the e of BC and the first quick relaxation is followed by a very slow decay to a second plateau. We find the curves to be well fitted by a summation of two decay modes,

$$Q(t) = Q_M + \Delta Q_S \exp\left\{-(t/\tau_S)^{\alpha}\right\} + \frac{\Delta Q_L}{1 + \log(1 + t/\tau_L)},$$

ΔQ_S and ΔQ_L are, respectively, the amplitudes of the fast stretched-exponential relaxation and of the slow decay. The slow decay process could also be fitted by a stretched exponential but it seems to be better fitted with the functional shape in eqn (8), which crossovers from a Lorentzian decay at small t to a logarithmic decay for $t > \tau_L$.

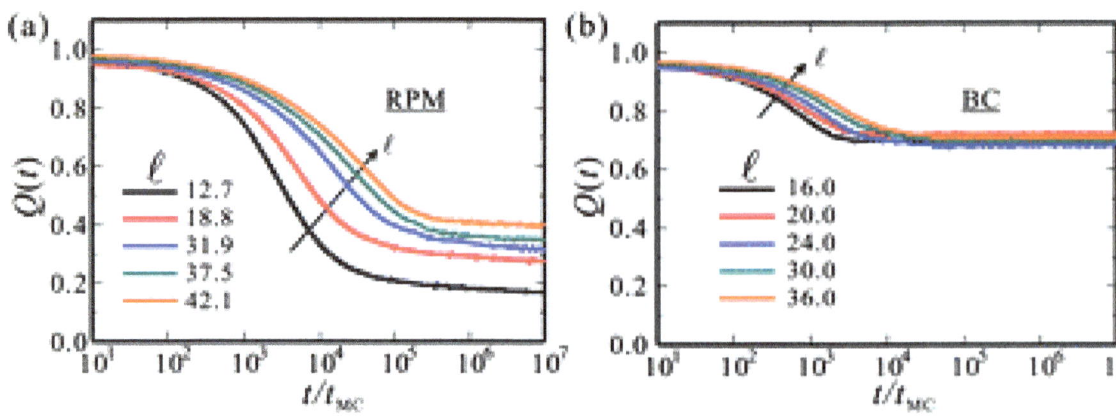

Figure 9 The long-time relaxation process of the global nematic order after the removal of a strong field for (a) RPM and (b) BC with various pore sizes ℓ. In RPM, a very slow logarithmic decay mode is observed (see eqn (8)).

The relaxation times of the fast stretched exponential decay τ_S for RPM and BC geometries are shown in Fig. 10(a). The values obtained in the two geometries match, indicating that the fast decay is a relaxation process common to the two systems and not associated with the change in defect configurations. τ_S grows as ℓ^2, a scaling generally found for the viscoelastic relaxation of nematics in conventional liquid crystal cells. Indeed, in these systems viscoelastic relaxation is expected to govern the evolution of the system from a field-induced ordered state to the configuration of the minimum elastic energy attainable without topological rearrangements of defects. In the e of BC, such a configuration is characterized by DLs stabilized by the surface geometry, so that no further relaxation takes place at all ℓ values. Our simulations indicate that the regular pattern consisting of loops shown in Fig. 8 is the global energy minimum state for BC.

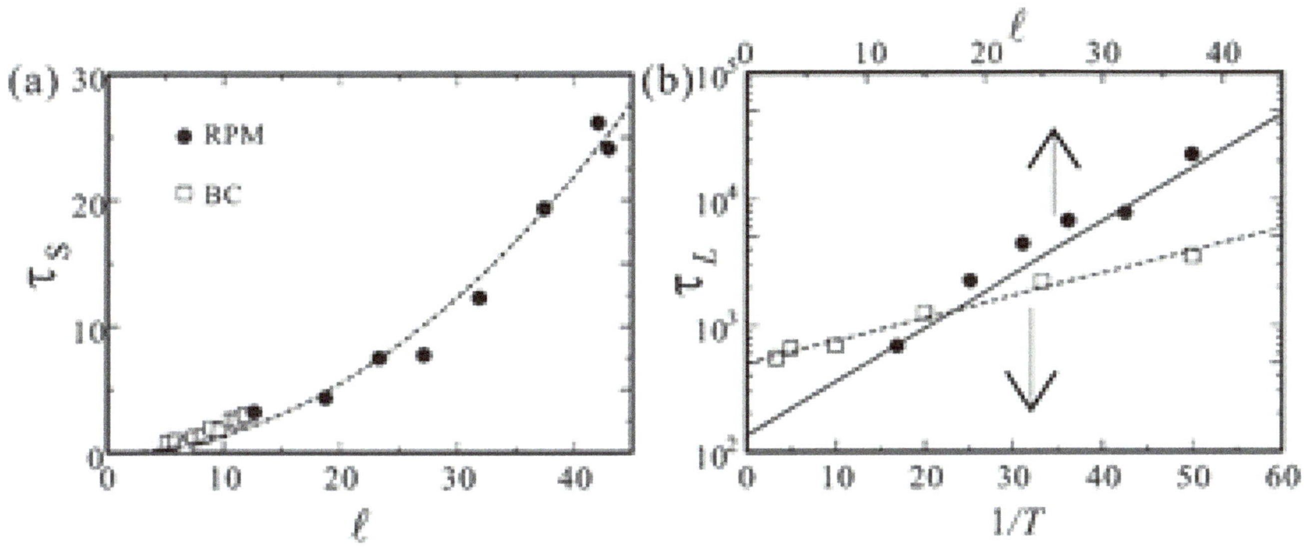

Figure 10 (a) The relaxation times for the fast stretched exponential decay τ_S for RPM and BC geometries as functions of the pore size ℓ. (b) Plots of τ_L in RPM against the mean pore size ℓ and the inverse temperature 1/T.

When instead the topology attained after the fast relaxation is not stabilized by a surface geometry in a unique manner, rearrangements of DLs leading to a state with a lower energy are possible while accompanying a further relaxation, as for RPM. The slow relaxation involves topological rearrangements of defects. After the system has reached one of the local minima of its elastic energy states, that is, for $t \approx (2 - 3) \times \tau_S$, DLs move only by thermal excitations. These continued motions lead DLs to transiently adopt conformations less energetically favoured, and thus foster DL collisions which may result in topological changes of defects, if energetically more favoured configurations can be accessed by the defect collisions. Such events in RPM can be directly seen in Fig. 11. A simple scaling analysis confirms that the activation barrier for such a DL collision is proportional to ℓ, leading to the relationship $\ln(\tau_L/t_{MC}) \propto \ell/T$. Indeed, if a porous structure is uniformly enlarged by a factor of b while keeping the elastic constants and the DL connectivity unchanged, the free energy required for bringing two given DLs together grows linearly with the expansion factor b, as confirmed in Fig. 10(b). Inspection of the simulations also provides us with a clue to understand the logarithmic nature of the slow glassy relaxation (eqn (8)). Each DL collision event leading to a topological rearrangement is followed by a fast viscoelastic relaxation to a new state in which DLs experience a new energy landscape. This leads to a cade effect, in which the activation energies are modified as the relaxation proceeds. More specifically, a logarithmic decay may be a consequence of an increase of the activation energy proportional to $Q(t)^{-1}$. It is natural to expect that the DL reorganization requires higher energy barriers for further DL collisions as the time proceeds.[48]

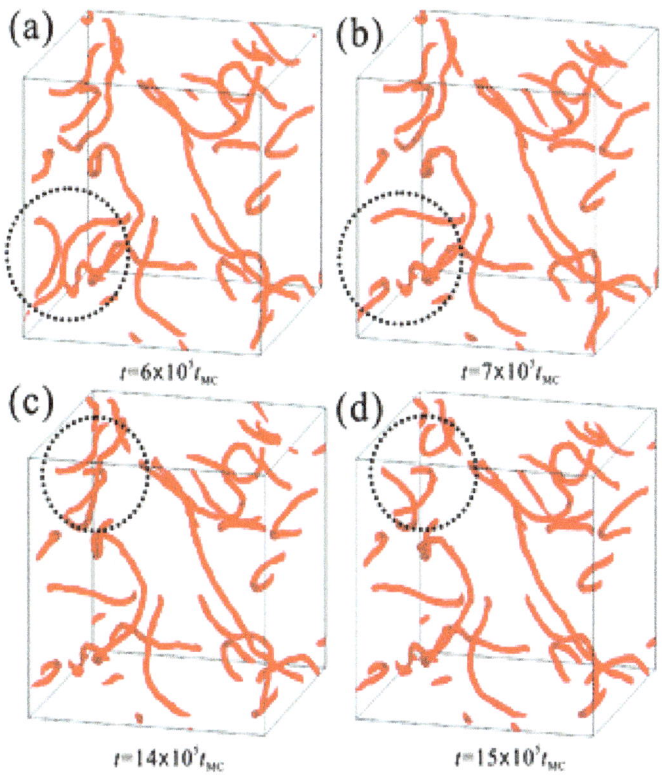

Figure 11 Snapshots of DLs in RPM at selected times during their relaxation process after the field removal. The comparisons between (a) and (b) as well as between (c) and (d) enable us to defect spontaneous topological changes in the DLs. The matrix pattern is not shown here to show the defect patterns better.

Robustness of Memory

For practical applications, crucial is the robustness of the memory effects. The topology-based non-ergodicity described here relies on a large energy necessary to 'reconfigure' irreducible DL loops. The bicontinuous nature of the matrix is essential for this form of multistability. This can be appreciated by studying the behaviour of DLs in an array of isolated spheres (SC_n, here). DLs compensate for the topological charge introduced by the particle with homeotropic anchoring. In the equilibration process, DLs are annealed into Saturn-rings, as shown in Fig. 2 and 8. Upon the field removal, rings can continuously rotate around the spheres without topological changes. Saturn-rings are held parallel to one another by mutual elastic interactions and their collective rotation is nearly a soft mode. Because the total surface area is small in SC_n, the amplitude of the fast viscoelastic mode becomes small, in comparison to that in BC. On turning off the field, we found that the nematic order starts rotating at a finite temperature, as shown in Fig. 12. Although the defect patterns in both BC and SC_n consist of loops, such spontaneous rotations of the global nematic order is prohibited for BC. In BC, half of the regularly aligned DLs are irreducible, encircling the porous network. So, they cannot change their orientation to other directions without topological rearrangements. As noted in the previous section, the energy cost necessary for rearrangements of DLs is much higher than the thermal energy. The topological irreducibility of DLs is the key for the robustness of the memory effects. We note that the local curvature of tubes forming BC stabilizes the locations of loops along the field axis, which also plays a crucial role in enhancing the memory effects. This mechanism of stabilization is weaker in a 'jungle gym' structure discussed below since it consists of straight cylinders (the zero curvature along the field axis).

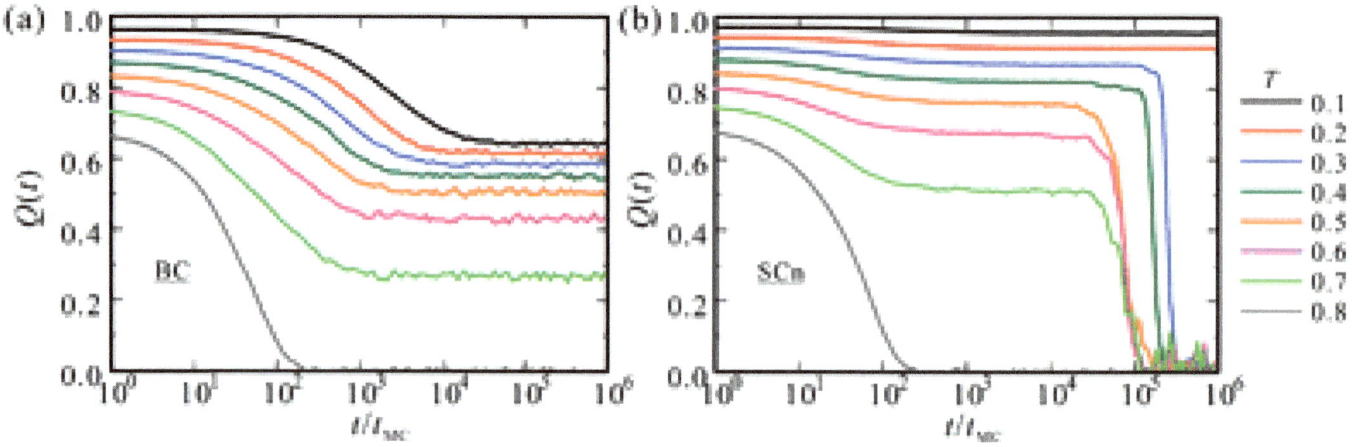

Figure 12 Thermal decay of the global nematic order along the z axis for various temperatures in porous media (a) with and (b) without topological constraint. The nematic liquid crystals are confined in (a) BC and (b) SC_n, respectively. The isotropic–nematic transition temperature in a bulk is $T_{IN} \cong 0.75$. Even below T_{IN}, the memory is spontaneously lost in the porous medium without topological constraint.

Towards experimental realization of high memory effects

In the above, we find that BC provides one of the best surface topologies for NLC memory. However, the formation of such a regular structure is not so easy with currently available technologies. Fabrication of a regular porous structure with two-photon polymerization is a promising method and very recently applied to NLC memory.[23] In this technique, the resolution of the laser is different in the x–y plane and along the z direction, and therefore all the channels do not have the same size, or the cross-section of the channels is not necessarily circular. Thus, it is important to evaluate the influence of imperfection and anisotropy on regular matrices by simulations for effective comparison with experiments. Hence, here we introduce anisotropy to 'jungle gym' matrices, which consist of cylindrical pillars oriented along the three orthogonal axes (see the inset of Fig. 13), and study its influence on the memory effect. The separation between the nearest parallel pillars is set to be $\ell = 16$.

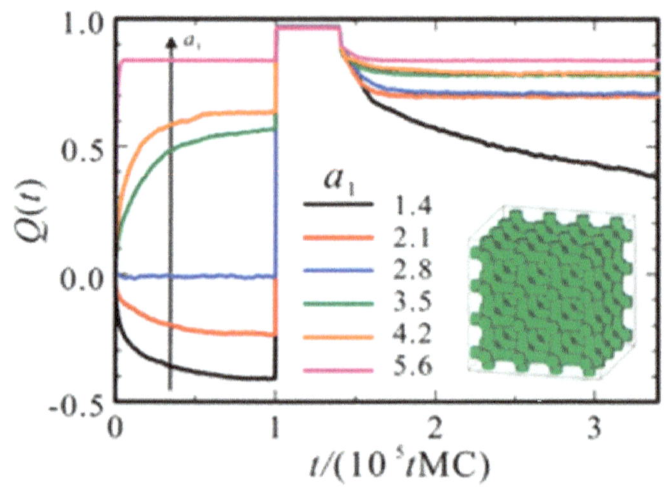

Figure 13 Evolutions of the global nematic order along the z axis in 'jungle gym' porous networks (see the inset). During $10^5 < t/t_{MC} < 1.5 \times 10^5$, an external field is applied along thez direction, $E_z = 0.5$. We vary the radii of the pillars along x and y axes with fixing the radius of the z-cylinders to $a_2 = 2.8$.

Fig. 13 shows temporal evolutions of the global orientations along the z axis, before, during and after the application of an external electric field. We vary the radius a_1, the radius of the channels in the x and y directions, while $a_2 = 2.8$, the radius of the channels in the z direction. In the symmetric matrix ($a_1 = a_2$) the spontaneous global order is nearly zero and remaining DLs are randomly entangled after zero-field cooling (not shown here), as observed in BC and RPM. In asymmetric networks, on the other hand, the NLCs show spontaneous alignment: the global order parameter Q in the direction of the anisotropy is much bigger or smaller than zero if the pillars in that direction are bigger ($a_1 > a_2$) or smaller ($a_1 < a_2$), respectively. In matrices with large a_1/a_2, in particular, remaining DLs are already spontaneously aligned in x–y planes (not shown) even without any external field. This intrinsic alignment is directly caused by the anchoring interaction, and thus its quantity depends on the difference between surface areas projected on the three orthogonal planes. The mechanism of these spontaneous alignments will be reported elsewhere. The memory effects after the application of external fields are also affected by the anisotropic structures. In particular, the NLC for $a_1 = 1.4$ shows a large decay mode as shown in Fig. 13. Pillar size and anisotropy are important parameters affecting the memory effects, as shown in Table 1, which lists the values of the memory parameter $M = Q_M - Q_{ZFC}$, i.e., the difference between the remnant order parameter Q_M after field application and the order parameter Q_{ZFC} after zero field cooling, for various es. It appears that the 'jungle gym' scaffolds of $a_1 \approx a_2$ show larger memory. However, the mechanisms of loss of memory are of two different kinds: in the structures with $a_1 \ll a_2$ the DLs order induced by the applied field is completely lost. Since they recover their negative intrinsic orientations, they will retain memory only if the applied field is in the x or y direction. On the other hand, the structures with very high anisotropy and $a_1 \gg a_2$ already have a very strong spontaneous alignment before the field is applied, therefore they do not 'gain' any more alignment from the application of an external field. The memory effects after the application of a field along the x direction is also important, in order to have a bistable (or multistable) system. Hence, symmetric porous structures are also preferred.[48]

Table 1 Occurrence of memory effects after the application of a field in the z direction in 'jungle gym' structures. The radii of the cylindrical pillars in the x and y directions are a_1 and that in the z axis is a_2. Memory is expressed by the parameter $M = (Q_M - Q_{ZFC})$. Cerulean ●: $M < 0.1$, green ▲: $0.1 < M < 0.3$, orange ■: $0.3 < M < 0.6$, red ♦: $0.6 < M < 0.8$, brown ▼: $M > 0.8$. The half-circles represent the es where memory is lost due to a 'saturation' effect because of high spontaneous ordering

a_2/a_1	$1/2a_2$	$3/4a_2$	a_2	$5/4a_2$	$3/2a_2$	$2a_2$
0.8	●	●	●	▲	▲	■
1.6	●	●	▼	■	■	●
2.0	▲	▼	▼	■	■	◗
2.4	▼	▼	♦	■	▲	◗
2.8	♦	▼	♦	■	▲	◗
3.2	♦	♦	♦	■	◗	◗
4.0	♦	♦	♦	▲	◗	◗

DYNAMICAL COUPLING BETWEEN NLC FLOW AND DEFECT PATTERNS

Dragging a colloidal particle accompanying a Saturn-ring defect in NLCs

Here we focus on a remarkable feature of NLCs, i.e., the coexistence of orientational order and fluidity. Hydrodynamics for NLCs is different from that for isotropic liquids due to the dynamical coupling between the orientational degree of freedom and the hydrodynamic one. The viscosity of a NLC depends on the local orientational field and the elastic stress, called Ericksen stress, acts on the anisotropic liquid itself.[1] Then, sedimenting or electrophoresing particles in nematic solvents show nonlinear flow behaviours, which are not seen in isotropic liquids, even in the Stokes regime. In particular, the coupling between topological defects and flow causes complex particle motions.[26,78–82] For example, Stark and Ventzki reported lift-up motions, the directions of which are not parallel to the applied force, for a particle having a hyperbolic point defect.[79,80] Furthermore, Lavrentovich et al. observed a drift motion of an electrophoresing particle in AC electric fields.[81] Interestingly, its velocity is proportional to the square of the field strength.

In ref. 68, we reported dynamics of a spherical particle accompanying a Saturn-ring defect. This is a quite interesting e where defect topology matters. We employed the same numerical scheme as that in Section 3. Fig. 14 shows representative snapshots of a particle moving in a NLC solvent. The background director field is oriented along the z axis. As shown inFig. 14(a) and (b), we apply body forces to a particle towards the z and x directions, respectively. In NLCs, defect motions are caused not only by the convectional flow, but also by the rotational motion of the director field. The dynamic behaviours are well characterized by the Ericksen number Er, which is the ratio between the elastic stress and the viscous stress. The Ericksen number is roughly proportional to the particle velocity. We note that the key process of defect motion is the director rotation. In our problem, thus, the relevant viscosity for the Ericksen number is the rotational viscosity, not the shear viscosity as often assumed.[68]

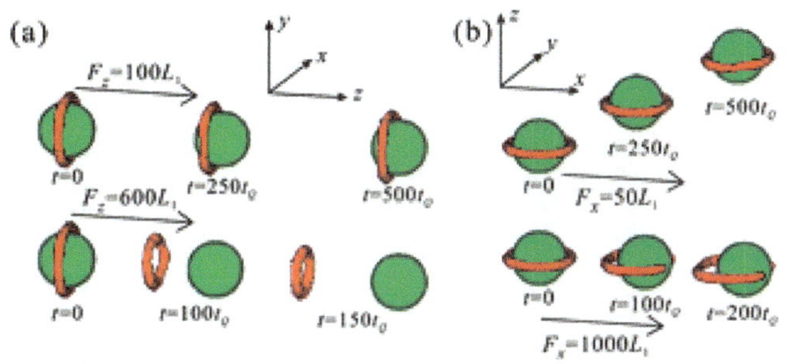

Figure 14 Representative snapshots of a particle moving in a NLC solvent. At rest, the background director is oriented along the z axis. We apply body forces for the particle towards the (a) z and (b) x directions. The radii of the particles are a = 8. The upper and lower panels in (a) correspond to regimes (a2) and (a3), respectively. The particle motions in (b) indicate regimes (b1) (upper) and (b2) (lower).

When we apply a force in parallel to the bulk director field selected by an external field, we observed three regimes for the particle motions. (a1) When the particle velocity is slow enough (low Er), the accompanied defect is capable of keeping its most preferred position relative to the particle. (a2) As the particle velocity is increased (moderate Er), the defect can follow the particle, but the relative defect position is shifted backward from the preferred equator position at a quiescent state. Here the amount of the shift from the equator position increases with an increase in the particle velocity. (a3) If the velocity is increased further (high Er), the defect cannot catch up with the particle and the relative separation then starts to increase with time, which results in the increase in the elastic energy. When the elastic energy cost exceeds the anchoring energy, the anchoring state on the particle surface is broken and the defect disappears eventually. In Fig. 14(a), we show these moderate and strong pulling es. In relation to this, we note that the following behaviour has been recently reported: for a fluid particle like an emulsion droplet, the behaviour is the same as that described above, whereas for a solid particle the defects shift forward.[82] The reason for this discrepancy is not clear at this moment. We also note that defect motion may be strongly influenced by the flow-alignment coupling in the nematohydrodynamic equations.

More interesting is the e when a force is applied perpendicular to the bulk director field selected by an external field. Again we found three regimes, depending on Er. (b1) We observed that the particle motion is not parallel to the force direction for low Er regime. Since a portion of the defect in the front of the particle motion behaves as an obstacle for the particle motion, the particle tends to circumvent the defect and it chooses one of the two possible detours with the equal statistical weight.[68] (b2) As the particle velocity is increased, the front part of the Saturn-ring defect penetrates to the particle, breaking the anchoring condition (see Fig. 14(b)). Upon this anchoring breaking, the particle can move just along the applied force. (b3) In a high Er regime, a full portion of the disclination loop disappears: flow-induced disanchoring transition. The degree of the off-axis motion in (b1) depends on the strength of the homeotropic anchoring condition: a particle with weaker anchoring surface can be driven more parallel to the field direction. This suggests an interesting possibility for separating particles in terms of their surface properties.[68] This may offer a new method for electrophoretic separation of colloids or proteins in terms of their surface properties.

Flow of NLCs in a porous matrix

Recently, flow behaviours of NLCs in porous networks or microfluidic cavities have attracted much attention.[49,83–85] We recently reported a lattice Boltzmann simulation on flow properties of a NLC in a BC matrix.[86,87] Fig. 15 shows a switching process of the defect pattern in a BC matrix induced by flow. Initially, we prepare a regular defect pattern by applying an electric field along the z axis, as shown in the second row of Fig. 8. Then, we start to flow the nematic solvent in BC towards the x direction at t = 0. When the flow speed is slow enough, the defects stay at their preferred positions. Under a moderate flow speed, the DL loops laying x–y plains change their orientation to a new direction. Because of the multistability of this pattern, the new orientation is stable even after the flow is stopped. So, we can say that flow direction can be recorded as the defect pattern in the porous network. This switching behaviour is similar to that caused by an electric or a magnetic field, as discussed in Section 4.[48] We found that the resistance coefficient relating the flow speed to the force depends on the director pattern. This leads to an abrupt increase of the flow speed upon this switching. We also expect that the flow path can be dynamically selected by manipulating local defect configurations. As the flow speed is increased further, cyclic motion of the trapped defects is observed, and its characteristic frequency depends on whether or not the defects are topologically concatenated to the network. These nonlinear behaviours are strongly influenced by the anchoring condition. The conventional argument based on the Ericksen number does not take the anchoring interaction into account. In microfluidic devices and porous networks, the coupling between flow and anchoring is not negligible, as in the above e of colloids driven in a NLC. This will open up an interesting possibility of applications to flow control.

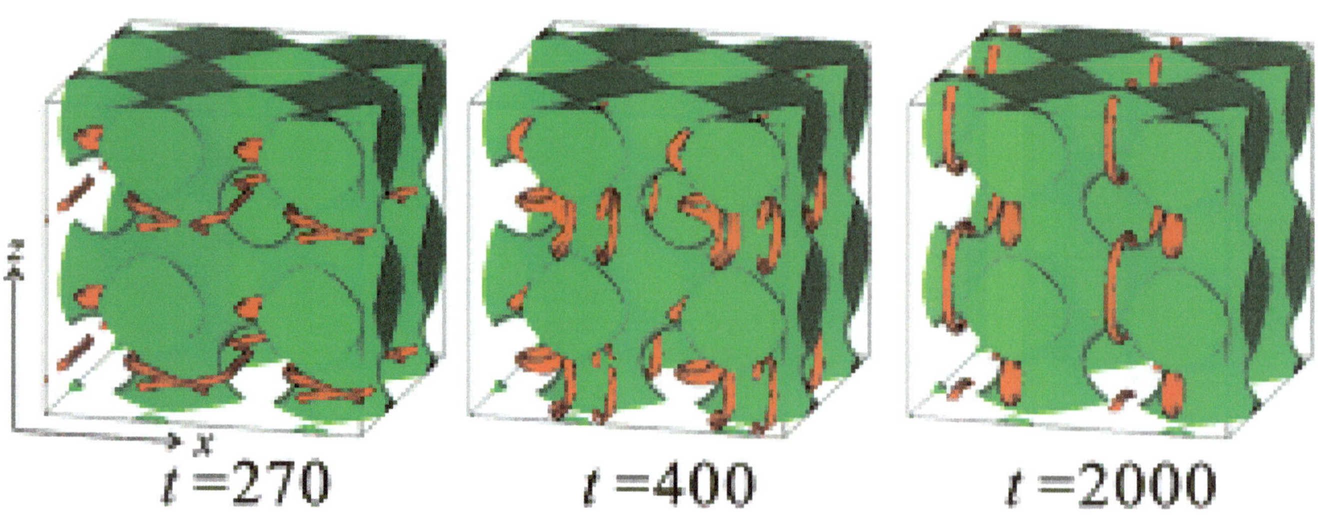

Figure 15 Switching process of the multistable defect pattern in a cubic porous medium. The NLC solvent is flowing towards the x direction

CONCLUSION AND OUTLOOK

In this article, we mainly review our numerical studies on topological defects of NLCs formed by geometrical frustration. In complex geometries with curved surfaces, frustrations between the elastic field favouring the undistorted nematic order and the anchoring effect distorting the order lead to the formation of stable defects, whose spatial configuration reflects how orientational order is organized in space. Since the energy barrier required for a change of the defect topology is far beyond the thermal energy, structures stabilized by defects are exceptionally robust among various structures in soft matter. The coexistence of this robustness of the defect structure and flowing ability of NLCs is a quite unique opportunity of defect engineering, which applies defects to create new functions. Defect engineering discussed here offers interesting applications of defects, which have been thought to be undesirable in conventional technology. Optical manipulation of defects is a quite promising way to control a defect configuration in liquid crystals.[10,39] While we are focussing here only on nematics, we note that defects in a smectic A phase can also be manipulated and patterned optically[88] and also geometrically.[89–92]

We show here mainly two types of NLC systems, in which the homeotropic anchoring to surfaces is assumed: isolated colloidal particles immersed in a NLC, where solids are mobile, and a NLC confined in porous media, where solids are immobile. For the former, we found a new type of metastable state, in which a ‘figure of eight’ disclination entangles around a pair of particles and strongly binds them. This effective interaction mediated by entangled DLs is also relevant for many particles, so that the particles can be irreversibly self-organized.

For the latter, stable defects formed in frustrated geometries have attracted many researchers for about two decades. However, global energy minimum states have been mainly considered so far. If only point defects remain in the system, they can move towards their most favourable positions without topological constraints, and then the system can reach the global energy minimum state. On the other hand, DLs are often trapped at high energy configurations because of their entanglement effects. However, such metastable states are not necessarily unfavourable. The entangled DLs now provide us with fascinating and promising tools to design systems with multistabilities. Indeed, in NLCs confined in porous media, remaining DLs wander through the channels of the solid network while keeping a large number of possible trajectories. Random porous media provide highly degenerate metastable states, leading to glassy slow dynamics, a topic of fundamental interest. After the removal of an external field, the system exhibits two relaxation modes in the global order parameter. The slower mode represents a thermally activated process, which accompanies the topological rearrangements of the DL pattern, and gives rise to the memory effect of topological origin. We demonstrated that the slow mode can be suppressed by designing the topology of matrix structures, more precisely, by removing the degeneracy of metastable states and realizing a unique topologically stable structure. This allows us to drastically enhance the capability of the memory effect. Here the bicontinuity of the matrix and the resulting irreducibility of DLs are the key to the robustness of the memory.

As emphasized in this review article, liquid crystals provide an ideal opportunity for investigating how topological constraint can affect not only self-organization of an ordered structure but also its dynamical behaviour. Geometrical frustration and the resulting glassy behaviour including memory effects are also important subjects. We would like to stress that the concept of defect science and engineering described in this paper may be relevant to strongly correlated electron systems and defect dynamics in Bose–Einstein condensates, including superconductors, liquid He, and laser cooled systems. What is common to these systems is the coexistence of order and fluidity. Here we consider only the topology of DLs but do not consider the degree of freedom of twist explicitly. Liquid crystalline phases such as nematic and smectic A do not have spontaneous topological defects, whereas some phases with chiral symmetry breaking, such as twisted grain boundary and cholesteric blue phases, have defects as their intrinsic structural elements. Thus, it is also interesting to consider the effects of chirality as well as effects of geometrical frustration on intrinsic defects.

Research on these topics has just started and is still immature. We hope that this review article would facilitate further intensive research that may lead us to the basic understanding of this fascinating physics on topology and its applications.

ACKNOWLEDGEMENTS

The authors are indebted to Tommaso Bellini and Marco Buscaglia for their collaboration and fruitful discussion on memory effects in porous media. T.A. and H.T. are grateful to the JSPS Core-to-Core Program “International research network for non-equilibrium dynamics of soft matter”. T.A. acknowledges KAKENHI (no. 24540433 and 23244088). F.S. acknowledges JSPS fellowship support for travel and stay in Kyoto University. H.T. acknowledges a grant-in-aid from the Ministry of Education, Culture, Sports, Science and Technology, Japan and Aihara Project, the FIRST program from JSPS, initiated by CSTP. A part of the computational work was performed using the facilities at the Supercomputer Center, Institute for Solid State Physics, University of Tokyo.

REFERENCES

1. P. G. de Gennes and J. Prost, The Physics of Liquid Crystals, Oxford University Press, Oxford, 2nd edn, 1993 .
2. P. M. Chaikin and T. C. Lubensky, Principles of Condensed Matter Physics, Cambridge University Press, Cambridge, 1995 .
3. M. Kleman and O. D. Lavrentovich, Soft Matter Physics: An Introduction, Springer-Verlag, 2003 .
4. G. P. Alexander, B. G. Chen, E. A. Matsumoto and R. D. Kamien, Rev. Mod. Phys., 2012, **84**, 497–514 .
5. A. Vilenkin and E. Shellard, Cosmic strings and other topological defects, Cambridge University Press, 2000 .
6. T. W. B. Kibble, J. Phys. A: Math. Gen., 2001, **9**, 1387 .
7. I. Chuang, N. Durrer, R. Turok and B. Yurke, Science, 1991, **251**, 1336–1342 .
8. H. Kikuchi, M. Yokota, Y. Hisakado, H. Yang and T.

Kajiyama, Nat. Mater., 2002, **1**, 64–68 .

9. P. Poulin, H. Stark, T. C. Lubensky and D. A. Weitz, Science, 1997, **275**, 1770–1773 .
10. I. Muševič, M. Škarabot, U. Tkalec, M. Ravnik and S. Žumer, Science, 2006, **2006**, 954–958 .
11. I. Chuang, N. Turok and B. Yurke, Phys. Rev. Lett., 1991, **66**, 2472–2475 .
12. K. Tojo, A. Furukawa, T. Araki and A. Onuki, Eur. Phys. J. E: Soft Matter Biol. Phys., 2009, **30**, 55–64 .
13. H. Tanaka and T. Araki, Phys. Rev. Lett., 2000, **85**, 1338–1341 .
14. C. Lapointe, A. Hultgren, D. M. Silevitch, E. J. Felton, D. H. Reich and R. L. Leheny, Science, 2004, **303**, 652 .
15. F. R. Hung, Phys. Rev. E: Stat., Nonlinear, Soft Matter Phys., 2009, **79**, 021705 .
16. J. Dontabhaktuni, M. Ravnik and S. Žumer, Soft Matter, 2012, **8**, 1657–1663 .
17. B. Senyuk, Q. Liu, S. He, R. D. Kamien, R. B. Kusner, T. C. Lubensky and I. I. Smalyukh, Nature, 2013, **493**, 200–205 .
18. S. C. Glotzer, Science, 2004, **306**, 419–420 .
19. A. van Blaaderen, Nature, 2006, **439**, 545–546 .
20. M. Conradi, M. Ravnik, M. Bele, M. Zorko, S. Žumer and I. Muševič, Soft Matter, 2009, **5**, 3905–3912 .
21. H. Tanaka, J. Phys.: Condens. Matter, 2000, **12**, R207 .
22. G. Gompper, M. Schick and S. Milner, Phys. Today, 1995, **48**, 91 .
23. F. Serra, S. M. Eaton, R. Cerbino, M. Buscaglia, G. Cerullo, R. Osellame and T. Bellini, Adv. Funct. Mater., 2013 DOI:10.1002/adfm.201203792.
24. E. M. Terentjev, Phys. Rev. E: Stat. Phys., Plasmas, Fluids, Relat. Interdiscip. Top., 1995, **51**, 1330–1337 .
25. O. V. Kuksenok, R. W. Ruhwandl, S. V. Shiyanovskii and E. M. Terentjev, Phys. Rev. E: Stat. Phys., Plasmas, Fluids, Relat. Interdiscip. Top., 1996, **54**, 5198–5203 .
26. H. Stark, Phys. Rep., 2001, **351**, 387–474 .
27. H. Mori and H. Nakanishi, J. Phys. Soc. Jpn., 1988, **57**, 1281–1286 .
28. J.-C. Loudet, P. Barois and P. Poulin, Nature, 2000, **407**, 611–613 .
29. O. Mondain-Monval, J. C. Dedieu, T. Gulik-Krzywicki and P. Poulin, Eur. Phys. J. E: Soft Matter Biol. Phys., 1999, **12**, 167–170 .
30. J.-C. Loudet, P. Barois, P. Auroy, P. Keller, H. Richard and P. Poulin, Langmuir, 2004, **20**, 11336–11347 .
31. K. Kita, M. Ichikawa and Y. Kimura, Phys. Rev. E: Stat., Nonlinear, Soft Matter Phys., 2008, **77**, 041702 .
32. R. W. Ruhwandl and E. M. Terentjev, Phys. Rev. E: Stat., Nonlinear, Soft Matter Phys., 1997, **55**, 2958 .
33. O. Guzmán, E. B. Kim, S. Grollau, N. L. Abbott and J. J. de Pablo, Phys. Rev. Lett., 2003, **91**, 235507 .
34. O. Guzmán, N. L. Abbott and J. J. de Pablo, J. Polym. Sci., Part B: Polym. Phys., 2005, **43**, 1033–1040 .
35. S. Žumer, Modeling of constrained nematic order: from defects to colloidal structures, Plenary talk at 21st International Liquid Crystal Conference, Keystone, Colorado, July 2–7, 2006 .
36. M. Ravnik, M. Škarabot, S. Žumer, U. Tkalec, I. Poberaj, D. Baič, N. Osterman and I. Muševič, Phys. Rev. Lett., 2007, **99**, 247801 .
37. M. Ravnik and S. Žumer, Soft Matter, 2009, **5**, 269–274 .
38. M. Ravnik and S. Žumer, Soft Matter, 2009, **5**, 4520–4525 .
39. U. Tkalec, M. Ravnik, S. Čopar, S. Žumer and I. Muševič, Science, 2011, **333**, 62–65 .
40. V. S. R. Jampani, M. Škarabot, M. Ravnik, S. Čopar, S. Žumer and I. Muševič, Phys. Rev. E: Stat., Nonlinear, Soft Matter Phys., 2011, **84**, 031703 .
41. T. A. Wood, J. S. Lintuvuori, A. B. Schofield, D. Marenduzzo and W. C. K. Poon, Science, 2011, **334**, 79–83 .
42. T. Araki and H. Tanaka, Phys. Rev. Lett., 2006, **97**, 127801 .
43. G. P. Crawford and S. Žumer, Liquid Crystals in Complex Geometries Formed by Polymer and Porous Networks, Taylor and Francis, London, 1996 .
44. I. Muševič and S. Žumer, Nat. Mater., 2011, **10**, 266–268 .
45. Z. Bradač, S. Kralj and S. Žumer, Phys. Rev. E: Stat. Phys., Plasmas, Fluids, Relat. Interdiscip. Top., 1998, **58**, 7447–7454 .
46. S. Kralj and S. Žumer, Phys. Rev. E: Stat. Phys., Plasmas, Fluids, Relat. Interdiscip. Top., 1995, **51**, 366–379 .
47. A. Shams, X. Yao, J. O. Park, M. Srinivasarao and A. D. Rey, Soft Matter, 2012, **8**, 11135 .
48. T. Araki, M. Buscaglia, T. Bellini and H. Tanaka, Nat. Mater., 2011, **10**, 303–309 .
49. F. Serra, K. C. Vishnubhatla, M. Buscaglia, R. Cerbino, R. Osellame, G. Cerullo and T. Bellini, Soft Matter, 2011, **7**, 10945–11032 .
50. D. L. Stein, Phys. Rev. A: At., Mol., Opt. Phys., 1979, **19**, 1708–1711 .
51. X. Wu, W. I. Goldburg, M. X. Liu and J. Z. Xue, Phys. Rev. Lett., 1992, **69**, 470–473 .
52. T. Bellini, N. Clark, C. Muzny, L. Wu, C. Garland, D. Schaefer and B. Oliver, Phys. Rev. Lett., 1992, **69**, 788–791 .
53. G. S. Iannacchione, G. P. Crawford, S. Žumer, J. W. Doane and D. Finotello, Phys. Rev. Lett., 1993, **71**, 2595–2598 .
54. T. Bellini, M. Buscaglia, C. Chiccoli, F. Mantegazza, P. Pasini and C. Zannoni, Phys. Rev. Lett., 2002, **88**, 245506 .
55. D. Kang, J. E. Maclennan, N. A. Clark, A. A. Zakhidov and R. H. Baughman, Phys. Rev. Lett., 2001, **86**, 4052–4055 .
56. M. Buscaglia, T. Bellini, C. Chiccoli, F. Mantegazza, P. Pasini, M. Rotunno and C. Zannoni, Phys. Rev. E: Stat., Nonlinear, Soft Matter Phys., 2006, **74**, 011706 .

57. D. E. Feldman, Phys. Rev. Lett., 2000, **84**, 4886–4889 .
58. F. Serra, M. Buscaglia and T. Bellini, Mater. Today, 2011, **14**, 488–494 .
59. P. A. Lebwohl and G. Lasher, Phys. Rev. A: At., Mol., Opt. Phys., 1972, **6**, 426–429 .
60. A. Maritan, M. Cieplak, T. Bellini and J. Banavar, Phys. Rev. Lett., 1994, **72**, 4113–4116 .
61. J. Ilnytskyi, S. Sokołowski and O. Pizio, Phys. Rev. E: Stat. Phys., Plasmas, Fluids, Relat. Interdiscip. Top., 1999, **59**, 4161–4168 .
62. M. Rotunno, M. Buscaglia, C. Chiccoli, F. Mantegazza, P. Panisi, T. Bellini and C. Zannoni, Phys. Rev. Lett., 2005, **94**, 097802 .
63. L. Petridis and E. M. Terentjev, Phys. Rev. E: Stat., Nonlinear, Soft Matter Phys., 2006, **74**, 051707 .
64. J. M. Fish and R. L. C. Vink, Phys. Rev. Lett., 2010, **105**, 147801 .
65. J. Yamamoto and H. Tanaka, Nature, 2001, **409**, 321–325 .
66. J. Fukuda and H. Yokoyama, Eur. Phys. J. E: Soft Matter Biol. Phys., 2001, **4**, 389–396 .
67. R. Yamamoto, Phys. Rev. Lett., 2001, **87**, 075502 .
68. T. Araki and H. Tanaka, J. Phys.: Condens. Matter, 2006, **18**, L193–L203 .
69. H. Tanaka and T. Araki, Chem. Eng. Sci., 2006, **61**, 2108–2141 .
70. C. Völtz, Y. Maeda, Y. Tabe and H. Yokoyama, Phys. Rev. Lett., 2006, **97**, 227801 .
71. E. B. Kim, O. Guzmán, S. Grollau, N. L. Abbott and J. J. de Pablo, J. Chem. Phys., 2004, **121**, 1949–1961 .
72. S. Čopar and S. Žumer, Phys. Rev. Lett., 2011, **106**, 177801 .
73. S. Čopar, T. Porenta and S. Žumer, Phys. Rev. E: Stat., Nonlinear, Soft Matter Phys., 2011, **84**, 051702 .
74. R. D. Kamien, Rev. Mod. Phys., 2002, **74**, 953–971 .
75. K. Binder and W. Kob, Glassy materials and disordered solid, World Scientific Pub., 2005
76. A. T. Ogielski, Phys. Rev. B: Condens. Matter Mater. Phys., 1985, **32**, 7384–7398 .
77. P. E. Lammert, D. S. Rokhsar and J. Toner, Phys. Rev. E: Stat. Phys., Plasmas, Fluids, Relat. Interdiscip. Top., 1995, **52**, 1778–1800 .
78. J. Fukuda, H. Stark, M. Yoneya and H. Yokoyama, J. Phys.: Condens. Matter, 2004, **16**, S1957 .
79. H. Stark and D. Ventzki, Phys. Rev. E: Stat., Nonlinear, Soft Matter Phys., 2001, **64**, 031711 .
80. H. Stark and D. Ventzki, Europhys. Lett., 2002, **57**, 60 .
81. O. D. Lavrentovich, I. Lazo and O. P. Pishnkyak, Nature, 2010, **467**, 947–950 .
82. B. T. Gettelfinger, J. A. Moreno-Razo, G. M. Koenig Jr, J. P. Hernández-Ortiz, N. L. Abbott and J. J. de Pablo, Soft Matter, 2010, **6**, 896–901 .
83. A. Sengupta, U. Tkalec and C. Bahr, Soft Matter, 2011, **7**, 6542–6459 .
84. A. Sengupta, C. Pieper, J. Enderlein, C. Bahr and S. Herminghaus, Soft Matter, 2013, **9**, 1937 .
85. A. Sengupta, U. Tkalec, M. Ravnik, J. M. Yeomans, C. Bahr and S. Herminghaus, Phys. Rev. Lett., 2013, **110**, 048303 .
86. T. Araki, Phys. Rev. Lett., 2012, **109**, 257801 .
87. T. Araki, Proceedings of the 4th International Symposium on "Slow Dynamics in Complex Systems", 2012 .
88. Y. Iwashita and H. Tanaka, Phys. Rev. Lett., 2003, **90**, 45501 .
89. F. Nallet and J. Prost, Europhys. Lett., 2007, **4**, 307 .
90. Y. Iwashita and H. Tanaka, Phys. Rev. E: Stat., Nonlinear, Soft Matter Phys., 2008, **77**, 041706 .
91. D. K. Yoon, M. C. Choi, Y. H. Kim, M. W. Kim, O. D. Lavrentovich and H.-T. Jung, Nat. Mater., 2007, **6**, 866–870 .
92. Y. H. Kim, D. K. Yoon, H. S. Jeong, O. D. Lavrentovich and H. T. Jung, Adv. Funct. Mater., 2011, **21**, 610–627

CHAPTER 5

MESOPOROUS INORGANIC SALTS WITH CRYSTAL DEFECTS: UNUSUAL CATALYSTS AND CATALYST SUPPORTS

Xinchen Kang [a], Wenting Shang [a], Qinggong Zhu [a], Jianling Zhang [a], Tao Jiang [a], Buxing Han [a], Zhonghua Wu [b], Zhihong Li [b] and Xueqing Xing [b]

[a]Beijing National Laboratory for Molecular Sciences, Key Laboratory of Colloid and Interface and Thermodynamics, Institute of Chemistry, Chinese Academy of Sciences, Beijing 100190, China.

[b]Beijing Synchrotron Radiation Facility, Institute of High Energy Physics, Chinese Academy of Sciences, Beijing 100049, China

ABTRACT

We proposed a strategy to synthesize mesoporous inorganic salt particles using the special properties of ionic liquid (IL) mixtures, and hollow mesoporous LaF_3, NdF_3, and YF_3 particles were synthesized and characterized using different techniques. The size of the mesopores in the salt particles was about 4 nm, and the materials were full of crystal defects. The LaF_3, NdF_3 and YF_3 particles were used as the catalysts for the cyanosilylation reaction of benzaldehyde using trimethylsilyl cyanide, and Ru/LaF_3 and Ru/NdF_3, in which Ru nanocatalysts were supported on the LaF_3 and NdF_3 particles with mesopores, were used to catalyze hydrogenations of benzene to cyclohexane and levulinic acid (LA) to γ-valerolactone (GVL). It was discovered that the activities of these catalysts were unprecedentedly high for these reactions. Detailed study showed that both the crystal defects and the mesopores in the salt particles played crucial roles for the extremely high catalytic activity.

INTRODUCTION

Porous materials have wide-ranging applications due to their admirable and intriguing structures. Common porous materials include porous carbon materials,[1] porous polymers,[2] zeolite materials,[3] metal–organic frameworks (MOFs),[4] and covalent organic frameworks (COFs),[5] and so on. Porous materials have a variety of applications or potential applications, such as catalysis,[6] gas separation, capture and storage,[7] and electrode materials.[8] Inorganic salts are composed of cations and anions of small size. The strong interaction among the ions leads to a close-packed structure of the salts, which is more feasible to form a perfect crystal. The reports about preparation of porous salts are very limited. Cong *et al.* synthesized NaCl and KCl with macropores (380 nm in diameter) using copolymer colloids as the template.[9] Chu *et al.* prepared porous FeF_3 particles consisting of small nanoparticles through two steps: $Fe(OH)_3$synthesis and FeF_3 fabrication.[10]

Ionic liquids (ILs) have received much interest due to their special properties such as negligible vapor pressure, good thermal stability, high conductivity, and strong solvent power for both organic and inorganic substances.[11] Especially, ILs have a designable nature.[12] These unique properties make them have great potential in applications in different fields, such as material synthesis,[13] chemical reactions,[14] extraction and fractionation,[15] and gas absorption.[16]

Catalysis is quite prevalent in chemical reactions and chemical industry.[17] Supported metal catalysts are widely

Citation: Xinchen Kang, Wenting Shang, Qinggong Zhu, Jianling Zhang, Tao Jiang, Buxing Han, Zhonghua Wu, Zhihong Li and Xueqing Xing. Mesoporous inorganic salts with crystal defects: unusual catalysts and catalyst supports. Chem. Sci., 2015,6, 1668-1675, DOI: 10.1039/C4SC03736G used because the metal components are active for various reactions.[18] It is well known that for supported catalysts, catalyst supports often have great influence on the activity, selectivity, and stability of the catalysts.[19] Mesoporous inorganic salt particles may have many advantages in catalysis which are not known because their synthesis is challenging.

Herein, we proposed a strategy to synthesize porous inorganic salts using IL–organic salt mixtures. LaF_3, NdF_3, and YF_3 particles with mesopores were successfully synthesized. As examples of applications, the mesoporous salt particles were used both as the catalysts directly and supports of Ru nanocatalysts. It was discovered that the salts and the salt supported catalysts exhibited unprecedented activity for different reactions. Systematic study indicated that both the crystal defects and mesopores contributed significantly to the very high catalytic activity. To the best our knowledge, this is the first work to use mesoporous inorganic salts as catalysts and catalyst supports.

Results and discussion

We synthesized the inorganic salts in tetrabutylammonium hexafluorophosphate ($N(Bu)_4PF_6$) and IL 1-octyl-3-methylimidazolium chloride (OmimCl) mixtures. $N(Bu)_4PF_6$ is an organic salt with a melting point of 246–247 °C.[20] OmimCl is liquid at room temperature. Fig. S1† shows the solubility of $N(Bu)_4PF_6$ (x_2, in mole fraction) in OmimCl determined in this work. The solubility of $N(Bu)_4PF_6$ in OmimCl decreases with decreasing temperature. Therefore, some $N(Bu)_4PF_6$ will precipitate as its content in the OmimCl–$N(Bu)_4PF_6$ mixture exceeds the solubility (Fig. S1†).

We synthesized LaF_3 in OmimCl–$N(Bu)_4PF_6$ mixtures at x_2 (mole fraction of $N(Bu)_4PF_6$ in OmimCl–$N(Bu)_4PF_6$ mixture) of 0.50 and 0.33 using $La(NO_3)_3$ as the metal precursor. Some precipitated $N(Bu)_4PF_6$ particles existed in the OmimCl–$N(Bu)_4PF_6$ mixtures at these conditions. The LaF_3 was generated from $La(NO_3)_3$ and $N(Bu)_4PF_6$, and the reaction equation is shown in Scheme S1.†[21] The X-ray diffraction (XRD) patterns of the as-prepared materials demonstrate clearly the formation of LaF_3 (Fig. 1). The half-peak width in the XRD patterns was much larger than the commercial LaF_3 (Fig. 1), indicating the poor crystallinity and existence of significant amounts of crystal defects in the LaF_3 prepared in this work. The scanning electron microscopy (SEM) and transmission electron microscopy (TEM) images of the LaF_3 materials are given in Fig. 2. The size of the mesopores in the particles was about 4 nm (Fig. 2e), which are intracrystal pores. The size of the particles was about 400 nm and each particle had a hollow core of about 80 nm.

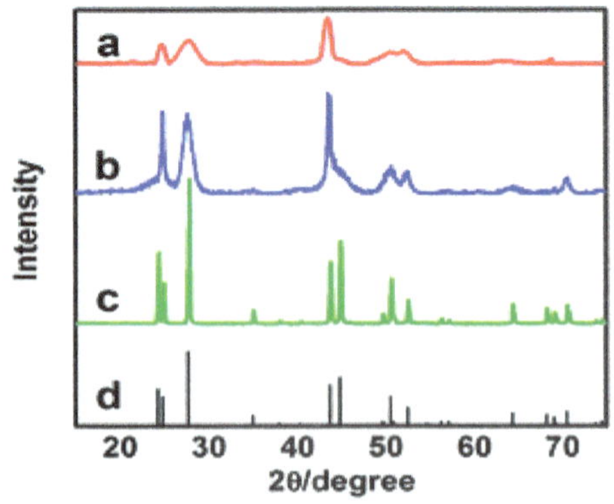

Figure 1 XRD patterns of LaF_3: (a) LaF_3 synthesized at $x_2 = 0.5$; (b) LaF_3 synthesized at $x_2 = 0.33$; (c) commercial LaF_3; and (d) standard pattern.

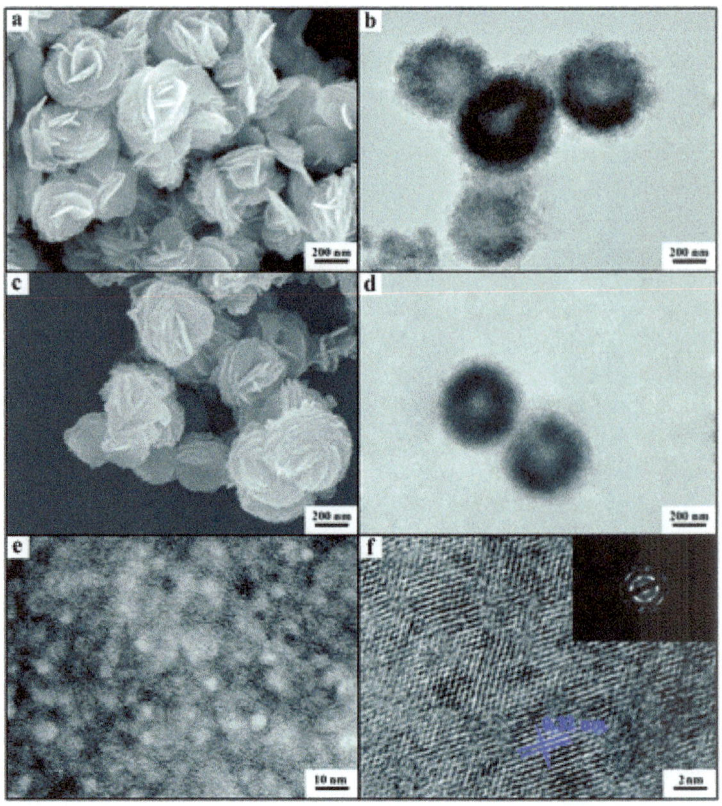

Figure 2 SEM and TEM images of mesoporous LaF_3 synthesized in the OmimCl–$N(Bu)_4PF_6$ mixtures: (a) SEM image of the LaF_3 synthesized at $x_2 = 0.5$; (b) TEM image of the LaF_3 synthesized at $x_2 = 0.5$; (c) SEM image of the LaF_3 synthesized at $x_2 = 0.33$; (d)

TEM image of the LaF_3 synthesized at $x_2 = 0.33$; (e) HRTEM image to show the pores on the walls of the LaF_3 synthesized at $x_2 = 0.5$; and (f) HRTEM image to show the crystal structure and the SAED pattern (the inset pattern) of the LaF_3 synthesized at $x_2 = 0.5$./

The porous LaF_3 particles prepared were further characterized using the N_2 adsorption/desorption method, and the adsorption/desorption isotherms and the pore size distributions of the LaF_3 particles are given in Fig. 3. The pore size distribution curve of the LaF_3 particles prepared had two peaks, which centered at about 4 nm and 55 nm, corresponding to the mesopores in the walls and hollow cores of the particles, respectively. The size of the pores determined by the N_2 adsorption/desorption method was roughly consistent with that observed from the TEM images (Fig. 2). The surface areas obtained from the BET (Brunauer, Emmett, and Teller) method and the pore volumes calculated from the Barrett–Joyner–Halenda method are presented inTable 1. The results further demonstrated the mesoporous character.

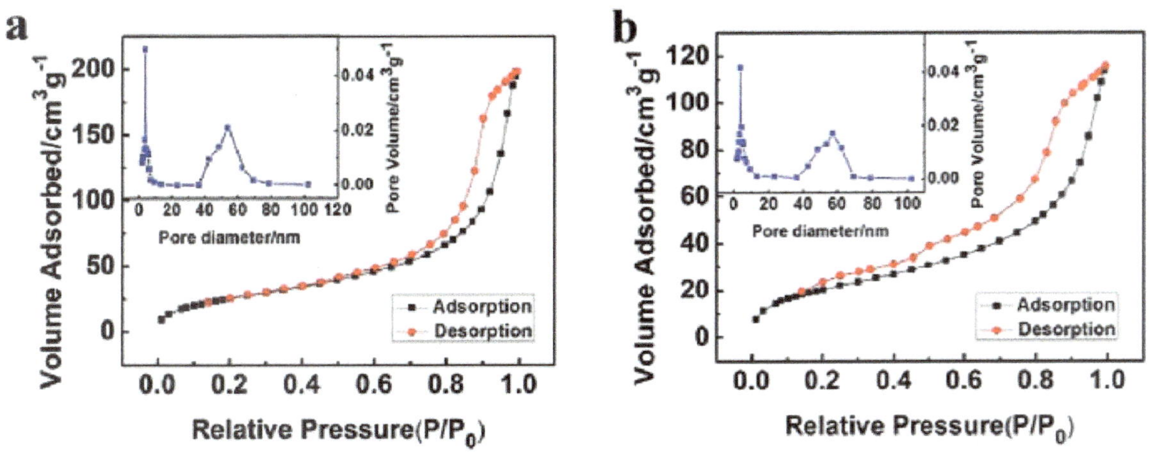

Figure 3 N_2 adsorption/desorption isotherms and pore size distribution (the insets) of the LaF_3particles synthesized at $x_2 = 0.5$ (a) and 0.33 (b).

Table 1 The BET surface area (S) and total pore volume (V) of different materials

Entry	Samples	S (m^2 g^{-1})	V (cm^3 g^{-1})
a Porous LaF_3 synthesized at $x_2 = 0.5$ (Fig. 2a and b). b Porous LaF_3 synthesized at $x_2 = 0.33$ (Fig. 2c and d). c Porous LaF_3 synthesized at $x_2 = 0.17$ (Fig. S3a and b). d Porous NdF_3 synthesized at $x_2 = 0.5$ (Fig. 8a and b). e Porous YF_3 synthesized at $x_2 = 0.5$ (Fig. 8c and d). f Commercial LaF_3 (Fig. S4a and b). g La-MOF prepared in this work (Fig. S4c and d).			
1	LaF_3[a]	97.9	0.32
2	LaF_3[b]	77.6	0.26
3	LaF_3[c]	59.1	0.16
4	NdF_3[d]	98.7	0.34
5	YF_3[e]	95.7	0.28
6	LaF_3[f]	0.48	0.002
7	La-MOF[g]	28.5	0.05

The compactness of the materials can be characterized by the small angle X-ray scattering (SAXS) technique *via* analyzing the fractal dimension D.[22] In this work we conducted a SAXS study (Fig. 4). According to the SAXS data, it can be known that the mass fractal existed in the LaF_3 materials, indicating that the materials had a loose structure with lots of crystal defects,[22] which agrees with the fact that the salt particles had poor crystallinity as shown by the XRD patterns (Fig. 1). Moreover, the LaF_3 synthesized had plenty of micropores ranging from 0.5 nm to 2 nm, and the pore size distribution is shown in Fig. S2.† The surface area of the micropores was 1.0 m^2 g^{-1}, which further confirms the existence of crystal defects in the LaF_3 materials. The crystal defects in the LaF_3 materials originate mainly from the mesopores, edges, and tips in the salts.

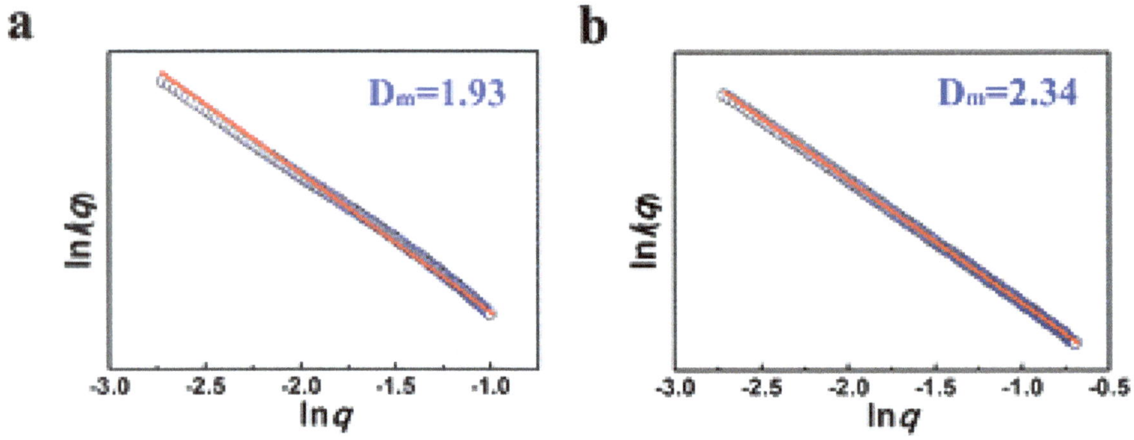

Figure 4 Mass fractal dimension (D_m) from SAXS curves of the as-prepared LaF_3 materials synthesized at $x_2 = 0.5$ (a), 0.33 (b).

In order to study the reason for the formation of the hollow cores and mesopores in the LaF_3 particles synthesized, we studied the size and shape of the precipitated $N(Bu)_4PF_6$ particles in the OmimCl–$N(Bu)_4PF_6$ mixtures using the SAXS technique, which is commonly used for characterizing the size and shape of particles dispersed in liquids.[23] The SAXS curves of the mixtures with different compositions are presented in Fig. 5a. The data in the small-angle region ($q < 1$) were used to obtain the size of the precipitated $N(Bu)_4PF_6$ particles. The generalized indirect Fourier transformation gives the pair-distance distribution function, $p(\mathrm{r})$, which is shown in Fig. 5b. The curves were nearly symmetric, suggesting that the precipitated $N(Bu)_4PF_6$particles in the OmimCl–$N(Bu)_4PF_6$ mixtures were spherical.[23] The size of the $N(Bu)_4PF_6$ particles calculated from the SAXS data agreed roughly with that of the hollow cores of the LaF_3particles shown in Fig. 2. SAXS is also a useful technique to study the microstructure of ILs. It has highlighted the existence of the nano-scale aggregates, and the size can be estimated using Bragg's diffraction law $d = 2\pi/q$.[24] The size of the IL aggregates calculated using this method was 3 nm. These could act as the templates to form the mesopores of the LaF_3 particles.

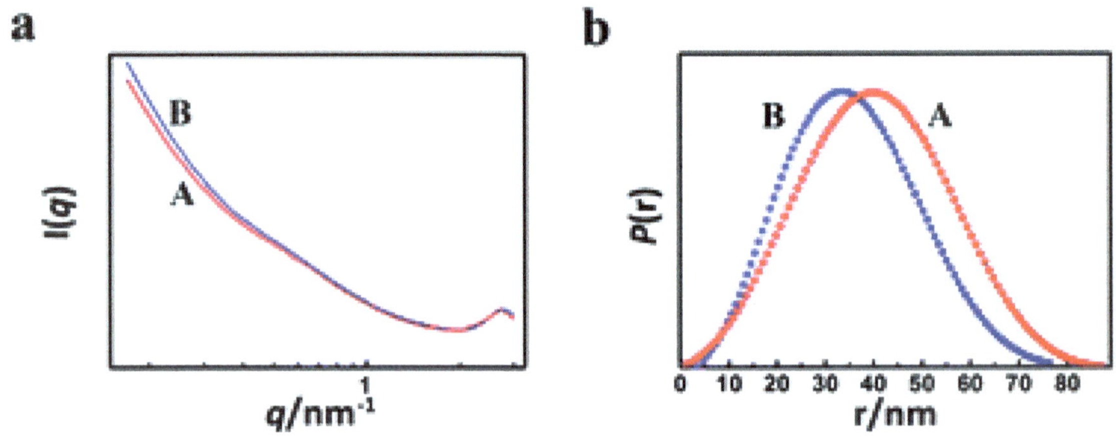

Figure 5 SAXS curves (a) and normalized pair-distance distribution function curves (b) of the OmimCl–$N(Bu)_4PF_6$ mixtures at $x_2 = 0.5$ (A) and 0.33 (B).

On the basis of the experimental results we can discuss the formation processes of the mesoporous particles, which are shown schematically in Fig. 6. In the OmimCl–$N(Bu)_4PF_6$mixtures, some $N(Bu)_4PF_6$ exists as the precipitated particles at $x_2 = 0.5$ and 0.33, and the rest of the $N(Bu)_4PF_6$ is dissolved in the solution, as discussed above. The $La(NO_3)_3$ reacts with the $N(Bu)_4PF_6$ particles at the surface to generate LaF_3. At the same time, the $La(NO_3)_3$ also reacts with the $N(Bu)_4PF_6$ in the solution. The LaF_3 generated in the solution aggregates with the LaF_3 nanocrystals on the surface of the $N(Bu)_4PF_6$ particles. Some IL aggregates are entrained when the LaF_3 particles aggregate around the $N(Bu)_4PF_6$ particles. The regular hollow LaF_3 particles with mesopores are formed after removing the OmimCl and $N(Bu)_4PF_6$ by washing with acetone. In this process, the $N(Bu)_4PF_6$ particles and the entrained IL aggregates act as the templates for the core and the mesopores, respectively (Fig. 2 and 6). In order to get the evidence that the $N(Bu)_4PF_6$ particles act as the template for the formation of the hollow core, we also synthesized the LaF_3 particles in the OmimCl–$N(Bu)_4PF_6$ mixture at $x_2 = 0.17$, and the SEM and TEM images, XRD patterns, and N_2 adsorption/desorption isotherms are given in Fig. S3.† In this e, all the $N(Bu)_4PF_6$ dissolves in OmimCl (Fig. S1†). $La(NO_3)_3$ reacts with the $N(Bu)_4PF_6$ in the solution due to the absence of the precipitated $N(Bu)_4PF_6$ particles, and the generated LaF_3 aggregates arbitrarily and irregular LaF_3 particles are formed (Fig. S3†). The particles with mesopores were also obtained after removing the entrained OmimCl and $N(Bu)_4PF_6$, but there was no hollow core (Fig. S3†). The results support the argument that the precipitated $N(Bu)_4PF_6$ particles act as the template for the formation of the hollow core of the salt particles at higher $N(Bu)_4PF_6$ concentrations ($x_2 = 0.5$ and 0.33).

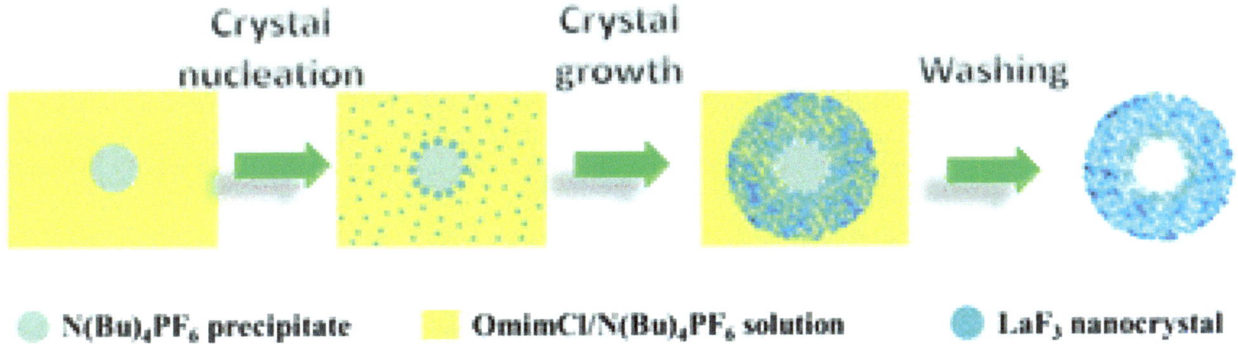

Figure 6 The schematic diagram for the formation mechanism of mesoporous LaF_3 particles with a hollow core.

We also synthesized NdF_3 and YF_3 particles in the OmimCl–$N(Bu)_4PF_6$ mixture of x_2 = 0.50 using this method. The XRD patterns, SEM and TEM images are given in Fig. 7 and 8, respectively. The porosity properties determined using the N_2 adsorption/desorption method are given in Table 1. The XRD patterns confirmed the formation of NdF_3 and YF_3 (Fig. 7). All the characterizations indicate that the morphology, structure, and porosity properties of the NdF_3 and YF_3 particles were similar to those of LaF_3 prepared, which indicates that the strategy proposed in this work is versatile.

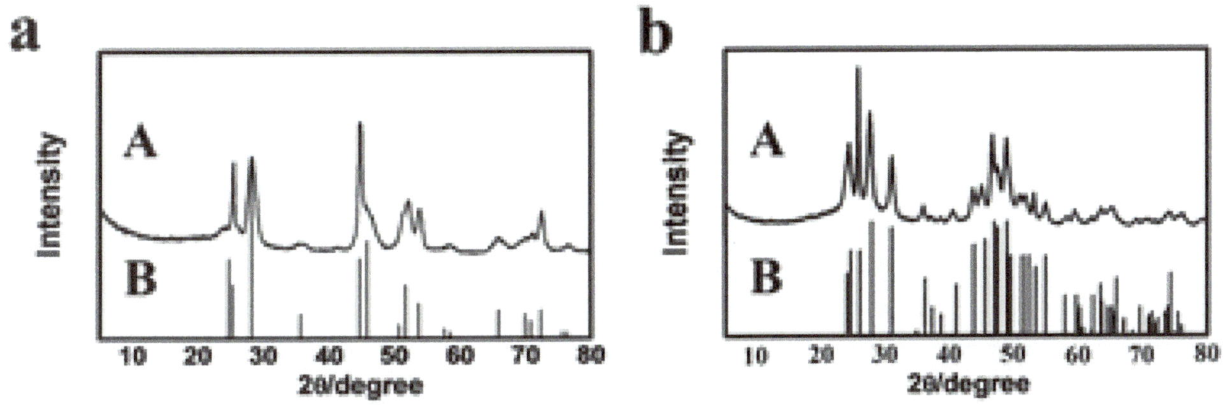

Figure 7 XRD patterns of the as-synthesized fluoride salts (A) synthesized at x_2 = 0.5 and the corresponding standard patterns (B): (a) NdF_3; and (b) YF_3.

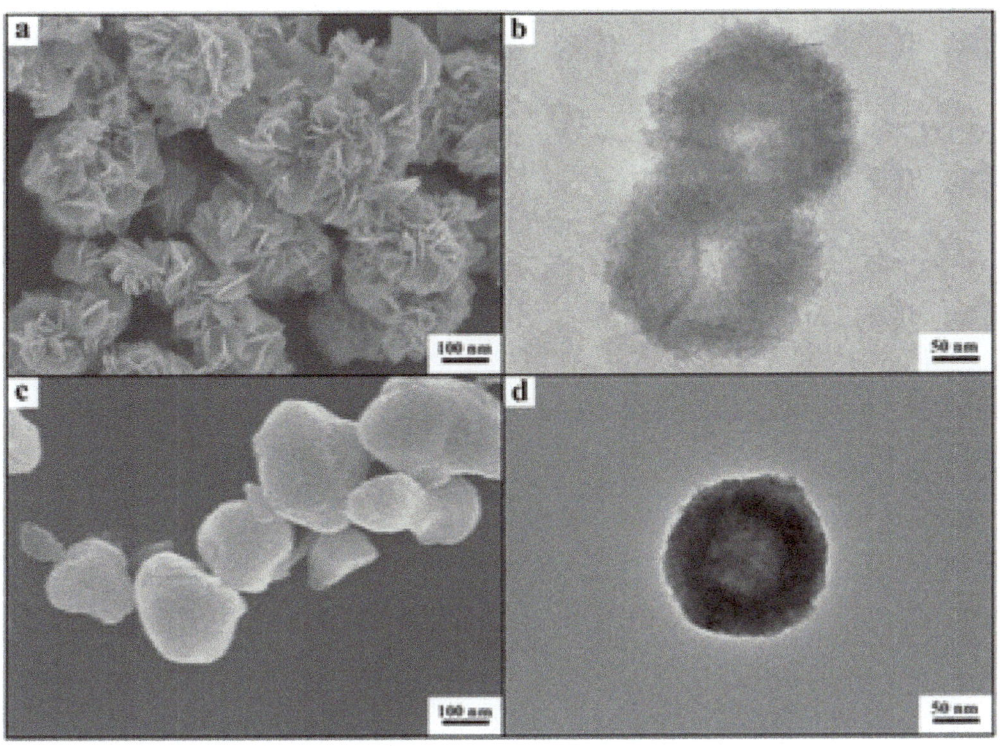

Figure 8 SEM and TEM images of NdF_3 (a and b) and YF_3 (c and d) materials synthesized at x_2= 0.5.

The mesoporous salt particles may find wide-ranging applications in different fields. They are very stable inorganic salts and are not soluble in water and common organic solvents, which are excellent characteristics of heterogeneous catalysts and catalyst supports. As examples of applications, we used the salt particles both as a catalyst directly and as the support of metal nanocatalysts for different reactions.

Cyanosilylation of carbonyl compounds provides a convenient route to produce cyanohydrins,[25] which are key intermediates in the synthesis of fine chemicals and pharmaceuticals.[26]The mesoporous LaF_3 shown in Fig. 2a and b was used as a heterogeneous catalyst in the cyanosilylation reaction of benzaldehyde and trimethylsilyl cyanide (TMSCN) to produce cyanohydrin, and the results are given in Table 2. Surprisingly, the LaF_3 particles with mesopores were extremely active and selective for the reaction (entry 1). In addition, the mesoporous NdF_3 (Fig. 8a and b) and YF_3 (Fig. 8c and d) had similar activity for this reaction (entries 2–3). We also studied the activity of commercial LaF_3 for this reaction (entry 4), and the activity was much lower (Fig. S4a and b† and Table 1). It has been reported that La metal–organic framework (La-MOF) was an efficient catalyst for this reaction.[27] In this work, we synthesized the La-BTC MOF (La-MOF hereafter). The as-prepared La-MOF was characterized by SEM and TEM (Fig. S4c and d†), XRD (Fig. S5†), and N_2 adsorption/desorption (Table 1) methods. The XRD pattern was the same as that reported previously.[28] The La-MOF was also used to catalyze the reaction, and it was demonstrated that the activity of the mesoporous LaF_3 particles was much higher than the La-MOF (entries 1 and 5) for the reaction. In order to study the effect of the defects in the mesoporous LaF_3, NdF_3 and YF_3 particles on the reaction, we calculated the turnover frequencies (TOFs) per unit surface area of all the catalysts in Table 2, which were calculated from the reaction data in Table 2 and the surface area data in Table 1. Surprisingly, the TOFs per unit surface area of the as-prepared mesoporous LaF_3, NdF_3 and YF_3 were much higher than the commercial LaF_3 that had no pores, and the La-MOF. It is well known that the mass transfer in mesopores is poorer than that at the surface of nonporous materials. Therefore, it can be concluded that the crystal defects of the mesoporous LaF_3, NdF_3 and YF_3contributed significantly to their extremely high activity. The reaction mechanism is shown in Scheme 1. The reaction is catalyzed by Lewis acid.[27] The crystal defects result in Lewis acid sites,[29] which enhances the acidity and is favourable for accelerating the reaction. The commercial LaF_3 has limited Lewis acid sites due to the high crystallinity. Thus, both the mesopores and the crystal defects in the salt particles played crucial roles for the very high catalytic activity.

Table 2 Cyanosilylation of benzaldehyde using TMSCN[a,i]

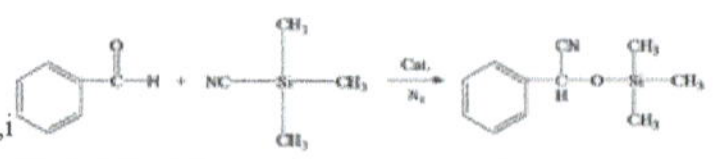

Entry	Catalysts	t (min)	C (%)	TOF_1[b] (h^{-1})	TOF_2[c] ($g\ h^{-1}\ m^{-2}$)
1	LaF_3[d]	1	100	1200	12.3
2	NdF_3[e]	1	100	1200	12.2
3	YF_3[f]	1	100	1200	12.5
4	LaF_3[g]	10	2.2	2.6	5.4
5	La-MOF[h]	10	6.4	7.7	0.27

a Reaction conditions: 2 mL benzaldehyde, 5 mol% catalyst, benzaldehyde : TMSCN molar ratio of 1 : 1.5, 50 °C, solvent free. b TOF_1 is the turnover frequency, which was calculated as moles of converted benzaldehyde per mole of catalyst per hour. c TOF_2 is the turnover frequency, which was calculated as moles of converted benzaldehyde per mole of catalyst per hour per unit area, and surface area data determined in this work (Table 1) were used in the calculation. d Porous LaF_3 (Fig. 2a and b). e Porous NdF_3 (Fig. 8a and b). f Porous YF_3 (Fig. 8c and d). g Commercial LaF_3 (Fig. S4a and b). h La-MOF prepared in this work (Fig. S4c and d). The quantitative analysis of the product was carried out using GC. i C = conversion of benzaldehyde.

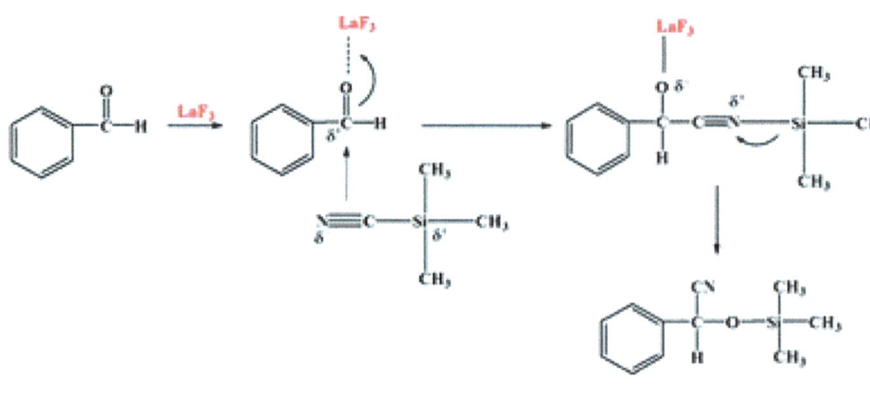

Scheme 1 The reaction mechanism for cyanosilylation of benzaldehyde to cyanohydrin

We also used the mesoporous LaF_3 and NdF_3 particles (Fig. 2a and b and 8a and b) as catalyst supports. The Ru/LaF_3 and Ru/NdF_3 were synthesized using water as the solvent, $RuCl_3$ as the precursor, and $NaBH_4$ as the reductant. The loading of Ru in the catalysts was 1.0 wt% as determined using the ICP-AES (VISTA-MPX) method. The TEM image of the Ru/LaF_3 and the size distribution of the Ru particles are shown in Fig. S6.† The Ru nanoparticles of about 1.2 nm were immobilized uniformly on the supports. The Ru nanoparticles in the Ru/NdF_3 had a similar size and size distribution.

Benzene hydrogenation to cyclohexane is a very important reaction in chemical industry.[30] We conducted the reaction using the Ru/LaF_3, Ru/NdF_3, and commercial Ru/C (TEM and SEM images are given in Fig. S6b and c†) as the catalysts at the same conditions, and the results are shown in Table 3. The catalytic activity of the Ru/LaF_3 and Ru/NdF_3 was amazingly high compared with that of the commercial Ru/C. It is obvious that the mesoporous morphology of the LaF_3 and NdF_3 particles played an important role in the reaction. It has also been reported that Lewis acids can activate aromatic rings.[31] LaF_3 and NdF_3, which are Lewis acids, activated the benzene ring and the Ru nanoparticles activated the H_2, which made the reaction more feasible to take place (Scheme 2). The crystal defects in the LaF_3 and NdF_3 lead to the existence of partial charge, which enhances the Lewis acidity and therefore the crystal defects can promote the reaction. The reusability of the Ru/LaF_3 catalyst was also studied. It was shown that the catalyst could be reused at least five times without reducing the activity, as shown in Fig. 9.

Table 3 Catalytic performances of different catalysts for benzene hydrogenation to cyclohexane[a,f] 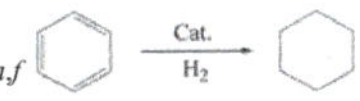

Entry	Catalysts	t (h)	T (°C)	Y (%)	TOF[b] (h^{-1})
1	Ru/LaF_3[c]	0.22	50	>99	11 364
2	Ru/LaF_3[c]	0.77	25	>99	3247
3	Ru/NdF_3[d]	0.33	50	>99	7576
4	Ru/C[e]	8	50	95.5	298

a Reaction conditions: 2 mL benzene, benzene/Ru (mol mol^{-1}) = 2500, 4 MPa H_2. b TOF was calculated as moles of converted benzene per mole of Ru per hour. c Porous LaF_3 (Fig. 2a and b) as support. d Porous NdF_3 (Fig. 8a and b) as support. e The commercial Ru/C catalyst (Fig. S6b and c). The quantitative analysis of the product was carried out using GC. f Y = yield of cyclohexane.

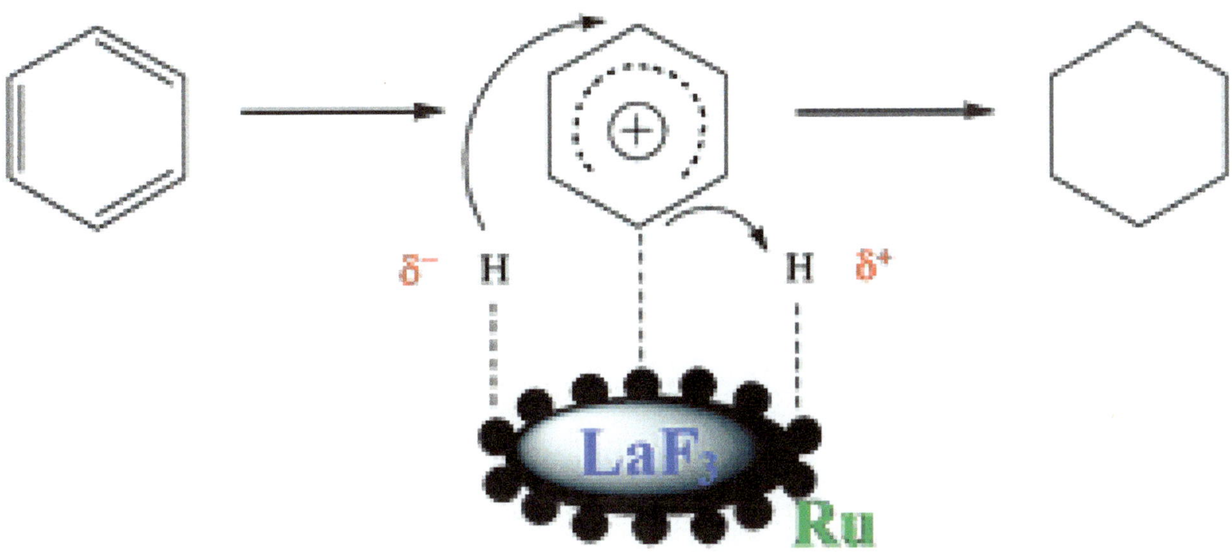

Scheme 2 The reaction mechanism for benzene hydrogenation to cyclohexane.

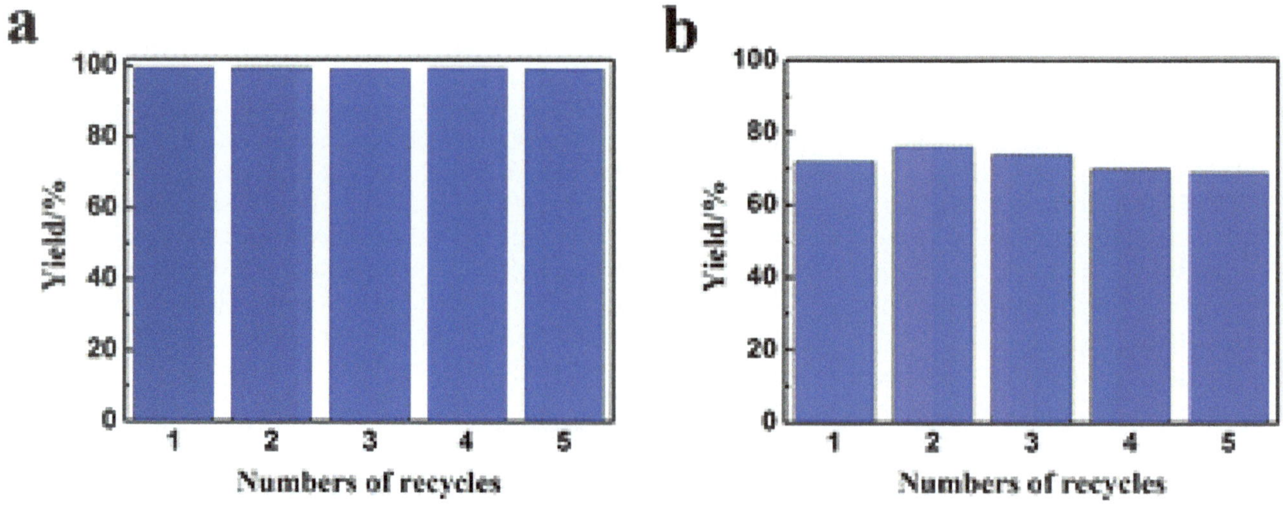

Figure.9 The reusability of the Ru/LaF_3 catalyst for benzene hydrogenation to cyclohexane at 50 °C, 0.22 h (a) and 25 °C, 0.5 h (b). The other conditions were the same as in Table 1.

LA hydrogenation to GVL is an important reaction in biomass utilization, and Ru is a very efficient catalyst for the reaction.[32] In this work, we carried out the reaction catalyzed by the Ru/LaF_3 prepared in this work and commercial Ru/C catalyst under solvent-free conditions, and the results are listed in Table 4. The dependence of the conversion on reaction time at some typical conditions is shown in Fig. S7.† The Ru/LaF_3 had much higher activity than the commercial Ru/C and the supported Ru nanocatalysts reported by other researchers. The LaF_3interacts strongly with the carbonyl group by the Lewis acid–base interaction, which may be favourable to the reaction. In addition, an esterification reaction is feasible under acid and alkali conditions and the LaF_3 as a Lewis acid can promote the esterification process. Therefore, the special morphology and structure, Lewis acid nature, and enhanced Lewis acidity by the defects of the LaF_3 support are the main reasons for the extremely high activity and selectivity of the Ru/LaF_3 catalyst for the reaction (Scheme 3).[33]

Table 4 Catalytic performances of the Ru/LaF_3 and commercial Ru/C catalysts for LA hydrogenation to GVL[a,f]

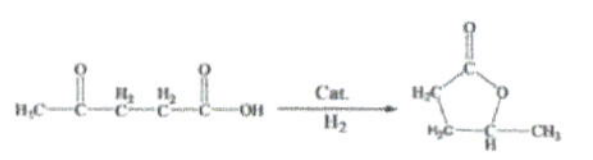

Entry	Catalysts	T (°C)	t (h)	C (%)	S (%)	TOF[b] (h^{-1})
1	Ru/LaF_3[c]	130	0.5	>99	>99	4000
2	Ru/LaF_3[c]	100	1.25	>99	>99	1600
3	Ru/LaF_3[c]	70	3.5	>99	>99	571
4	Ru/LaF_3[c]	40	9	>99	>99	222
5	Ru/C[d]	130	7	>99	>99	286
6	Ru/C[d]	100	18	>99	>99	111
7	Ru/Al_2O_3[e]	70	3	24	96	137

a Reaction conditions: 2 mL LA, LA/Ru (mol mol^{-1}) = 2000, initial pressure of H_2 was 6 MPa, solvent free. b TOF was calculated as moles of converted LA per mole of Ru per hour. c Porous LaF_3 (Fig. 2a and b) as support. dThe commercial Ru/C catalyst (Fig. S6b and c). e The catalyst reported in ref. 32. The quantitative analysis of the product was carried out by GC. f C = conversion of LA, S = selectivity of GVL.

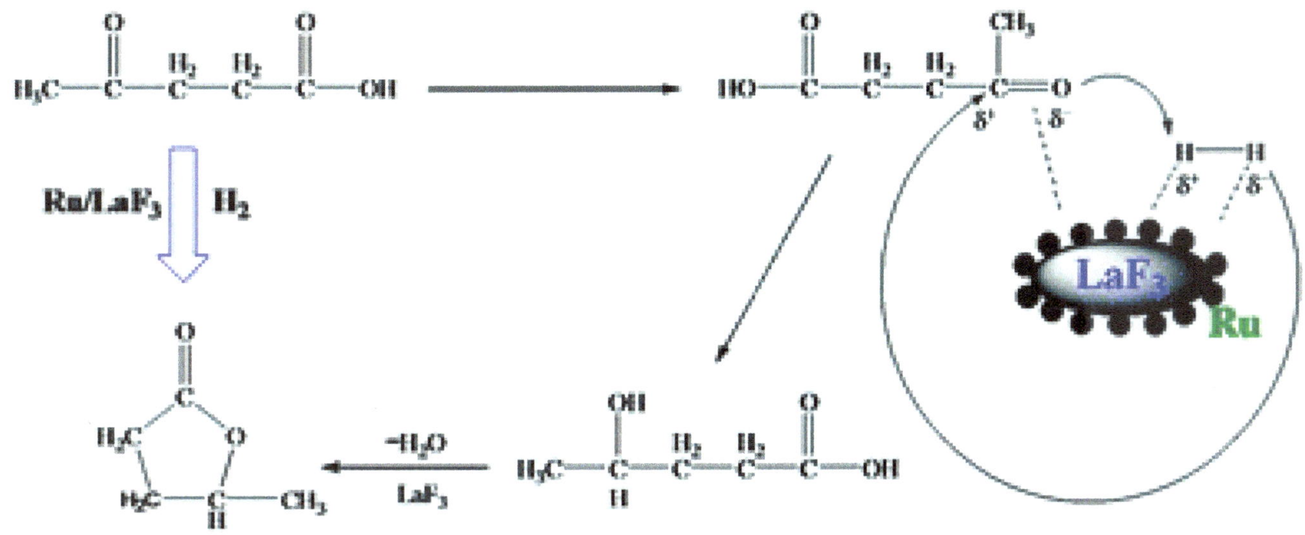

Scheme 3 The reaction mechanism for LA hydrogenation to GVL.

CONCLUSIONS

In summary, a protocol to synthesize mesoporous inorganic salts was proposed in this work, and mesoporous LaF_3, NdF_3 and YF_3 were prepared successfully and were full of crystal defects. The catalytic activity of the mesoporous LaF_3, NdF_3 and YF_3 particles could be more than 800 times higher than that of the commercial LaF_3 without pores for the cyanosilylation reaction of benzaldehyde using trimethylsilyl cyanide. The catalytic activity of the supported catalysts Ru/LaF_3 and Ru/NdF_3, in which Ru nanocatalysts were supported on the mesoporous salt particles, was also extremely high for the hydrogenations of benzene to cyclohexane and LA to GVL. The abundant crystal defects in the mesoporous salt particles contributed significantly to the excellent catalytic performance of the catalysts. We believe that this method can also be used to fabricate some other mesoporous inorganic salts, which may have unique features in catalysis, extraction and fractionation, gas absorption, and other fields.

EXPERIMENTAL

Materials

OmimCl (purity > 99%) and $N(Bu)_4PF_6$ (purity > 99%) were purchased from the Centre of Green Chemistry and Catalysis, LICP, . LaF_3 (purity > 99.99%), $La(NO_3)_3{\cdot}6H_2O$ (A. R. grade), $Nd(NO_3)_3{\cdot}6H_2O$ (A. R. grade), $Y(NO_3)_3{\cdot}6H_2O$ (A. R. grade), $RuCl_3{\cdot}3H_2O$ (Ru ≥ 37%), benzene (A. R. grade), ethanol (A. R. grade), cyclohexane (A. R. grade), toluene (A. R. grade), and *n*-butyl alcohol (A. R. grade) were provided by Sinopharm Chemical Reagent Co., Ltd. Benzaldehyde (purity > 98%), trimethylsilyl cyanide (purity > 97%), H_3BTC (purity > 99%), LA (purity > 99%) and GVL (purity > 99%) were obtained from J & K Scientific Ltd. $NaBH_4$ (purity ≥ 98%) was provided by Alfa Aesar China (Tianjin) Co., Ltd. The commercial Ru/C catalyst was purchased from Baoji Ruike Corporation, China.

The solubility of $N(Bu)_4PF_6$ in OmimCl

In the experiment, known amounts of $N(Bu)_4PF_6$ and OmimCl were mixed in a glass tube with an inner diameter of 10 mm, which were immersed in an oil bath of desired temperature. The total mass of the mixture was 5.0 g. The mixture was heated to 160 °C to obtain a homogeneous solution. Then the temperature was decreased slowly until the solution became cloudy, indicating that $N(Bu)_4PF_6$ began to precipitate. At this temperature, the mixture became clear again as the temperature was increased. The process was repeated five times to determine the saturation temperature.

SAXS study

The apparatus and the procedures were similar to that used to study the microstructure of IL gels.[13b] SAXS experiments were carried out at Beamline 1W2A at the Beijing Synchrotron Radiation Facility. The data were collected using a CCD detector (MAR) with a maximum resolution of 3450 × 3450 pixels. The wavelength of the X-ray was 1.54 Å, and the distance of the sample to detector was 1.596 m. For OmimCl/$N(Bu)_4PF_6$ samples, desired amounts of OmimCl and $N(Bu)_4PF_6$ were mixed in a glass tube at 160 °C to obtain a homogeneous solution. Then the glass tube was cooled to 85 °C. In a typical experiment, some sample was added into the sample cell with a heating unit around the cell, and the X-ray scattering data were recorded. For the LaF_3 samples, the sample was smeared on the sample cell at room temperature, and the other procedures were similar. The 2-D SAXS images were obtained from the detector and then transformed into the profiles of intensity (I) *vs.* wavevector (q) using the software SAXS Data Pre-process V2.0.0. The pair-distance distribution function p(r) was obtained from SAXS data by using an Irena tool suite within the Igor pro and Gnom application software. The fractal dimension D of LaF_3 materials was calculated and analyzed using the method reported.[22]

Synthesis of the Porous Salt Particles

We describe mainly the procedures to synthesize LaF_3 because those to synthesize NdF_3 and YF_3 were similar. In a typical experiment, desired amounts of OmimCl and $N(Bu)_4PF_6$ were mixed in a conical flask of 50 mL at 160 °C until the mixture became a homogeneous solution. The total mass of the mixture was 18 g. Then 1.5 mmol of $La(NO_3)_3$ was added to the system under vigorous stirring, and then the mixture was cooled to 85

°C in 10 min. The system was maintained at this temperature for 72 h. The obtained mixture containing the materials was mixed with 50 mL acetone and then centrifuged with a centrifugal speed of 5000 rpm. The obtained LaF_3 was washed with acetone 10 times and dried in a vacuum oven at 40 °C for 24 h. To synthesize NdF_3 or YF_3, the procedures were the same. The main difference was that $Nd(NO_3)_3$ or $Y(NO_3)_3$ were used, instead of $La(NO_3)_3$.

Synthesis of the La-MOF

The procedure for synthesizing the La-MOF was similar to that used previously.[28] In the experiments, 18 g of OmimCl and water were mixed in a conical flask of 50 mL, and the molar fraction of water in the solution was 0.5. Then 1.5 mmol of $La(NO_3)_3{\cdot}nH_2O$ and 1.5 mmol H_3BTC were added to the conical flask. The system was maintained at 85 °C for 72 h under stirring. The obtained mixture containing the materials was centrifuged with a centrifugal speed of 5000 rpm. The obtained La-MOF was washed with acetone 10 times and dried in a vacuum oven at 40 °C for 24 h.

Ru/LaF_3 and Ru/NdF_3 synthesis

We describe the procedure to synthesize Ru/LaF_3 because that for the synthesis of Ru/NdF_3 was similar. In a typical experiment, 20 mL $RuCl_3$ aqueous solution (1 mmol) was added into a 50 mL flask. Then 0.2 g LaF_3 was introduced into the flask and dispersed by ultrasound. The solution was stirred for 0.5 h. Then, 10 mL $NaBH_4$ aqueous solution (20 mmol) was added into the mixture slowly until the Ru^{3+} was completely reduced. The Ru/LaF_3 was washed with water five times and dried at 40 °C for 24 h under vacuum.

Characterization

X-Ray diffraction (XRD) analysis of the samples was performed on an X-ray diffractometer (Model D/MAX2500, Rigaka) with Cu-Kα radiation, and the scan speed was 5° min^{-1}. The morphologies of the products were characterized using a HITACHI S-4800 scanning electron microscope (SEM) and a JEOL-1010 transmission electron microscope operated at 100 kV. The crystal structure and SAED pattern were characterized using a TEM JEOL-2100F. The porosity properties of the materials were obtained from nitrogen adsorption–desorption isotherms using a Micromeritics ASAP 2020M system, and then calculated from the Barrett–Joyner–Halenda method. The Ru loading in the Ru/LaF_3 catalyst and Ru/NdF_3 was determined using the ICP-AES method (VISTA-MPX).

Cyanosilylation of Benzaldehyde

To carry out the reaction, 2 mL benzaldehyde and a certain amount of LaF_3 were placed into a 25 mL glass vial, which was immersed in a constant-temperature water bath of 50 °C. The mixture was stirred (800 rpm) under a N_2 atmosphere, and then TMSCN (another reactant) was added and the reaction began.[27] After a suitable reaction time, the reaction mixture was centrifuged to precipitate the LaF_3 particles. The reaction mixture was analyzed using a gas chromatograph (GC, HP 4890) equipped with a flame ionization detector (FID), and toluene was used as the internal standard. The identification of the products and reactants was done using a GC-MS (SHIMADZU-QP2010) and the ^1H NMR spectra were recorded on a Bruker Avance III 400 HD spectrometer in $CDCl_3$ with TMS as an internal standard, as well as by comparing the retention times to respective standards in GC traces.

Benzene Hydrogenation

The solvent free hydrogenation of benzene to cyclohexane was carried out in a 20 mL stainless steel batch reactor, which was similar to that used previously.[31b] In the experiment, 2 mL benzene and a desired amount of catalyst were added into the reactor. The reactor was immersed in a water bath of desired temperature, and the reaction was started. Then H_2 was introduced into the system at 4 MPa and the pressure was maintained during the reaction. The reaction mixture was cooled in ice-water and H_2 was released after 10 minutes. The procedures for products analysis were similar to those described above. In the reuse experiments, the catalyst was washed with ethanol (3 × 5 mL) and was used for the next run after drying at 40 °C under vacuum.

LA Hydrogenation

The solvent free hydrogenation of LA to GVL was carried out in a 20 mL stainless steel batch reactor, which was the same as that for the hydrogenation of benzene described above. In the experiment, 2 mL LA and a certain amount of catalyst were added into the reactor. The reactor was immersed in an oil bath of desired temperature. Then H_2 was introduced into the system at 6 MPa and the stirrer was started. The reaction mixture was cooled in ice-water for 10 minutes and H_2 was released. The procedures for product analysis were similar to those described above. Differently, the internal standard for this reaction was *n*-butyl alcohol.

ACKNOWLEDGEMENTS

The authors thank the National Natural Science Foundation of China (21133009, U1232203, 21321063), Chinese Academy of Sciences (KJCX2.YW.H30).

REFERENCES

1. (*a*) L. Qie, W. M. Chen, Z. H. Wang, Q. G. Shao, X. Li, L. X. Yuan, X. L. Hu, W. X. Zhang and Y. H. Huang, *Adv. Mater.*, 2012, 24, 2047 ; (*b*) J. Liang, R. F. Zhou, X. M. Chen, Y. H. Tang and S. Z. Qiao, *Adv. Mater.*, 2014, 26, 6074 .
2. (*a*) W. G. Lu, D. Q. Yuan, J. Sculley, D. Zhao, R. Krishna and H. C. Zhou, *J. Am. Chem. Soc.*, 2011, 133, 18126 ; (*b*) D. W. Johnson, C. Sherborne, M. P. Didsbury, C. Pateman, N. R. Cameron and F. Claeyssens, *Adv. Mater.*, 2013, 25, 3178 .
3. (*a*) Z. P. Wang, J. H. Yu and R. R. Xu, *Chem. Soc. Rev.*, 2012, 41, 1729 ; (*b*) L. L. Zhang, Z. Q. Hu and J. W. Jiang, *J. Am. Chem. Soc.*, 2013, 135, 3722 ; (*c*)Q. Wen, J. C. Di, L. Jiang, J. H. Yu and R. R. Xu, *Chem. Sci.*, 2013, 4, 591 .
4. (*a*) T. R. Cook, Y. R. Zheng and P. J. Stang, *Chem. Rev.*, 2013, 113, 734 ; (*b*) S. Cao, G. Gody, W. Zhao, S. Perrier, X. Y. Peng, C. Ducati, D. Y. Zhao and A. K. Cheetham, *Chem. Sci.*, 2013, 4, 3573 ; (*c*) M. Li, D. Li,

M. O'Keeffe and O. M. Yaghi, *Chem. Rev.*, 2014, 114, 1343 .

5. (*a*) T. Tozawa, J. T. A. Jones, S. I. Swamy, S. Jiang, D. J. Adams, S. Shakespeare, R. Clowes, D. Bradshaw, T. Hasell, S. Y. Chong, C. Tang, S. Thompson, J. Parker, A. Trewin, J. Bacsa, A. M. Z. Slawin, A. Steiner and A. I. Cooper, *Nat. Mater.*, 2009, 8, 973 ; (*b*) J. W. Colson, A. R. Woll, A. Mukherjee, M. P. Levendorf, E. L. Spitler, V. B. Shields, M. G. Spencer, J. Park and W. R. Dichtel, *Science*, 2011, 332, 228 .
6. (*a*) C. M. A. Parlett, K. Wilson and A. F. Lee, *Chem. Soc. Rev.*, 2013, 42, 3876 ; (*b*) C. H. Cui, H. H. Li and S. H. Yu, *Chem. Sci.*, 2011, 2, 1611 .
7. (*a*) T. A. Makal, J. R. Li, W. G. Lu and H. C. Zhou, *Chem. Soc. Rev.*, 2012, 41, 7761 ; (*b*) P. Nugent, Y. Belmabkhout, S. D. Burd, A. J. Cairns, R. Luebke, K. Forrest, T. Pham, S. Q. Ma, B. Space, L. Wojtas, M. Eddaoudi and M. J. Zaworotko, *Nature*, 2013, 495, 80 ; (*c*) Y. L. Zhu and W. Zhang, *Chem. Sci.*, 2014, 5, 4957 .
8. (*a*) N. Liu, K. Lee and P. Schmuki, *Angew. Chem., Int. Ed.*, 2013, 52, 12381 ; (*b*) X. Huang, H. Yu, J. Chen, Z. Y. Lu, R. Yazami and H. H. Hng, *Adv. Mater.*, 2014, 26, 1296 .
9. H. L. Cong and W. X. Cao, *Solid State Sci.*, 2006, 8, 1056 .
10. Q. X. Chu, Z. C. Xing, J. Q. Tian, X. B. Ren, A. M. Asiri, A. O. Al-Youbi, K. A. Alamry and X. P. Sun, *J. Power Sources*, 2013, 236, 188 .
11. (*a*) R. D. Rogers and K. R. Seddon, *Science*, 2003, 302, 792 ; (*b*) J. Dupont, *Acc. Chem. Res.*, 2011, 44, 1223 ; (*c*) M. Y. Lui, L. Crowhurst, J. P. Hallett, P. A. Hunt, H. Niedermeyer and T. Welton, *Chem. Sci.*, 2011, 2, 1491 .
12. (*a*) M. Smiglak, W. M. Reichert, J. D. Holbrey, J. S. Wilkes, L. Y. Sun, J. S. Thrasher, K. Kirichenko, S. Singh, A. R. Katritzky and R. D. Rogers, *Chem. Commun.*, 2006, 2554 ; (*b*) Z. F. Fei, T. J. Geldbach, D. B. Zhao and P. J. Dyson, *Chem.–Eur. J.*, 2006, 12, 2122 .
13. (*a*) Y. Zhou and M. Antonietti, *J. Am. Chem. Soc.*, 2003, 125, 14960 ; (*b*) P. S. Wheatley, P. K. Allan, S. J. Teat, S. E. Ashbrook and R. E. Morris, *Chem. Sci.*, 2010, 1, 483 ; (*c*) X. C. Kang, J. L. Zhang, W. T. Shang, T. B. Wu, P. Zhang, B. X. Han, Z. H. Wu, G. Mo and X. Q. Xing, *J. Am. Chem. Soc.*, 2014, 136, 3768 .
14. (*a*) J. P. Hallett and T. Welton, *Chem. Rev.*, 2011, 111, 3508 ; (*b*) Y. F. Zhao, B. Yu, Z. Z. Yang, H. Y. Zhang, L. D. Hao, X. Gao and Z. M. Liu, *Angew. Chem., Int. Ed.*, 2014, 53, 5922 .
15. S. P. M. da Silva, A. M. D. Lopes, L. B. Roseiro and R. Bogel-Lukasik, *Adv.*, 2013, 3, 16040 .
16. (*a*) G. K. Cui, J. J. Zheng, X. Y. Luo, W. J. Lin, F. Ding, H. R. Li and C. M. Wang, *Angew. Chem., Int. Ed.*, 2013, 52, 10620 ; (*b*) D. Z. Yang, M. Q. Hou, H. Ning, J. Ma, X. C. Kang, J. L. Zhang and B. X. Han, *ChemSusChem*, 2013, 6, 1191 .
17. (*a*) S. Schauermann, N. Nilius, S. Shaikhutdinov and H. J. Freund, *Acc. Chem. Res.*, 2013, 46, 1673 ; (*b*) C. K. Prier and D. W. C. MacMillan, *Chem. Sci.*, 2014,5, 4173 ; (*c*) G. X. Chen, Y. Zhao, G. Fu, P. N. Duchesne, L. Gu, Y. P. Zheng, X. F. Weng, M. S. Chen, P. Zhang, C. W. Pao, J. F. Lee and N. F. Zheng, *Science*, 2014, 344, 495 .
18. (*a*) P. F. Xu, R. B. Yu, H. Ren, L. B. Zong, J. Chen and X. R. Xing, *Chem. Sci.*, 2014, 5, 4221 ; (*b*) A. F. Lee, J. A. Bennett, J. C. Manayil and K. Wilson, *Chem. Soc. Rev.*, 2014,43, 7887 .
19. (*a*) Z. W. Seh, S. H. Liu, M. Low, S. Y. Zhang, Z. L. Liu, A. Mlayah and M. Y. Han, *Adv. Mater.*, 2012, 24, 2310 ; (*b*) L. Y. Chen, H. R. Chen, R. Luque and Y. W. Li, *Chem. Sci.*, 2014, 5, 3708 .
20. J. Y. Jeon, J. K. Varghese, J. H. Park, S. H. Lee and B. Y. Lee, *Eur. J. Org. Chem.*, 2012, 2012, 3566 .
21. D. S. Jacob, L. Bitton, J. Grinblat, I. Felner, Y. Koltypin and A. Gedanken, *Chem. Mater.*, 2006, 18, 3162 .
22. W. D. Cheng, X. Q. Xing, D. H. Wang, K. H. Zhang, Q. Cai, G. Mo, Z. J. Chen and Z. H. Wu, *J. Wood Sci.*, 2011, 57, 470 .
23. Z. M. Xue, J. L. Zhang, L. Peng, J. S. Li, T. C. Mu, B. X. Han and G. Y. Yang, *Angew. Chem., Int. Ed.*, 2012, 51, 12325 .
24. T. L. Greaves, D. F. Kennedy, S. T. Mudie and C. J. Drummond, *J. Phys. Chem. B*, 2010, 114, 10022 .
25. (*a*) N. Kurono, K. Arai, M. Uemura and T. Ohkuma, *Angew. Chem., Int. Ed.*, 2008, 47, 6643 ; (*b*) T. Yang, A. Bartoszewicz, J. Ju, J. L. Sun, Z. Liu, X. D. Zou, Y. X. Wang, G. B. Li, F. H. Liao, B. Martín-Matute and J. H. Lin, *Angew. Chem., Int. Ed.*, 2011, 50, 12555 .
26. R. J. H. Gregory, *Chem. Rev.*, 1999, 99, 3649 .
27. R. F. D›Vries, M. Iglesias, N. Snejko, E. Gutiérrez-Puebla and M. A. Monge, *Inorg. Chem.*, 2012, 51, 11349 .
28. W. T. Shang, X. C. Kang, H. Ning, J. L. Zhang, X. G. Zhang, Z. H. Wu, G. Mo, X. Q. Xing and B. X. Han, *Langmuir*, 2013, 29, 13168 .
29. R. Wischert, P. Laurent, C. Copéret, F. Delbecq and P. Sautet, *J. Am. Chem. Soc.*, 2012, 134, 14430 .
30. N. Kuhl, M. N. Hopkinson, J. Wencel-Delord and F. Glorius, *Angew. Chem., Int. Ed.*, 2012, 51, 10236 .
31. (*a*) R. R. Deshmukh, J. W. Lee, U. S. Shin, J. Y. Lee and C. E. Song, *Angew. Chem., Int. Ed.*, 2008, 47, 8615 ; (*b*) H. Z. Liu, T. Jiang, B. X. Han, S. G. Liang and Y. X. Zhou, *Science*, 2009, 326, 1250 .
32. A. M. R. Galletti, C. Antonetti, V. De Luise and M. Martinelli, *Green Chem.*, 2012, 14, 688 .
33. Z. P. Yan, L. Lin and S. J. Liu, *Energy Fuels*, 2009, 23, 3853

CHAPTER 6

VIEDMA RIPENING: A RELIABLE CRYSTALLISATION METHOD TO REACH SINGLE CHIRALITY

Leyla-Cann Sögütoglu , René R. E. Steendam , Hugo Meekes , Elias Vlieg and Floris P. J. T. Rutjes

Radboud University, Institute for Molecules and Materials, Heyendaalseweg 135, 6525 AJ Nijmegen, The Netherlands.

ABSTRACT

Crystallisation processes have evolved to practical methods that allow isolation of an enantiopure product in high yield. Viedma ripening in particular enables access to enantiopure products in a reliable way, simply through grinding of crystals in a solution. This tutorial review covers the basic principles behind asymmetric crystallisation processes, with an emphasis on Viedma ripening, and shows that to date many novel organic molecules can be obtained in enantiopure solid form.

INTRODUCTION

Mirror symmetry is widely present throughout Nature and has fascinated and inspired mankind to date. Although any object has a mirror image, not every mirror image is identical to the original. Such objects are coined "chiral", derived from the ancient Greek word χειρ (kheir) which means "hand". Chirality is also present at the molecular level: a molecule is chiral – and consists of two mirror-image isomers or enantiomers – if it has no internal plane of symmetry. Such a situation occurs for example if a molecule contains a tetrahedral carbon atom bearing four different substituents. Thermodynamically, the enantiomers (denoted as *S* or *R*) are equally likely to exist, yet most chiral molecules in living organisms are overrepresented in one enantiomeric form. How this molecular asymmetry emerged under prebiotic conditions remains a much studied topic in science.[1] The single handedness in living systems furthermore implies that the two enantiomers of a molecule interact differently with chiral biomolecular receptors. This has enormous consequences for the pharmaceutical industry as generally only one of the two enantiomers exerts the desired medicinal effect. Due to the effect that active pharmaceutical ingredients tend to become more selective with increasing complexity, progressively more single enantiomeric drugs are being launched as compared to racemic- or achiral drugs.[2] Therefore, strategies to create single enantiomeric compounds are of pivotal importance to the fine chemical and pharmaceutical industry. Over the past ten years, amongst a variety of chemical and enzymatic methods, Viedma ripening (Fig. 1) has been established as a reliable solid-state method for the deracemisation of racemic mixtures of crystalline compounds into single enantiomers, simply by continuously grinding a suspension.[3] The process is not yet used on an industrial scale.

Citation: Leyla-Cann Sögütoglu , René R. E. Steendam, Hugo Meekes , Elias Vlieg and Floris P. J. T. Rutjes. Viedma ripening: a reliable crystallisation method to reach single chirality. Chem. Soc. Rev., 2015, 44, 6723-6732, DOI: 10.1039/C5CS00196J

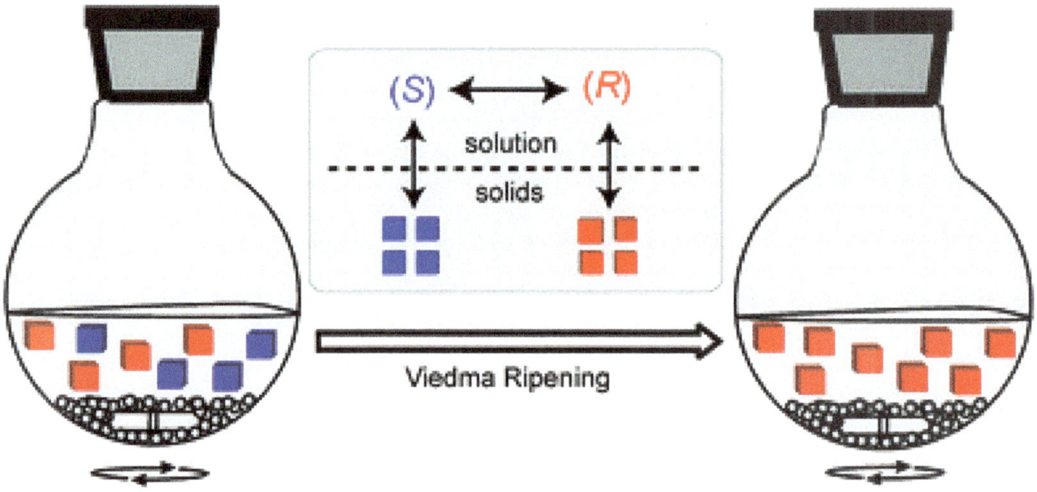

Figure 1 Schematic representation of Viedma ripening.

After its discovery in 2005, an increasing number of reports demonstrate the broad applicability of Viedma ripening which enabled the complete deracemisation of a large variety of crystalline chiral compounds.[3] Section 2 of this tutorial review covers the basic principles behind asymmetric crystallisation processes with a brief history on the discovery of Viedma ripening. The principles underlying Viedma ripening are addressed in Section 3. Finally, a number of recent applications involving crystallisation-induced deracemisation methods are highlighted in Section 4.

ASYMMETRIC CRYSTALLIZATION

Nucleation

The very first aggregate of growth units (molecules, ions, atoms) large enough to grow to a macroscopic crystal, is called a nucleus. Nucleation in solution starts when the compound is not soluble any more. Common salt (NaCl) for example, will become insoluble in water (*i.e.* the solution becomes saturated) at room temperature at a concentration of 357 g L^{-1} and will start to nucleate if the solubility is decreased further, typically by cooling or through evaporation of the solvent. The solution or liquid from which crystals precipitate is called the mother liquor. Two types of nucleation are generally distinguished, called primary and secondary nucleation (Fig. 2). Primary nucleation is the initial formation of a crystal when there are no other crystals present. Secondary nucleation is the formation of nuclei from an existing macroscopic crystal when this crystal is crushed into smaller pieces, for example as the result of convection,[4]crystal–crystal collisions or mechanical breakage by a stirrer bar. Primary nucleation can proceed homogeneously or heterogeneously. The nucleation is called homogeneous in e the compound nucleates without interplay of any media other than the solution. The term heterogeneous nucleation, on the other hand, is used when the compound nucleates onto another compound or dust particles in the solution. Crystal growth starts nearly always *via* heterogeneous nucleation.

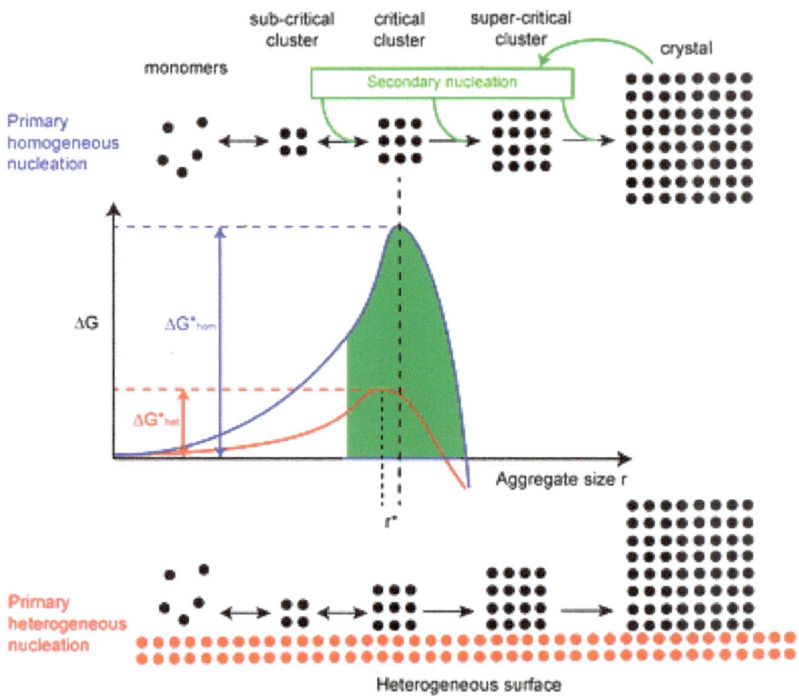

Figure 2 The formation of crystals from solution proceeds either through primary-(homogeneous or heterogeneous) or through secondary nucleation.

In a simplified view, crystal growth encounters a barrier as a result of unfavourable crystal-solution surface energy *versus* favourable bulk energies. The critical nucleus size (r^*) is reached when the favourable and unfavourable energy contributions just counterbalance. Note that the critical nucleus size for primary and secondary nucleation does not need to be the same as it depends on an interfacial energy contribution. Once an aggregate is larger than the critical nucleus it can grow 'downhill' spontaneously. In other words, the crystal as thermodynamic end product has a kinetic barrier called nucleation barrier. The barrier can be reduced by primary heterogeneous nucleation (red curve) or even 'disregarded' by using secondary nucleation (green area), which supplies clusters, usually larger than the critical nucleus.

For further reading on the crystallisation of molecules from solution, the reader is advised to consult a previously reported tutorial review.[5]

Crystallisation of Enantiomers

Chiral discrimination and enantiomeric separation is extremely difficult to achieve in solution without chiral reagents. The synthesis of chiral molecules typically results in a racemic mixture in which both enantiomers are present in approximately equal amounts. The enantiomeric excess (ee) is used to define the ratio between enantiomers:

$$\text{ee}\ (\%) = (([R] - [S])/([R] + [S])) \times 100 \tag{1}$$

A racemic mixture corresponds to an ee of 0% whereas enantiopure products have an overall 100% ee or −100% for the opposite enantiomer.

In the solid state, chiral discrimination is easier to realise. This especially applies to molecules which, in the crystal structure, have a greater affinity for the same enantiomer than for the opposite enantiomer (Fig. 3). Such a racemic conglomerate is in fact a mechanical mixture of enantiomerically pure crystals of one enantiomer and its opposite. This is the e for approximately 5–10% of all chiral crystalline molecules.[6] When molecules have a greater affinity for the opposite enantiomer than for the same enantiomer, the molecules form a single crystalline phase in which the two enantiomers are present in an ordered 1 : 1 ratio. The crystallographic unit cell contains both enantiomers and the solid is called a racemic compound or true racemate. Approximately 90–95% of all crystalline racemic mixtures form racemic compounds. Finally, in less than 1% of the es a racemic mixture crystallises as a solid solution, containing molecules of each enantiomer in a random arrangement. A unit cell can still be assigned, but the molecular filling of this unit cell is not as well-defined as in the e of a true crystal.

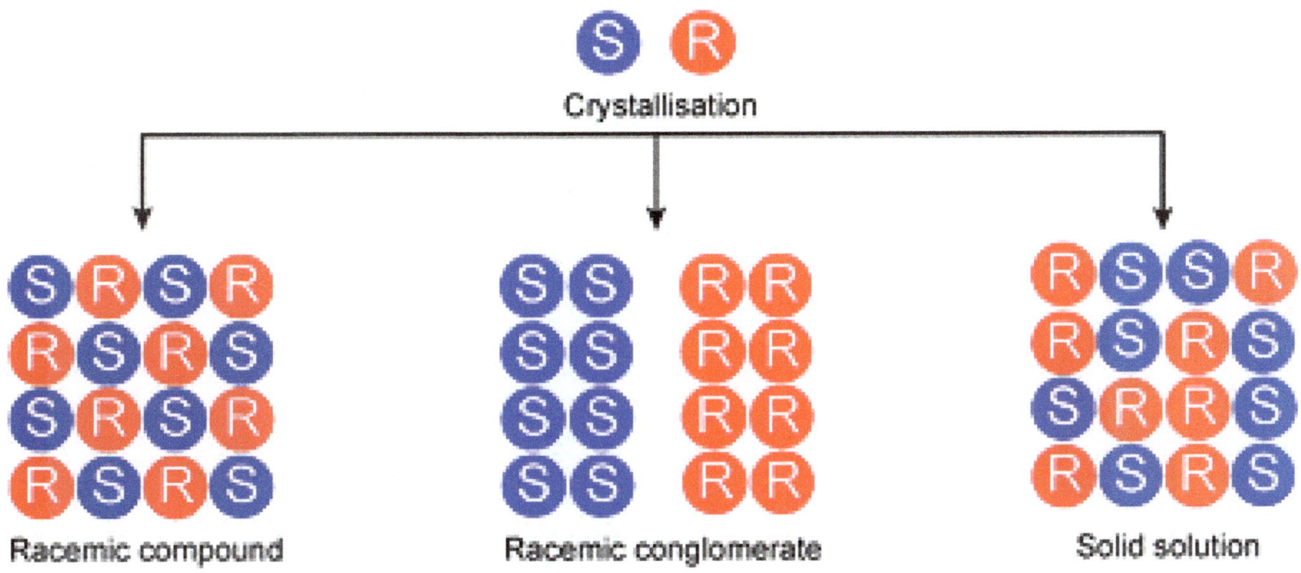

Figure 3 Enantiomers crystallise either as a racemic compound, racemic conglomerate or a solid solution.

Total Spontaneous Resolution

Chiral discrimination in the solid state can only be achieved with molecules that form racemic conglomerate crystals. Nucleation of these molecules from a homogeneous solution generally proceeds to give a crystal of single chirality (Fig. 4a). In this way, the solution becomes enriched in the opposite enantiomer that will also crystallise upon further increase of the supersaturation. Eventually, the overall solid state is racemic yet each crystal is enantiopure. If the crystals can be separated from each other, one can obtain enantiopure products and in this way perform spontaneous resolution. Louis Pasteur obtained enantiopure conglomerate crystals through the manual separation of crystals of sodium ammonium tartrate.[7] Although Pasteur was able to separate the crystals by hand based on morphological chirality, many crystals do not display such a chiral morphology. This separation method is still being used, albeit in a more practical way.[8]

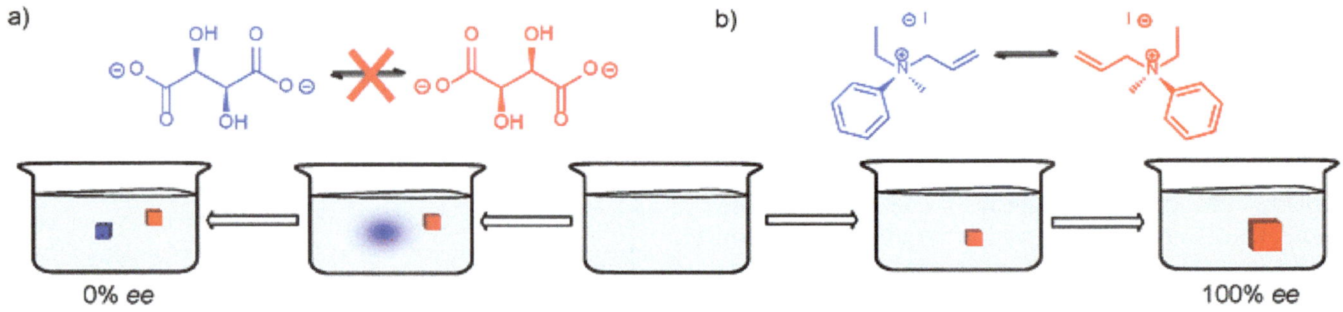

Figure 4 Crystallisation of conglomerate-forming molecules in (a) the absence and (b) presence of solution-phase racemisation.

To prevent the nucleation of the other enantiomer, Egbert Havinga used conglomerate-forming quaternary ammonium iodide (Fig. 4b) that racemises in solution as a spontaneous process toward maximal entropy.[9,10] Because of the decrease in supersaturation due to the first crystal formed (Fig. 2), primary nucleation of either enantiomer will less likely take place. As the mother crystal grows larger, it retains its chirality by taking up only the monomers with matching chirality. The supply of these monomers is maintained through racemisation in solution. This way enantiopure crystals spontaneously form without the need for separation. This process is also known as total spontaneous resolution.[6]

Havinga postulated in 1941, at the Dutch organic chemistry conference, the three requirements for a racemic mixture to have a high probability to undergo total spontaneous resolution:[9]

1) The compound must form separate *R* and *S* crystals (*i.e.* the compound is a racemic conglomerate).
2) The compound must be able to racemise in solution, possibly aided by a catalyst speeding up the racemisation.
3) The rate of crystal nucleation is low, while the rate of crystal growth is high. The rate of racemisation should be high as well.

At first Havinga was unaware of the impact of his research, which was published in the Chemisch Weekblad, a journal written in Dutch. Soon after a publication of a model on spontaneous asymmetric synthesis by the theoretical physicist Sir Charles Frank,[11] Havinga published his experimental findings again, but this time in English in an international journal.[10] Interestingly, an often overlooked detail in the experimental section of the Dutch paper is that Havinga mentions that in some experiments several crystals were formed as the result of agitation. When combined, the mixture of crystals still displayed optical activity. Havinga proposed that these crystals were grown from seeds originating from the initial mother crystal (secondary nucleation) and therefore were of the same handedness (Fig. 5).

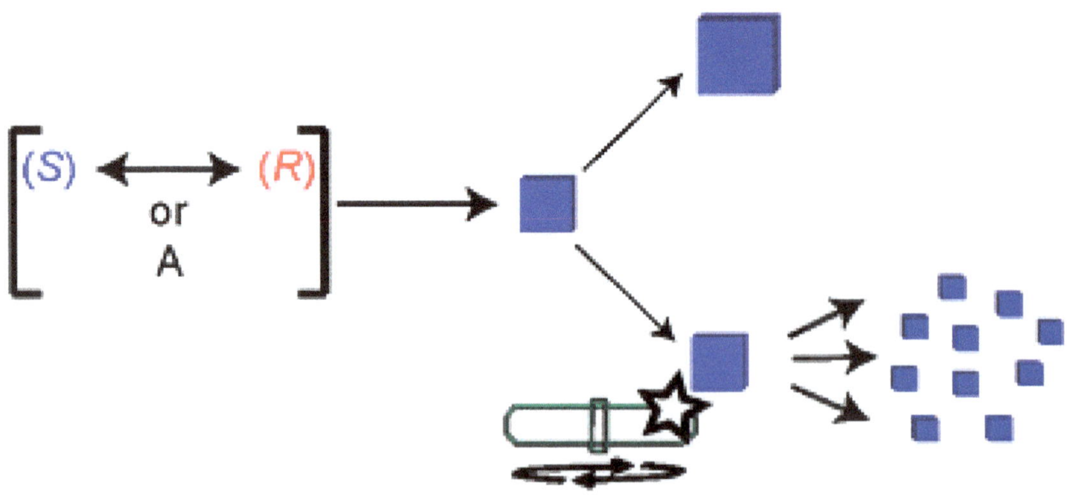

Figure 5 Primary nucleation to give an enantiopure crystal which can grow larger (top) or will undergo secondary nucleation (bottom).

This effect of secondary nucleation was studied in 1990 in more detail by Kondepudi *et al.* who found that crystallisation from a stirred solution[12] or melt[13,14] leads to many small crystals which are nearly all of the same handedness. In their solution-phase experiments, the compound used was sodium chlorate which is achiral as a molecule in solution but in the solid state the arrangement is such that the crystals become chiral (Fig. 6).

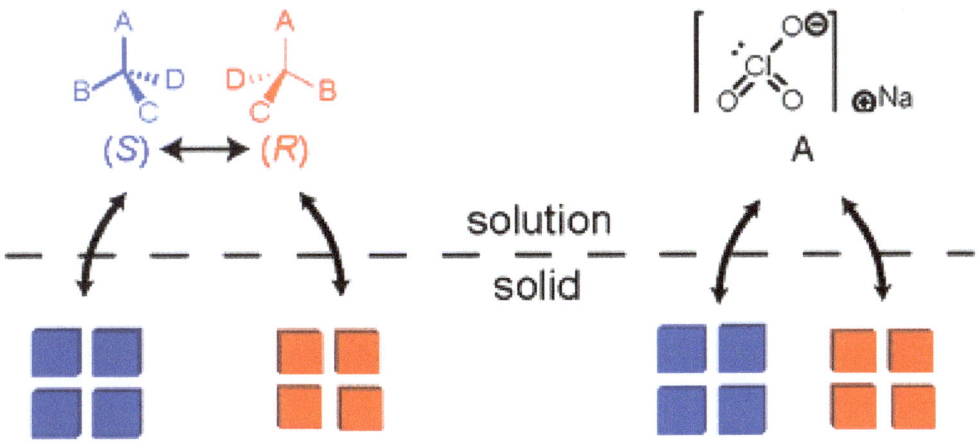

Figure 6 Solution-solid equilibria of a chiral molecule (left) and sodium chlorate (right).

Kondepudi *et al.* showed that nearly all crystals (99.7% of all crystals formed in an experiment) had the same chirality when an aqueous solution of sodium chlorate was stirred during crystallisation. He stated that "in order to produce total asymmetry of close to 100% in the product crystals in every try, autocatalysis and competition between the L- and D-crystals are needed." In Kondepudi›s experiment too, all requirements for total spontaneous resolution as stated by Havinga are met: (1) $NaClO_3$ crystallises as chiral crystals (thus being a conglomerate in the solid state), (2) racemisation in solution is not even needed, because the building block is achiral and (3) the rate of primary nucleation was kept low, while the rate of crystal growth was high. The latter condition was possible, because Kondepudi made use of the rapid generation of secondary nuclei which reduced the concentration to a level at which the rate of primary nucleation is virtually zero under stirring conditions. Primary nucleation is a cumbersome process, with a kinetic barrier of reaching the critical nucleus, whereas secondary nuclei larger than the critical nucleus grow as soon as they are formed. As such, the left- and right-handed nuclei generated through secondary nucleation compete for the solute and the rapid crystal growth suppresses primary nucleation and therefore the formation of nuclei of the opposite handedness. In addition, Kondepudi noticed that when the solution is not stirred, there is no preference for one chiral form over the other. This means that all of the nuclei are produced through primary nucleation, homogeneous or heterogeneous, and their handedness is at random. In this e too, the depletion of the solute due to crystal growth may eventually stop the primary nucleation, but no chiral resolution occurred in these trials.

Later, Durand *et al.* showed that the asymmetry obtained through stirring-induced crystallisation of a conglomerate-forming achiral molecule can be converted to molecular chirality.[15]Stirring during the crystallisation of conglomerate-forming molecules which racemise in solution provide an attractive route to obtain molecules in high ee and high yield.[16] However, it is not the most reliable approach in reaching an enantiopure product as nucleation of the unwanted enantiomer still can take place.

VIEDMA RIPENING

In an attempt to test the hypothesis of secondary nucleation, Cristobal Viedma repeated the Kondepudi experiments in 2004, but under conditions which facilitate a much higher nucleation rate leading to many primary nucleation events of sodium chlorate.[17] Intriguingly, the solid state was still found to be of single chirality despite the fact that the high nucleation rate led to the nucleation of both enantiomers. In a follow-up experiment, Viedma prepared a saturated solution already containing crystals of both chiral forms which he then subjected to intensive grinding using glass beads in a closed system under isothermal conditions (Fig. 7).[18]

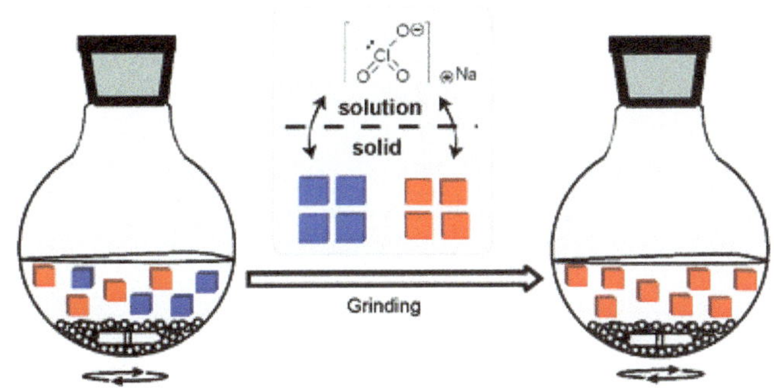

Figure 7 Schematic representation of Viedma's experiment.

Viedma discovered that the initially racemic mixture of sodium chlorate crystals was, over a period of several days, completely transformed into an end state in which all of the crystals were of one chiral form. This transformation, which is now called Viedma ripening, involves solid-to-solid deracemisation instead of solution-to-solid deracemisation and is therefore significantly different as compared to total spontaneous resolution.

Like Kondepudi *et al.*, Viedma used the achiral molecule sodium chlorate in his experiments. In 2008, Noorduin *et al.* extended Viedma ripening to intrinsically chiral molecules.[19] They showed that amino acid derivatives, such as the one depicted in Fig. 8, can also undergo solid-to-solid deracemisation. These molecules also form conglomerate crystals and undergo racemisation in solution using a strong base. Noorduin *et al.* showed that Viedma ripening of this compound can readily be scaled up to larger volumes of up to 320 mL of solvent, thereby demonstrating the industrial viability of Viedma ripening.[20] In addition, other pharmaceutically-relevant molecules, such as Naproxen[21] and a Clopidogrel intermediate,[22] were found to undergo complete deracemisation through Viedma ripening. The final configuration of the product can easily be controlled using additives,[19] difference in crystal size between the enantiomers[23] and even the order of process steps.[24]

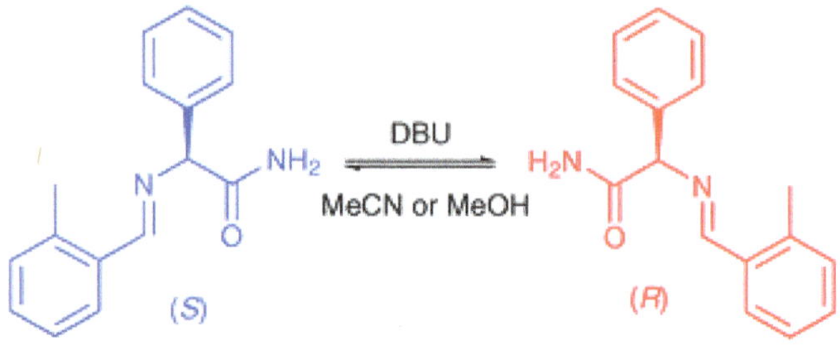

Figure 8 Extension of Viedma ripening to intrinsically chiral molecules that form racemic conglomerate crystals.

During the past decade, several research groups have studied the mechanism behind this intriguing transformation. Although "the practical execution of the process is remarkably simple"[25] the underlying mechanism is more complicated. After symmetry breaking of the racemic mixture of crystals due to random local fluctuations in ee or local fluctuations in the crystal size distribution (CSD) difference between the enantiomers, the system undergoes complete solid state deracemisation through an autocatalytic feedback mechanism in which the initial ee is amplified exponentially to an enantiopure end state. To date, many computational studies have been carried out to explain the mechanism behind Viedma ripening. To account for all these models, however, is beyond the scope of this review. Still, most modeling studies describe the Viedma ripening process along the four factors described in the following paragraphs (also indicated with the numbers in Fig. 9).[25]

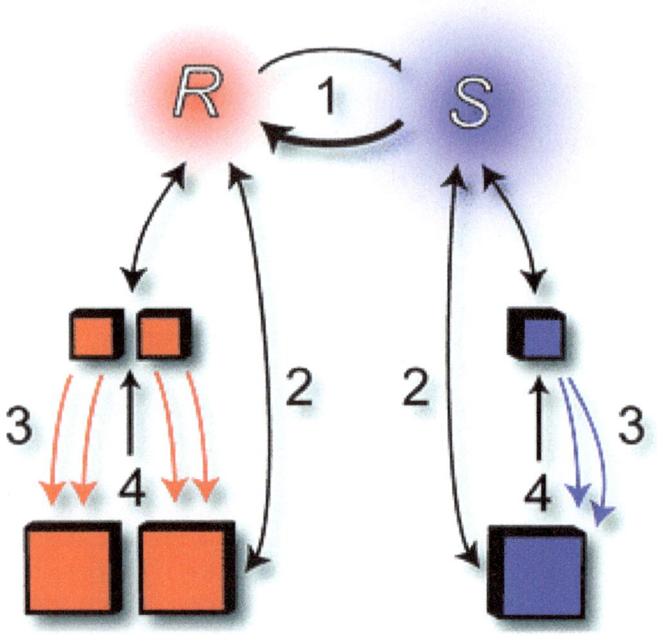

Figure 9 Schematic representation of a proposed mechanism behind Viedma ripening. The numbers are explained in the main text.

Racemisation in solution (which is not needed for achiral molecules that crystallise as a racemic conglomerate)

In order to convert all enantiomers in the solid phase to one form, the molecule must continuously racemise in solution. Furthermore, chiral clusters of the (*S*)-enantiomer have a lower probability to encounter a crystal of the same handedness, and thus dissolve more frequently. As a result, the solution becomes enriched with the enantiomer that is opposite to the one that is enriched in the solid phase. The enantiomer that is enriched in solution undergoes racemisation leading to the supply of more of the enantiomer that corresponds to the one that is enriched in the solid phase.

Ostwald ripening

Ostwald ripening, first described by Wilhelm Ostwald in 1896, is the continued dissolution and growth of crystals to reach a thermodynamically stable state wherein the surface area to volume ratio of the solid system is minimised. That is, larger crystals grow at the expense of smaller crystals. The latter is implicitly shown in Fig. 2, as the extrapolated Gibbs free energy is still lowering for larger particle size. There is a tendency towards the largest aggregate or crystal. When crystals of one enantiomer are sufficiently larger than the other, Viedma ripening leads to complete deracemisation in which the final product corresponds to the initially larger crystals (survival of the fittest).[23]

Enantioselective incorporation of clusters into larger crystals[26]

Fig. 9 shows the incorporation of both solute molecules as well as clusters into larger crystals of the same handedness. During Viedma ripening, crystals are continuously broken into clusters. The reincorporation of these clusters into larger crystallites occurs more often for the enantiomer which is in excess in the solid phase. This process, together with attrition, leads to the autocatalytic amplification effect typical for Viedma ripening. Experimental proof for enantioselective incorporation of chiral clusters remains to be found, although some indirect evidence relating to $NaClO_3$, the archetypical compound for Viedma ripening, has been reported.[27,28]

Attrition

While enantiopure clusters and molecules of *R* and *S* incorporate enantioselectively in the bulk crystals *R* and *S* respectively, the bulk crystals are ground, producing chiral fragments as well as monomers. The steady attrition also maintains overall small crystal sizes which in turn enhance the Ostwald ripening effect. A higher attrition intensity leads to shorter deracemisation times.[29]

As the following section will show, novel applications involving Viedma ripening and total spontaneous resolution are still being discovered.

RECENT EXAMPLES

This section highlights some examples in recent literature in which an enantiopure product can be obtained either through total spontaneous resolution or Viedma ripening. Noorduin *et al.*were able to apply Viedma ripening to the deracemisation of an amino acid derivative (Fig. 8) in which the final configuration of the product could be controlled using the rotation sense of circularly polarised light (CPL).[30] After the solution was subjected to r-CPL radiation, Viedma ripening reproducibly led to an enantiopure product with the (*R*)-configuration. An enantiopure product with the (*S*)-configuration could be obtained when l-CPL was used. The CPL radiation induces the formation of an enantioenriched (*i.e.* >0% ee) mixture of side products which stereoselectively hamper the crystal growth of one enantiomer.

Another contribution to the discussion on the origin of single chirality was proposed by Frank, already in 1953. Frank envisaged that an enantiopure product can in principle be formed from achiral starting materials provided that the product enantiomer catalyses its own formation and at the same time suppresses the formation of the other enantiomer.[11] He concluded his article with the sentence "*a laboratory demonstration is not necessarily impossible*". It took nevertheless more than 40 years until Frank's concept of asymmetric autocatalysis was experimentally realised. Asakura *et al.* found that an optically active cobalt complex can be formed through asymmetric autocatalysis.[31] In the same year, Soai *et al.* showed that an initial small amount of chiral product can be amplified to single chirality.[32] Without the addition of enantioenriched material from the beginning, however, enantiopure products could not be obtained.[33] These results underline the fact that the synthesis of an enantiopure product from achiral reactants without pre-existing enantioenrichment still is extremely difficult to achieve in solution. More recently, some of the present authors successfully extended Frank›s concept to a synthetic organic construction reaction[34] in which crystals of the product were used as the asymmetric autocatalytic driving force (Fig. 10).[35]

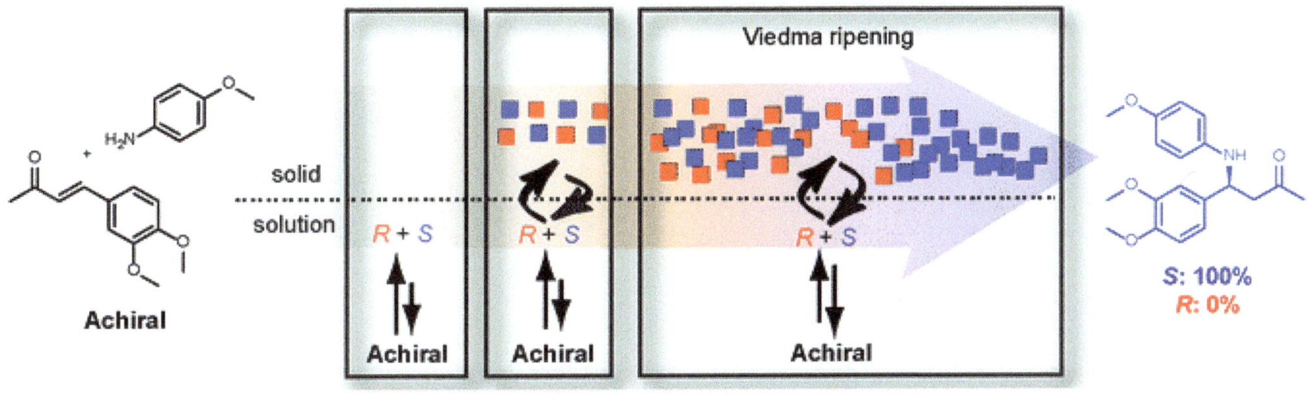

Figure 10 The synthesis of an enantiopure product from achiral conditions.

It was found that the product amine forms racemic conglomerate crystals while in solution it can racemise through a reversible aza-Michael reaction. Starting at a high concentration of achiral reactants, both enantiomers of the product are rapidly formed using DBU as an achiral catalyst in solution. Due to the poor solubility, both enantiomers of the product crystallise to give a crystal-solution system. The latter system subsequently undergoes deracemisation through Viedma ripening due to the applied grinding conditions. This way, an enantiopure product was reproducibly obtained in high yield from achiral reactants in a single reaction. The enantiopure product was formed either in the (*R*)- or (*S*)-configuration.

The reversible reaction can also be used to enhance the ee of other conglomerate-forming molecules, as was shown for a reversible Mannich reaction[36] and a reversible aldol reaction.[37]In addition to a reversible reaction, racemisation can also be induced in different ways as some recent reports have shown (Fig. 11). The group of Hakånsson significantly extended the list of metal complexes that can be resolved through total spontaneous resolution (Fig. 11a).[38,39] These complexes have achiral ligands but are chiral (denoted or instead of *R* and *S*) as a complex. On the other hand, pyrimidine-derivatives studied by Yagishita *et al.* exhibit axial chirality and these molecules as such were found to undergo racemisation through an achiral transition state at elevated temperatures without the need for a catalyst (Fig. 11b).[40] The combination of seeding and stirring gave the product with an ee of up to 91% through total spontaneous resolution. Another class of compounds that can be obtained in enantioenriched form through total spontaneous resolution are isoindolinones (Fig. 11c). These compounds undergo rapid racemisation through a ring-opened achiral intermediate in the presence of a strong base (DBU). Starting from a clear solution, evaporation of solvent led to the crystallisation of enantioenriched isoindolinones in quantitative yields through total spontaneous resolution.[41] Some of the present authors later reported that the isoindolinones can be deracemised completely through Viedma ripening provided that (1) there is sufficient attrition, (2) a suitable solvent is used in which the compounds have a high enough solubility and (3) that racemization is avoided during the ee analysis.[42] Moreover, the reported isoindolinones readily racemise in ethanol without a catalyst and this enabled the complete deracemisation of these isoindolinones through Viedma ripening in ethanol.

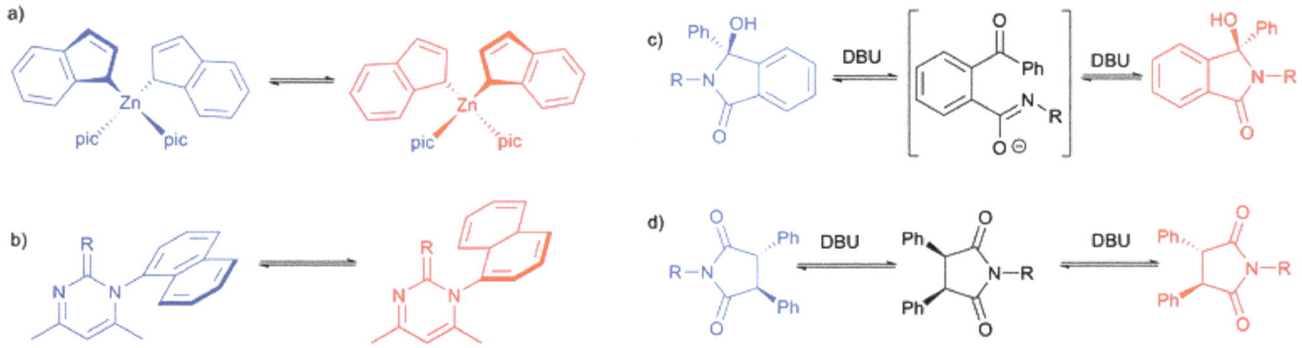

Fig. 11 Selected conglomerate-forming chiral molecules that racemise differently in solution. pic is 3-picoline.

Trans-Succinimides undergo racemisation through the achiral *cis*-isomers which possess an intrinsic mirror plane (Fig. 11d).[43] Starting with the achiral *cis*-isomer, desymmetrisation facilitated by DBU provides both enantiomers which rapidly racemise through the *cis*-isomer. In combination with conglomerate crystallisation, this resulted in total spontaneous resolution to give the products in quantitative yields and 85–98% ee.

In addition to chiral molecules, achiral molecules were recently obtained in enantiopure form through Viedma ripening. Of all achiral compounds, between 8 and 13% crystallise as conglomerate crystals. McLaughlin *et al.* applied Viedma ripening to deracemise ten achiral organic molecules that crystallise in a chiral fashion (three examples are shown in Fig. 12).[44]Solid-state circular dichroism (CD) spectroscopy was used to determine the solid state.

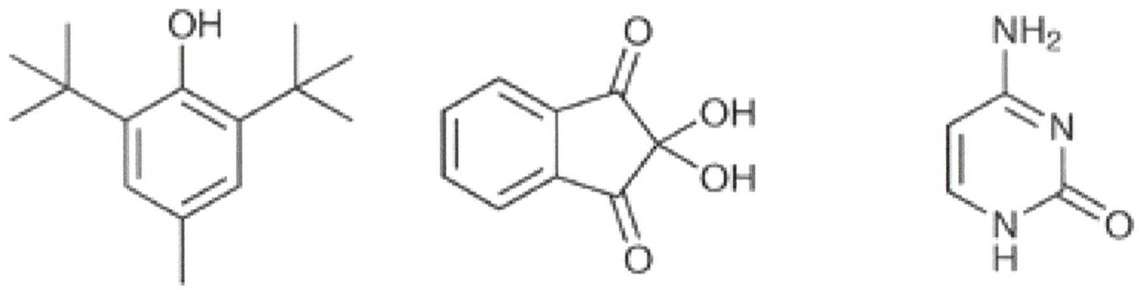

Figure 12 Selected achiral molecules which form conglomerate crystals.

Besides racemisation in solution, the formation of racemic conglomerate crystals is another prerequisite for reaching single chirality using the interplay between crystals and a solution. The synthesis of a library of derivatives is one approach to find conglomerate crystals. On the other hand, tailoring crystallisation processes can also open up possibilities that could enable the formation of additional examples of conglomerate crystals. This way, Spix *et al.* showed that a metastable conglomerate of glutamic acid can be deracemised through Viedma ripening (Fig. 13).[45] Although glutamic acid forms a racemic compound in the form of a hydrate, it was found that a metastable conglomerate form can be obtained as the kinetic product in the presence of acetic acid. Racemisation of glutamic acid proceeds at elevated temperatures which also facilitates formation of two other unwanted racemic forms of the product. Therefore, experimental conditions were tweaked which ultimately enabled access to enantiopure glutamic acid in 80% yield through Viedma ripening starting from low initial ee›s of 14%.

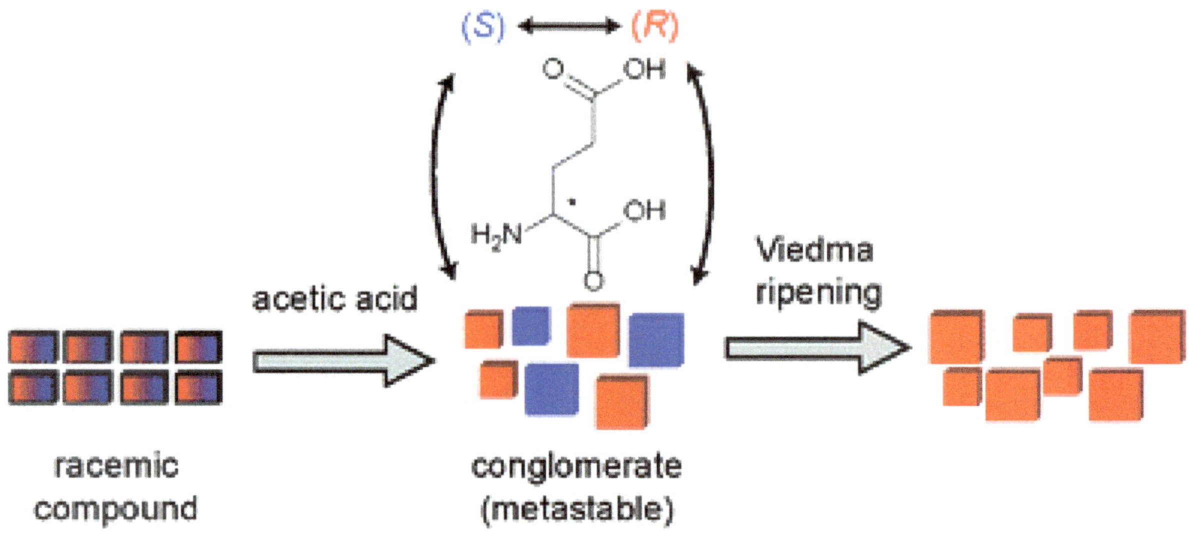

Figure 13 Deracemisation of a metastable conglomerate through Viedma ripening.

Alternatively, salt formation can be used as a tool to prepare a library of compounds which in turn could provide new conglomerate candidates. This way, Spix *et al.* combined six different amino acids with four different sulfonates resulting in twenty-four different salts.[46] It was found that three out of the twenty-four salts (13%) were racemic conglomerate crystals. Although leucine-2,5-xylenesulfonate was unstable under the applied Viedma ripening conditions which involves the use of acetic acid, the two other conglomerates (*i.e.* alanine-4-chlorobenzenesulfonate in 35–42% yield and phenylalanine-2,5-xylenesulfonate in 60–63% yield) could be deracemised through Viedma ripening (Fig. 14). In all these es, the yield is somewhat limited due to the solubility of the compounds.

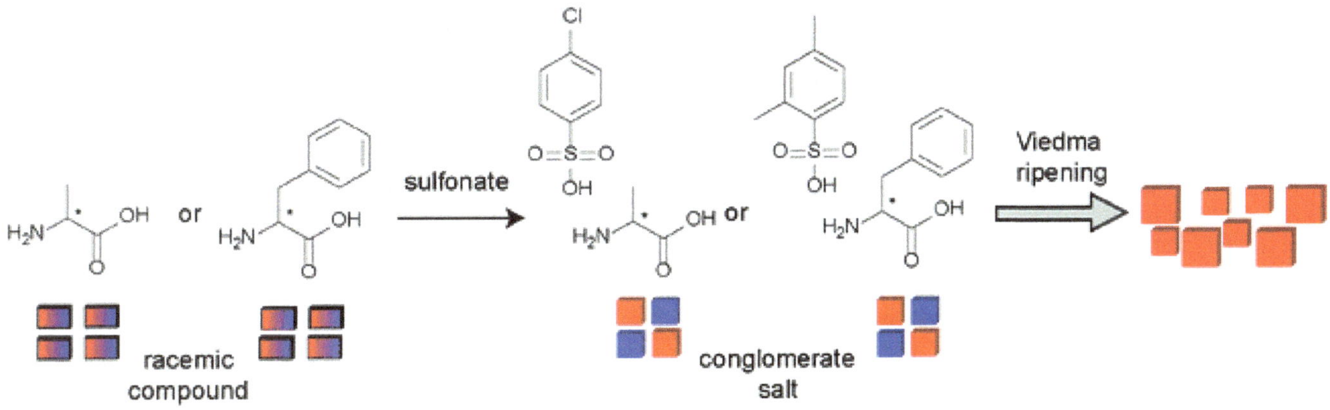

Figure 14 Amino acid salts that can be deracemised through Viedma ripening.

While Viedma ripening found broad applicability, Viedma himself in collaboration with other groups found that deracemisation can be achieved through the use of a temperature gradient at high temperatures and without grinding.[47,48] Coquerel *et al.* extended this observation to the use of deliberate 'temperature cycling' in order to gain a clearer picture of the mechanism behind this form of deracemisation, "of which the mechanism is still matter of debate".[49,50] With temperature cycling, all crystals start to dissolve simultaneously during a heating period, while the crystals remaining after the heating cycle are subsequently grown during the cooling period by consuming the excess of the solute molecules in the supersaturated solution (Fig. 15). A typical temperature program for temperature cycling involves rapid heating but slow cooling to avoid primary nucleation of the unwanted enantiomer. Such a program is repeated several times until an ee of 100% is reached.

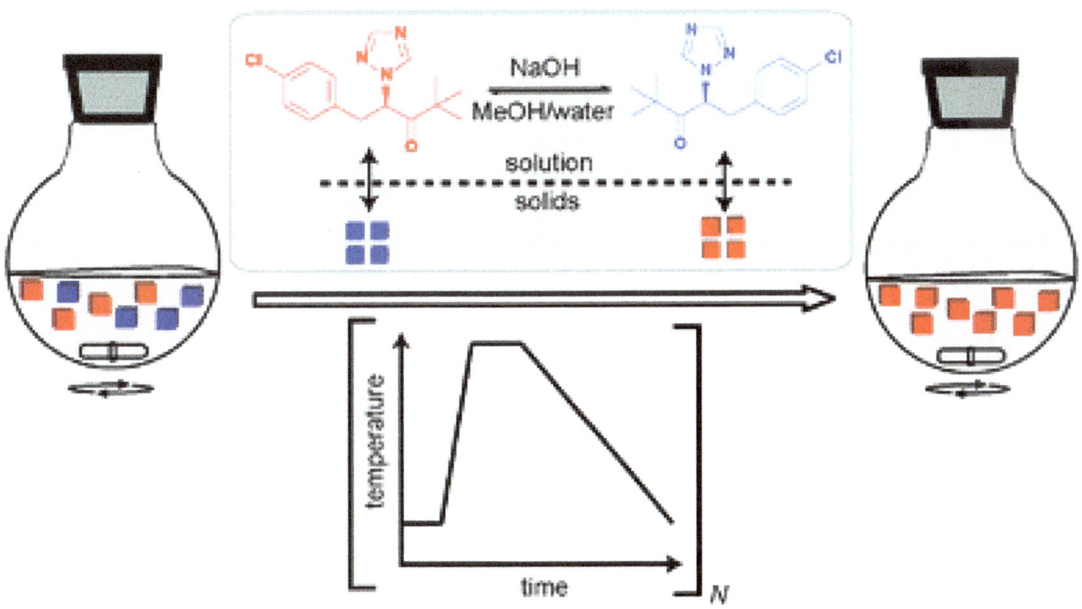

Figure 15 Deracemisation through repeated heating and cooling of a suspension.

CONCLUSIONS AND OUTLOOK

In this tutorial review we have shown that molecules that form conglomerate crystals and racemise in solution can easily be obtained in enantiopure form through total spontaneous resolution, Viedma ripening or variants thereof. The main bottleneck that prevents a wider application of these techniques remains the prerequisite of conglomerate crystallisation. Currently, it is unknown why most compounds crystallise as racemic compounds instead of racemic conglomerate crystals. Therefore, the most straightforward method to acquire conglomerates is to prepare a library of derivatives through synthesis or salt formation. Libraries of crystals could also be prepared through formation of co-crystals or solvates. Provided that the library is sufficiently large, a number of derivatives should turn out to be conglomerate crystals.

The second prerequisite is racemisation in solution, though this is not needed for achiral molecules that crystallise as a racemic conglomerate. Recent reports have shown that various racemisation approaches can be applied in combination with Viedma ripening and total spontaneous resolution, leading to even more examples of enantiopure products.

Even though there still is no absolute consensus, recent mechanistic studies have provided a deeper understanding of the driving forces behind Viedma ripening. Most studies have shown that a feedback mechanism must be involved to account for the exponential increase in ee. This feedback mechanism can be explained in terms of chiral clusters which selectively incorporate into crystals of the same handedness. Despite some indirect experimental observations, undisputed evidence for chiral clusters remains to be revealed.

After the discovery of the traditional Viedma ripening method, many novel applications have been developed which enable the use of Viedma ripening on a large industrial scale as well as in synthetic organic chemistry. Also, attrition is no longer required as gentle temperature fluctuations can lead to solid-state deracemisation. Crystal-solution systems are also very useful in asymmetric autocatalysis as enantiopure compounds can be obtained from achiral reactants.

From this review it appears that to date many compounds are still obtained in enantiopure form through total spontaneous resolution instead of Viedma ripening. However, it should be noted that Viedma ripening is a more robust and reliable method as crystals of the unwanted enantiomer, which often prevent total spontaneous resolution from reaching an enantiopure end state, are transformed into the desired enantiomer. Although seeding is used in total spontaneous resolution to obtain the best results in terms of chiral purity of the final product, Viedma ripening leads to complete deracemisation without seeding, even when starting from a racemic mixture of crystals.

REFERENCES

1. I. Myrgorodska, C. Meinert, Z. Martins, L. Le Sergeant d'Hendecourt and U. J. Meierhenrich, *Angew. Chem., Int. Ed.*, 2015, **54**, 1402–1412 .
2. V. Balzani, A. d. Meijere, K. N. Houk, H. Kessler, J. M. Lehn, S. V. Ley, S. L. Schreiber, J. Thiem, B. M. Trost, F. Vögtle and H. Yamamoto, *Top. Curr. Chem.*, 2007, **269**, 1–313 .
3. W. L. Noorduin, E. Vlieg, R. M. Kellogg and B. Kaptein, *Angew. Chem., Int. Ed.*, 2009, **48**, 9600–9606 .
4. T. Buhse, D. Durand, D. Kondepudi, J. Laudadio and S. Spilker, *Phys. Rev. Lett.*, 2000, **84**, 4405–4408 .
5. G. Coquerel, *Chem. Soc. Rev.*, 2014, **43**, 2286–2300 .
6. E. L. Eliel and S. H. Wilen, *Stereochemistry of Organic Compounds*, John Wiley & Sons, Inc., 1994
7. L. C. R. Pasteur, *Hebd. Séanc. Acad. Sci. Paris*, 1848, 535
8. J. E. Hein, B. Huynh Cao, C. Viedma, R. M. Kellogg and D. G. Blackmond, *J. Am. Chem. Soc.*, 2012, **134**, 12629–12636 .
9. E. Havinga, *Chem. Weekbl.*, 1941, **38**, 642–644 .
10. E. Havinga, *Biochim. Biophys. Acta*, 1954, **13**, 171–174 .
11. F. C. Frank, *Biochim. Biophys. Acta*, 1953, **11**, 459–463 .
12. D. K. Kondepudi, R. J. Kaufman and N. Singh, *Science*, 1990, **250**, 975–976 .
13. R. E. Pincock, R. R. Perkins, A. S. Ma and K. R. Wilson, *Science*, 1971, **174**, 1018–1020 .
14. D. K. Kondepudi, K. L. Bullock, J. A. Digits, J. K. Hall and J. M. Miller, *J. Am. Chem. Soc.*, 1993, **115**, 10211–10216 .
15. D. J. Durand, D. K. Kondepudi, P. F. Moreira Jr and F. H. Quina, *Chirality*, 2002, **14**, 284–287 .
16. D. K. Kondepudi and K. Asakura, *Acc. Chem. Res.*, 2001, **34**, 946–954 .
17. C. Viedma, *J. Cryst. Growth*, 2004, **261**, 118–121 .
18. C. Viedma, *Phys. Rev. Lett.*, 2005, **94**, 065504 .
19. W. L. Noorduin, T. Izumi, A. Millemaggi, M. Leeman, H. Meekes, W. J. P. Van Enckevort, R. M. Kellogg, B. Kaptein, E. Vlieg and D. G. Blackmond, *J. Am. Chem. Soc.*, 2008, **130**, 1158–1159 .
20. W. L. Noorduin, P. van der Asdonk, A. A. C. Bode, H. Meekes, W. J. P. van Enckevort, E. Vlieg, B. Kaptein, M. W. van der Meijden, R. M. Kellogg and G. Deroover, *Org. Process Res. Dev.*, 2010, **14**, 908–911 .
21. W. L. Noorduin, B. Kaptein, H. Meekes, W. J. P. van Enckevort, R. M. Kellogg and E. Vlieg, *Angew. Chem., Int. Ed.*, 2009, **48**, 4581–4583 .
22. M. W. van der Meijden, M. Leeman, E. Gelens, W. L. Noorduin, H. Meekes, W. J. P. van Enckevort, B. Kaptein, E. Vlieg and R. M. Kellogg, *Org. Process Res. Dev.*, 2009, **13**, 1195–1198 .
23. B. Kaptein, W. L. Noorduin, H. Meekes, W. J. P. van Enckevort, R. M. Kellogg and E. Vlieg, *Angew. Chem., Int. Ed.*, 2008, **47**, 7226–7229 .
24. W. L. Noorduin, H. Meekes, W. J. P. van Enckevort, B. Kaptein, R. M. Kellogg and E. Vlieg, *Angew. Chem., Int. Ed.*, 2010, **49**, 2539–2541 .
25. W. L. Noorduin, W. J. P. van Enckevort, H. Meekes, B. Kaptein, R. M. Kellogg, J. C. Tully, J. M. McBride and E. Vlieg, *Angew. Chem., Int. Ed.*, 2010, **49**, 8435–8438 .
26. M. Uwaha, *J. Phys. Soc. Jpn.*, 2004, **73**, 2601–2603 .
27. C. Viedma, J. M. McBride, B. Kahr and P. Cintas, *Angew. Chem., Int. Ed.*, 2013, **52**, 10545–10548 .
28. Z. El-Hachemi, J. Crusats, J. M. Ribo, J. M. McBride and S. Veintemillas-Verdaguer, *Angew. Chem., Int. Ed.*, 2011, **50**, 2359–2363 .
29. W. L. Noorduin, H. Meekes, W. J. P. van Enckevort, A. Millemaggi, M. Leeman, B. Kaptein, R. M. Kellogg and E. Vlieg, *Angew. Chem., Int. Ed.*, 2008, **47**, 6445–6447 .
30. W. L. Noorduin, A. A. C. Bode, M. van der Meijden, H. Meekes, A. F. van Etteger, W. J. P. van Enckevort, P. C. M. Christianen, B. Kaptein, R. M. Kellogg, T. Rasing and E. Vlieg,*Nat. Chem.*, 2009, **1**, 729–732 .
31. K. Asakura, K. Kobayashi, Y. Mizusawa, T. Ozawa, S. Osanai and S. Yoshikawa, *Phys. D*, 1995, **84**, 72–78 .

32. K. Soai, T. Shibata, H. Morioka and K. Choji, *Nature*, 1995, **378**, 767–768 .

33. K. Soai, I. Sato, T. Shibata, S. Komiya, M. Hayashi, Y. Matsueda, H. Imamura, T. Hayase, H. Morioka, H. Tabira, J. Yamamoto and Y. Kowata, *Tetrahedron: Asymmetry*, 2003, **14**, 185–188 .

34. J. B. Hendrickson, *J. Am. Chem. Soc.*, 1975, **97**, 5784–5800 .

35. R. R. E. Steendam, J. M. M. Verkade, T. J. B. van Benthem, H. Meekes, W. J. P. van Enckevort, J. Raap, F. P. J. T. Rutjes and E. Vlieg, *Nat. Commun.*, 2014, **5**, 6543

36. S. B. Tsogoeva, S. Wei, M. Freund and M. Mauksch, *Angew. Chem., Int. Ed.*, 2009, **48**, 590–594 .

37. A. M. Flock, C. M. M. Reucher and C. Bolm, *Chem. – Eur. J.*, 2010, **16**, 3918–3921 .

38. A. Lennartson, S. Olsson, J. Sundberg and M. Håkansson, *Angew. Chem., Int. Ed.*, 2009, **48**, 3137–3140 .

39. A. Lennartson and M. Håkansson, *Angew. Chem., Int. Ed.*, 2009, **48**, 5869–5871 .

40. M. Sakamoto, F. Yagishita, M. Ando, Y. Sasahara, N. Kamataki, M. Ohta, T. Mino, Y. Kasashima and T. Fujita, *Org. Biomol. Chem.*, 2010, **8**, 5418–5422 .

41. F. Yagishita, H. Ishikawa, T. Onuki, S. Hachiya, T. Mino and M. Sakamoto, *Angew. Chem., Int. Ed.*, 2012, **51**, 13023–13025 .

42. R. R. E. Steendam, M. C. T. Brouwer, E. M. E. Huijs, M. W. Kulka, H. Meekes, W. J. P. van Enckevort, J. Raap, F. P. J. T. Rutjes and E. Vlieg, *Chem. – Eur. J.*, 2014, **20**, 13527–13530 .

43. S. Hachiya, Y. Kasashima, F. Yagishita, T. Mino, H. Masu and M. Sakamoto, *Chem. Commun.*, 2013, **49**, 4776–4778 .

44. D. T. McLaughlin, T. P. T. Nguyen, L. Mengnjo, C. Bian, Y. H. Leung, E. Goodfellow, P. Ramrup, S. Woo and L. A. Cuccia, *Cryst. Growth Des.*, 2014, **14**, 1067–1076 .

45. L. Spix, H. Meekes, R. H. Blaauw, W. J. P. van Enckevort and E. Vlieg, *Cryst. Growth Des.*, 2012, **12**, 5796–5799

46. L. Spix, A. Alfring, H. Meekes, W. J. P. van Enckevort and E. Vlieg, *Cryst. Growth Des.*, 2014, **14**, 1744–1748 .

47. C. Viedma and P. Cintas, *Chem. Commun.*, 2011, **47**, 12786–12788 .

48. C. Viedma, J. E. Ortiz, T. de Torres, T. Izumi and D. G. Blackmond, *J. Am. Chem. Soc.*, 2008, **130**, 15274–15275 .

49. K. Suwannasang, G. Coquerel, C. Rougeot and A. E. Flood, *Chem. Eng. Technol.*, 2014, **37**, 1329–1339 .

50. K. Suwannasang, A. E. Flood, C. Rougeot and G. Coquerel, *Cryst. Growth Des.*, 2013, **13**, 3498–3504

CHAPTER 7

AN INVESTIGATION INTO THE EFFECT OF MIXING ON THE SECONDARY NUCLEATION OF SODIUM CHLORATE IN A STIRRED TANK AND AN OSCILLATORY BAFFLED CRYSTALLIZER

Craig J. Callahan and Xiong-Wei Ni

EPSRC Centre for Innovative Manufacturing in Continuous Manufacturing and Crystallisation, Centre for Oscillatory Baffled Reactor Applications (COBRA), School of Engineering and Physical Science, Heriot-Watt University, Edinburgh, EH14 4AS, UK.

ABTRACT

This paper reports the outcome of the examination of the effect of mixing intensity on the secondary nucleation mechanism of seeded crystallization of sodium chlorate in various configurations of stirred tank crystallizer (STC) and oscillatory baffled crystallizer (OBC). The results show that for the STC, an un-scraped system always yielded crystals of the same enantiomorphism as the seed crystal. The introduction of scraping to the STC resulted in product crystals of the opposite enantiomorphism to the seed being formed, however changing the rate of stirring (RPM) had no influence over the percentage similarity to the seed crystal. For the OBC, scraping always gave a product crystal crop with less than 100% similarity to the seed. Similarly, this was unaffected by altering the frequency of oscillation. Removing the scraping effect brought an increase of the similarity to the seed but the product crystals never reached 100% similarity to the seed enantiomorphism. For both scraped and un-scraped OBCs, reduction of the oscillation amplitude did increase the similarity of the products to the seed and 100% similarity to the seed was repeatedly achieved at the lowest operating amplitude for the un-scraped OBC. These findings provide further insight into the nucleation mechanism in the OBC and suggest that an alternative mechanism to that in the STC is observed.

INTRODUCTION

Over a hundred years ago, Young described how agitation would cause the onset of nucleation in a quiescent, supersaturated solution in which no crystallization would occur spontaneously.[1] This has since been studied in more detail and it is now recognized that mixing influences the metastable zone width (MSZW),[2] an important scientific parameter of solution crystallization. Some early works by Mullin and Raven showed that for the batch cooling crystallization of some aqueous salts, increasing the agitation rate raised the nucleation temperature; further increasing the stirrer speed reduced the nucleation temperature slightly and further intensifying the agitation rate resulted in enhanced nucleation with higher nucleation temperature once again. The suggested mechanism for this phenomenon was a highly complex combination of enhanced mass transfer due to reduced diffusion distance competing against disruption of pre-nuclei clusters due to fluid shear.[3,4]

Nývlt et al. investigated the relationship between stirrer speed and MSZW in order to probe the primary nucleation kinetics of aqueous solutions. They found that the higher the stirrer speed, the higher the nucleation rate constant, but the nucleation order was only dependent on the number of

Citation: Craig J. Callahan and Xiong-Wei Ni. An investigation into the effect of mixing on the secondary nucleation of sodium chlorate in a stirred tank and an oscillatory baffled crystallizer. CrystEngComm, 2014, 16, 690-697, DOI: 10.1039/C3CE41467A molecules required to form a critical nucleus, i.e. the nucleation order was independent of agitation rate.[5,6]

The relationship between nucleation and agitation rates investigated by Mullin and Nývlt were recently observed by Liang et al. in the study of L-glutamic acid crystallization. They hypothesised that reduction of boundary layer thickness was observed at higher agitation rates, which caused the change in nucleation rates,[7] although this parameter is difficult to observe and measure.

The mixing effect applies not only to primary nucleation, but also to secondary nucleation.[8] Melia and Moffitt presented data on the secondary nucleation of potassium chloride from aqueous solution and suggested two possible mechanisms for the secondary nucleation process.[9] The first possibility was that in the vicinity of the seed crystal, weakly attracted solute molecules were somewhat ordered in the seed crystal boundary layer without actually being incorporated into the crystal structure. These clusters remained in the boundary layer and, with increasing fluid shear, the boundary layer could be washed into the bulk solution and replaced with fresh solution. These clusters then developed into nuclei in the bulk solution away from the seed crystal. This type of nucleation was reported earlier by Powers in the study of sucrose crystallization.[10]

The second mechanism was that the solute molecules were incorporated into the seed crystal structure as dendrites. Through the action of fluid shear, or by crystallizer–crystal collisions, these dendrites could be ripped from the seed crystal and carried into the bulk solution to further grow. Melia and Moffitt supported this type of mechanism for the crystallization of ammonium chloride.[11] It would be reasonable to assume that these mechanisms are dependent on agitation rate.

Min and Goldburg investigated the influence of fluid shear on the formation of 2,6-lutidine droplets in water. In their experiments, the droplets appeared as temperature increased. Increasing the fluid shear resulted in a reduction of the embryo-type clusters and reduced the effective droplet nucleation rate.[12] This outcome is particularly relevant to the implementation of a two-step nucleation model where pre-nucleation clusters may exist as a dense fluid phase,[13] and relevant to what may occur in the boundary layer of a seed crystal when fluid dynamics of a crystallizer are altered.

The combined actions of collision breeding and the boundary layer models were discussed by Buhse et al. who concluded that secondary nucleation by either model was dependent on the crystallizer's hydrodynamics.[14] Chen et al. found that the thickness of sodium chlorate boundary layer was dependent on the fluid dynamic conditions.[15] It seems likely that the boundary layer bred nucleation process was competing with the collision bred one; the ratio of the two would depend on not only crystallizer hydrodynamics but also its fluid dynamics.

In general it has been shown that by altering the crystallizer›s fluid dynamics, it becomes possible to manipulate many important factors critical to nucleation. Increasing fluid agitation intensity and shear rates tends to reduce boundary layer thickness and would have measurable effects on what is happening in the vicinity of a seed crystal. Fluid shear may also influence any pre-nucleation clusters that may exist in the system[16] and allow some degree of impact over nucleation rates.

It has previously been stated that "crystallization may be the most difficult operation to scale-up – successfully" and highlighted the importance of mixing at all scales for control of nucleation and subsequent growth.[17] A drive to minimize differences in fluid dynamics at different scales of operation has led to the use of continuous crystallization by means of plug flow crystallizers such as the continuous oscillatory baffled crystallizer (COBC).[18,19] Due to the uniform mixing, plug flow characteristics,[20] enhanced mass[21] and heat transfer rates,[22] together with the ease and readiness of controlling temperature profiles along the length of the crystallizer, this provides a constant fluid mechanical condition for the nucleation and subsequent growth of crystals. This has led to uniform crystal sizes with significantly enhanced filtration rates.[23] The scale up of a lab scale COBC of typically 10–15 mm diameter to a pilot/full scale COBC of 40–100 mm diameter is a linear process based on the maintenance of the vessel geometry such as the baffle spacing and the orifice to tube cross sectional area ratio regardless of vessel length.[24] The ability to scale up in such a way means that there is little variation in fluid mechanical conditions between lab pilot and full modes of operation. This results in little variation of metastable zone widths and facilitates a direct and smooth scale up operation. In addition, analytical tools and monitoring techniques can be used at all scales without modification.

Typically, in order to successfully implement a continuous crystallizer, the process will be performed on the batch scale in order to confirm the kinetics (MSZW, nucleation mechanisms and growth mechanisms) and examine the effects of operating parameters on crystal specification. From those, the residence time for a COBC can be determined, which is roughly equal to the differences between the starting and the nucleation temperature and between the nucleation temperature and the end temperatures divided by suitable desirable cooling rates. It is therefore the scope of this work to understand how the fluid dynamics utilised in batch operation can assist the decisions to be made in continuous processing, and gain some insights as to how a crystallizing species may behave in a continuous crystallizer.

In recent studies of crystallization of active pharmaceutical ingredients (APIs) in the OBC, some interesting results were consistently observed. For example, higher nucleation temperatures and narrower MSZW was observed in an OBC than in a stirred tank crystallizer (STC), even though processes parameters, such as solution concentration, cooling rates and mixing intensity, i.e. power dissipated in the solution by the mixing device, remained constant.[25] Rapid growth of crystals in an OBC without the entrainment of impurities was possible[26] as was seeding being unnecessary in an OBC to obtain a certain crystal specification while it was essential in an STC for the same operation.[27] This current work aims to address the latter finding by utilizing a crystal compound whose enantiomorphism or morphology in the solid state could be related to the nucleation mechanism. Sodium chlorate is such a non-chiral substance that on crystallization produces crystalline solids possessing either a left (levorotatory) or right handed (dextrorotatory) enantiomorphism.[28] As such, the crystalline solid will display optical activity, a property that has been exploited in previous studies to determine the handedness of the crystals based on the direction of rotation of plane polarized light,[29,30] and in turn elucidate the origin of the product crystals.[31,32]

Sodium chlorate was previously utilised in two relevant investigations into nucleation mechanism by Denk and Botsaris,[33,34] where at a supercooling of 3 °C and 350 RPM stirring in a stirred tank crystallizer, a crop of product crystals had 100% right handedness when seeded with a single right handed crystal. The equivalent outcome was reported for seeding with a levorotatory crystal. This would indicate that secondary nucleation took place due to either dendrite coarsening or collision breeding within the crystallizer, i.e. the entire product crystals were originated directly from the seed crystal. At higher supercoolings of greater than 7 °C, primary nucleation was observed. This was shown from a mixture of 50 : 50 left and right handed product crystals obtained, suggesting that if spontaneous nucleation occurred, a mixture of both left and right handed crystals would be the evidence. This outcome was also supported by Martin et al.[35]

The hypothesis investigated here is that a different nucleation mechanism could be observed in an OBC compared to that in a STC due to different styles of mixing, or mixing mechanisms, while all operational conditions remain constant, such as supersaturation, temperature, mixing intensity and so on. In an earlier work of seeded sodium chlorate crystallization at a supercooling of 1 °C, it was observed that in the STC, all of the product crystals bore the same handedness as the seed crystal, indicating that secondary nucleation directly from the seed was the source of product crystals. This will be referred to as "seed-similar" nucleation hereafter. For the OBC, on the other hand, the product crystals were never more than 96% similar to the seed crystal, suggesting an alternative nucleation mechanism that facilitates a "seed-dissimilar" nucleation model. The scraping action of the baffle outer edge against the inner surface of the crystallizer wall was proposed as the cause of generation of the incorrect enantiomorph in the OBC. To test this hypothesis, the scraping was removed from the OBC, while it was introduced to the STC. The opposite results were obtained, i.e. the un-scraped OBC produced much more seed-similar crystals and the scraped STC produced seed-dissimilar crystals, which verified this hypothesis.[36] The objective of this work is to investigate how the identified nucleation mechanism would vary with mixing intensity and local fluid flow mechanics.

EXPERIMENTAL SETUP AND PROCEDURE

For this current work, two types of crystallizer were investigated, each had two possible configurations: with and without scraping. This allowed the nucleation mechanism to be further probed at various mixing intensities in both crystallizers, with the aims of reinforcing and consolidating these previous hypotheses and findings, as well as discovering new findings and extending the investigation through rigorous design of experiment.

The Stirred Tank Crystallizer

Fig. 1 shows the setup of the STC. The STC consisted of a jacketed round bottom glass vessel with a working volume of 500 mL. A curved, PTFE paddle type stirrer was driven by the overhead motor and was of diameter 80 mm and width 15 mm. In order to introduce scraping, the impeller can be pressed tightly against the vessel bottom due to the curvature of the paddle. Raising the impeller allowed the scraping effect to be removed. Stainless steel wall baffles were introduced via a stainless steel flange. Four stainless steel baffles were spaced equally 90° apart and extended approximately 30 mm into the fluid. This allowed the system to be well mixed,[37] yet provided the stirrer a reasonable range of clearances from the vessel bottom without colliding with the baffles. With a working volume of 500 mL, the height of the liquid in the vessel was approximately 80 mm. The rotation was adjustable from 50 to around 2000 RPM.

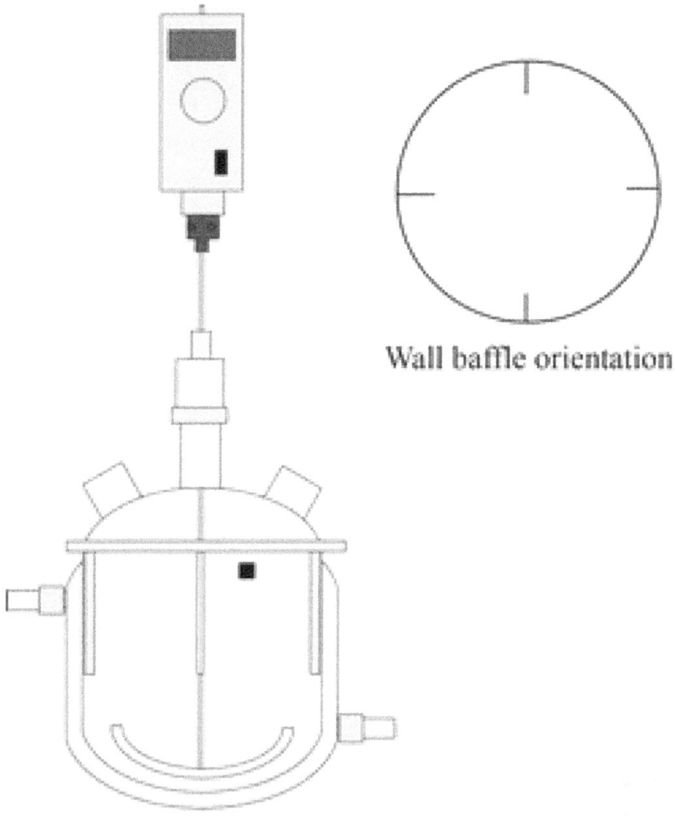

Figure 1 Image of the STC setup (not to scale). Jacket connectors allowed temperature controlvia water bath. The steel wall baffles were suspended by the flange at the top of the vessel and orientated according to the schematic at the top right of the figure. The stirrer height was set by clamping the shaft at the overhead stirrer. The seed location is indicated by the filled square.

The energy dissipation of a stirred tank crystallizer is defined by:[38]

$$\frac{P}{V} = \frac{P_O \rho N_s^3 D_S^5}{V_L}$$

where P/V is the energy dissipation (power per unit volume) (W m^{-3}), ρ the fluid density (1456 kg m^{-3} at 40 °C[39]), N_s the speed of the stirrer (rps), D_S the diameter of the stirrer (0.08 m), and V_L the volume of liquid in the STC (0.0005 m^3). P_O is the dimensionless power number of the agitator which was estimated at 1 based on data from Nagata.[40]

The Oscillatory Baffled Crystallizer

Fig. 2 shows the setup of the OBC. The OBC consisted of a jacketed glass column of an internal diameter of 50 mm and a height of approximately 500 mm. This gave a working volume of 500 mL. The vessel was held in position by a stainless steel flange. Two sets of baffle strings were used in order to create scraped and non-scraped systems. The first baffle string comprised 4 PTFE baffles with an outer diameter of 50 mm, an orifice diameter of 24 mm and a thickness of 3 mm. The baffles were spaced at 65 mm by stainless steel spacers and tightly fit against the wall of the column, allowing the actions of scraping. The second baffle string consisted of 4 PTFE baffles with an outer diameter of 40 mm, an orifice diameter of 17 mm and a thickness of 3 mm. Again, the baffles were spaced at 65 mm with stainless steel spacers. In this e, there was a gap between the outer edges of the baffles and the inner surface of the column, leading to no scraping against the vessel wall when oscillated by means of a linear motor (Copley Controls Corp.), which was held by an aluminium/plastic frame above the OBC. Frequencies of 0–10 Hz and amplitudes of 0–50 mm were possible. The selected trough to peak amplitude for the OBC was 30 mm (roughly half the baffle spacing and allowable by the hardware).

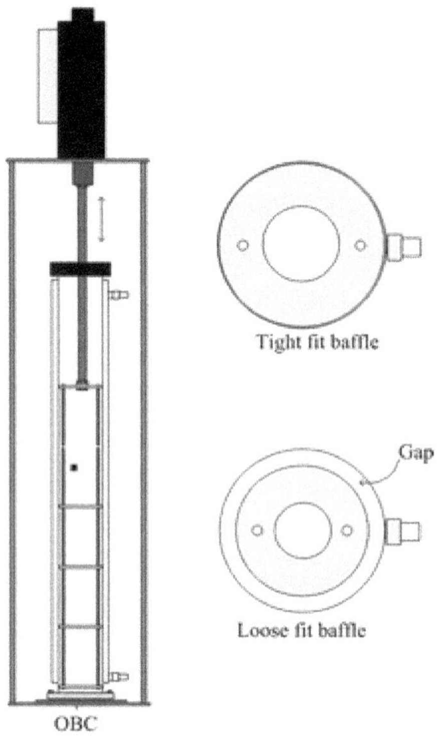

Figure 2 Image of the OBC setup (not to scale). The operating configuration of the OBC is shown to the left. A top view of the cross sectional area of the tight fitting baffle is shown top right and the loose fitting baffle shown bottom right. The seed location is indicated by the filled square.

The energy dissipation of the OBC is estimated using the quasi steady flow model proposed by Baird and Stonestreet:[41]

$$\frac{P}{V} = \frac{2\rho N_b}{3\pi C_D^2}\left(\frac{1-\alpha^2}{\alpha^2}\right)x_o^3(2\pi f)^3$$

where N_b is the number of baffles per unit length of OBC (15.38 m^{-1}), C_D the discharge coefficient of the baffles (taken as 0.7), x_o the centre to peak amplitude of oscillation (0.015 m), f the oscillation frequency (Hz) and α the baffle free area ratio (0.23 for the tight fit baffles, 0.48 for the loose fit baffles). The free area ratio, α is defined by the area of the orifice divided by the tube cross-sectional area and also takes into account the gap around the baffle edge (if present).

Three energy dissipation values (low, medium and high) were achieved by changing the oscillation frequencies in the OBC and the stirrer speed in the STC, as given in Table 1. For the purpose of comparison, similar energy dissipation was used as the basis for each of the devices.[26]

Table 1 Mixing conditions for each crystallizer. The mixing speed is given to the left and the approximate energy dissipation (W m^{-3}) given in brackets to the right

Mixing condition	STC (RPM)	OBC tight baffles (Hz)	OBC loose baffles (Hz)
Low	65 (12)	0.4 (9)	0.8 (13)
Medium	160 (180)	1.0 (145)	2.0 (216)
High	260 (776)	1.6 (595)	3.0 (732)
		2 (1162)	

Experimental Procedure

As with the previous work,[36] the crystallizers were temperature controlled by a Grant GP200R2 water bath and the temperature was monitored using stainless steel T-type thermocouples. The general procedure involved preparing an aqueous solution of laboratory grade sodium chlorate (sourced from Fisher Scientific UK, 99+% purity) saturated at 31 °C, c = 1.05 g $NaClO_3$per g H_2O. The solution was made using distilled water to the working volume of the crystallizer, and cured at 40 °C for one hour before being filtered using an oven heated, Whatman glass fiber filter (grade GF/C nominal pore size 1.2 µm). Solutions were held again at 40 °C for a further 30 minutes before being cooled, with agitation, to 30 °C, $c^* = 1.03$ g $NaClO_3$ per g H_2O, giving a calculated supersaturation, S, of 1.02. On reaching 30 °C, a single seed crystal of known handedness was washed with cold distilled water and dried using filter paper before being suspended by means of a wire in the crystallizing solution for three minutes whilst maintaining agitation. The seed crystal was then removed from the solution and the agitation stopped to allow the product crystals to grow quiescently overnight at 30 °C to form larger crystals. The product crystals were separated, dried and analysed by polarimetry. Both the left handed (levorotatory) and right handed (dextrorotatory) crystals were separated from an average of 3000 individual crystals for each run and the similarity to the seed crystals examined and calculated. Each experiment was repeated a further two times to establish the repeatability of the data. The apparatus was thoroughly cleaned between trials using household detergent at 60 °C followed by three rinses with hot water. The apparatus was allowed to dry prior to use.

RESULTS AND DISCUSSION

Before presenting the main results, a number of benchmark experiments were performed. In order to verify that no spontaneous nucleation could occur as a result of surface phenomena, the first set of experiments were conducted according to the procedure described earlier, except no seeds were used. Only the blank wire that was used for holding the single seed crystal was suspended in the solution. The crystallizers containing the supersaturated solutions were left overnight and it was found that no crystals were seen in any system (both scraped and un-scraped devices). The same outcome was observed three times in each device, indicating that the possibility of either the presence of a foreign body (the blank wire) or entraining foreign particles to initiate heterogeneous nucleation was not observable under the tested conditions. This suggests that for these experiments, the nucleation of product crystals stemmed entirely from the introduction of a single sodium chlorate seed crystal.

The next set of benchmark tests were conducted in the previously described manner, but no mixing at all was applied to the crystallizers. This was to verify the hypothesis that it is the mixing mechanism that has an effect on the secondary nucleation. When no mixing was applied before, during and after the seeding, it was seen that all product crystals bore the same handedness as the seed crystal in both devices. Again, three repeats were done in both the OBC and STC systems. This outcome would suggest that in quiescent solutions, where the seed boundary layer has not been disturbed, the solute molecules would adhere to the seed in the bulk solution, thus providing seed-similar nucleation of products. This is in contrast to the findings of Denk and Botsaris, who found that the percentage similarity to the seed crystal was around 60% at 1 °C supercooling.[33] The reason for the difference between their report and this current work may lie in the fact that while Denk and Botsaris were adding the wetted seed crystal to the solution, it "was rotated slightly to dissipate whatever water film might be adhering to its surfaces".[33] The seed was added dry to the solution in the experiments reported here.

The final set of benchmarking experiments involved halting the agitation prior to seeding, rather than after seeding. This was to further verify the above hypothesis. As with the previous trials, it was found that all of the seeds were of the same handedness as the seed crystal. It would indicate that the mixing mechanism would be the only likely cause of deviation from 100% similarity to the seed either through affecting the seed boundary layer, or via the previously identified scraping action.

The results from the seeded crystallizations in the STC are presented in Table 2.

Table 2 Percentage similarity to the seed crystal for the seeded crystallization in both the un-scraped (left) and scraped (right) STC

Un-scraped sample	Mixing intensity			Scraped sample	Mixing intensity		
	Low	Medium	High		Low	Medium	High
1	100.00	100.00	100.00	1	92.62	100.00	93.62
2	100.00	100.00	100.00	2	94.61	99.02	96.33
3	100.00	100.00	100.00	3	98.77	92.24	97.35
Average	100.00	100.00	100.00	Average	95.33	97.09	95.77
Standard error	0.00	0.00	0.00	Standard error	1.18	2.44	1.11

From the data of the un-scraped STC on the left of Table 2, it can be seen that the similarity to seed is consistently 100% for all power dissipations. While no mechanism was expected in the un-scraped STC to promote nucleation of the opposite-to-seed enantiomorph, increasing the power dissipated into the system would be expected to increase the rate of secondary nucleation by either crystal–crystal or crystal–crystallizer collisions,[42,43] without influencing the product crystal handedness.

When scraping was introduced to the STC by simply having the impeller tightly pressed against the bottom of the vessel, seed-dissimilar nucleation did occur (the right side of Table 2). This outcome was expected from the previous work.[36] It would seem reasonable to infer that as the mixing intensity increased in a system where a seed-dissimilar nucleation mechanism exists, one of two outcomes could be seen. The first possible scenario could be that the rate of seed-dissimilar nucleation because of the scraping mechanism would increase due to more scrapes per unit time (in this e rotations per second) resulting in a decrease in the percentage similarity to the seed of the crystals produced. The second possible outcome that may be observed could be the rate of seed-similar nucleation being enhanced due to either crystal–crystal or crystal–crystallizer collisions occurring more frequently because of the increased mixing. However the data in Table 2 suggest that neither scenario was observed. In fact, analysis of the data (Kruskal–Wallis[44]) suggests that there is no statistical significance in the differences between the results at various mixing intensities (P = 0.7326). These experiments suggest that although scraping did cause the formation of product crystals of the opposite enantiomorphism to the seed crystal, the energy dissipation at which scraping is applied had little or no influence on the rate of this seed-dissimilar nucleation. This could perhaps be due to both seed-similar and seed-dissimilar nucleation rates increasing with mixing intensity, resulting in no net change in the average similarity.

The experiments conducted in the STC were then repeated in the OBC. The results of these tests are given in Table 3.

Table 3 Percentage similarity to the seed crystal for the seeded crystallization in both the un-scraped (left) and scraped (right) OBC at a fixed centre to peak amplitude of 15 mm

Un-scraped sample	**Mixing intensity**			**Scraped sample**	**Mixing intensity**		
	Low	Medium	High		Low	Medium	High
1	100.00	98.86	99.58	1	94.02	99.00	93.09
2	96.99	99.49	99.83	2	94.89	93.43	93.20
3	99.88	98.69	99.01	3	96.31	92.90	90.49
Average	98.96	99.01	99.47	Average	95.07	95.11	92.26
Standard error	0.98	0.24	0.24	Standard error	0.67	1.95	0.89

Considering the scraped OBC first (the right side of Table 3), seed-dissimilar nucleation did occur, leading to the percentage of similarity to the seed being lower than 100%. It appears, however, that there is no clear trend with the change in mixing intensity (frequency). Indeed, no statistical significance can be found in the difference between the outcomes under the three conditions (P = 0.1479). Similarly, with the un-scraped OBC (the left side of Table 3), some seed-dissimilar nucleation also took place. Analysis of the data of the un-scraped set up suggests again that there is no significant difference between each mixing intensity (P = 0.4298). In terms of trends found in both the scraped STC and scraped OBC, the effect of mixing (stirring vs.oscillation) had little impact. Statistical analysis of these data shows there is no significant difference in the percentage similarity between the scraped OBC and the scraped STC (P = 0.2004).

The seed-dissimilar mechanism took place in the un-scraped OBC, while it was not observed in the un-scraped STC. The statistical difference in the percentage similarity between the two un-scraped set ups is significant (P = 0.0005). Why is this? Previous work on the numerical modelling of flow patterns in both tight fit and loose fit baffle arrangements in the OBC showed that more shear was predicted in the latter device than in the former.[45] The seed-dissimilar nucleation in the un-scraped OBC may be linked with the increased shear at the gap, even though it has previously been shown that the introduction of a gap in the OBC increased the mixing time, i.e. the larger the gap is, the poorer the mixing is.[46] This would suggest that the local fluid flow condition in the un-scraped OBC could have played a deciding role on the nucleation mechanism.

To put this hypothesis to the test, lower oscillation amplitudes of 7 and 15 mm were used at a fixed frequency, so that the amount of fluid shearing at the gap is effectively reduced as the shear (strain) rate is proportional to the change of oscillatory velocity ($x_o f$) at the gap as

$$\dot{\gamma} \propto \frac{x_0 f}{D}$$

where $\dot{\gamma}$ is the fluid shear rate (s^{-1}) and D is the baffle diameter (m).[47]

Clearly, a reduction of the amplitude at a fixed frequency reduces the value of the oscillatory velocity, x_0f, and therefore the shear (strain) rate of the system. The results of the varied amplitude experiments in both the scraped and un-scraped OBC are presented in Table 4.

Table 4 Percentage similarity to the seed crystal for the seeded crystallization in the scraped (left) and un-scraped (right) OBCs. These tests were conducted at a fixed frequency of 2 Hz, the approximate energy dissipation (W m^{-3}) is given in brackets next to the amplitude

Un-scraped sample	**Amplitude, x_0 (mm)**			**Scraped sample**	**Amplitude, x_0 (mm)**		
	3.5 (3)	7.5 (27)	15 (216)		3.5 (14)	7.5 (145)	15 (1162)
1	100.00	99.96	98.86	1	97.85	98.12	93.12
2	100.00	99.94	99.49	2	100.00	95.72	93.12
3	100.00	99.42	98.69	3	99.91	98.39	93.82
Average	100.00	99.77	99.01	Average	99.25	97.41	93.35
Standard error	0.00	0.18	0.24	Standard error	0.70	0.85	0.23

Examination of the data from the un-scraped OBC (the left side of Table 4) reveals that decreasing the amplitude, i.e. the shear (strain) effect, increased the similarity to the seed, and 100% similarity to the seed was achieved at the lowest amplitude of 7 mm, where the shear effect was at its lowest. Increasing the amplitude gave a significant difference in the outcome of the experiment ($P = 0.0349$). The data continue to suggest that the local fluid mechanical conditions could influence nucleation mechanisms. To further confirm this, the same conditions were tried in the scraped OBC as shown the right side of Table 4. A clear trend of the similarity to the seed increasing towards 100% with reducing oscillation amplitude is evident here, with a significant statistical difference ($P = 0.0496$). It seems that by reducing the effective scraped area via the length of the scraping in the system, the local shearing effect was effectively reduced, leading to the significant reduction of the amount of seed-dissimilar nucleation. The results suggest that both the overall mixing intensity and the local fluid mechanical conditions during the seeding within a crystallizer are equally important in influencing nucleation.

The question remains as how, exactly, the seed-dissimilar crystals were formed in the scraped and un-scraped systems. It could be speculated that the seed boundary layer could be the key factor in deciding whether or not nucleation would be seed-similar or seed-dissimilar. In the OBC, the fluid flow regime could disrupt this boundary layer differently, giving different pre-nucleation conditions[48] to those seen in the STC. For the scraped systems, the results presented here resemble an undergraduate chemistry lab project using a spatula scraping the surface of a beaker containing supersaturated solution, but the exact mechanism becomes harder to elucidate, perhaps with a combined scraping and fluid flow regime model in effect. As Davey et al. pointed out, determination of the crystal nucleus characteristics (such as size, shape, and, particularly, the enantiomorph) will require technology and data that are as yet unavailable.[49]

CONCLUSIONS

In conclusion, it was previously found that when a scraping mechanism is introduced to either a stirred tank or oscillatory baffled crystallizer containing a supersaturated solution of sodium chlorate, nuclei of opposite enantiomorphism to the seed crystal are found in the crop of product crystals.[36] This paper reported that no real change in the percentage of similarity has been seen in the two systems when the mixing intensity (in terms of frequency and stirring rate) was altered; further demonstrated that local fluid mechanical conditions (e.g. shearing effect due to the oscillatory amplitude) can also lead to the formation of seed-dissimilar crystals. By reducing such an effect, the un-scraped OBC gave 100% similarity to the seed at the lowest amplitude (shear effect). The results of this work indicate that the action of scraping generates nuclei of the opposite enantiomorphism to the seed, so does the local mixing/shearing conditions in a crystallizer.

ACKNOWLEDGEMENTS

The authors thank EPSRC and the EPSRC Centre for Innovative Manufacturing in Continuous Manufacturing and Crystallisation and Heriot-Watt University for the matching funding for this work. Thanks are expressed to: Chris Price and Alex Caillet of GSK; Gerry Steele and Amy Robertson of Astra Zeneca; Kevin Girard of Pfizer; Thomas Rammeloo of Johnson and Johnson; Jan Sefcik of The University of Strathclyde and the other Academics of the Continuous Crystallization Discussion Group (CCDG) for the most helpful monthly meetings. The authors thank Alison Nordon and Jaclyn Dunn of The University of Strathclyde for the most helpful discussion of statistical analysis.

REFERENCES

1. S. W. Young, J. Am. Chem. Soc., 1911, 33, 148–162 .
2. J. Ulrich and C. Strege, J. Cryst. Growth, 2002, 237–239, Part 3, 2130–2135.
3. J. W. Mullin and K. D. Raven, Nature, 1961, 190, 251–251 .
4. J. W. Mullin and K. D. Raven, Nature, 1962, 195, 35–38 .
5. J. Nývlt, J. Cryst. Growth, 1968, 3–4, 377–383 .
6. J. Nývlt, R. Rychlý, J. Gottfried and J. Wurzelová, J. Cryst. Growth, 1970, 6, 151–162 .
7. K. Liang, G. White, D. Wilkinson, L. J. Ford, K. J. Roberts and W. M. L. Wood, Ind. Eng. Chem. Res., 2004, 43, 1227–1234 .
8. A. Myerson, Handbook of Industrial Crystallization, Butterworth Heinemann, 2002.
9. T. P. Melia and W. P. Moffitt, Ind. Eng. Chem. Fundam., 1964, 3, 313–317 .
10. H. E. C. Powers, Ind. Chem., 1963, 39, 351 .
11. T. P. Melia and W. P. Moffitt, J. Colloid Sci., 1964, 19, 433–447 .
12. K. Y. Min and W. I. Goldburg, Phys. Rev. Lett., 1993, 71, 569 .
13. J. Chen, B. Sarma, J. M. B. Evans and A. S. Myerson, Cryst. Growth Des., 2011, 11, 887–895 .
14. T. Buhse, D. Durand, D. Kondepudi, J. Laudadio and S. Spilker, Phys. Rev. Lett., 2000, 84, 4405–4408 .
15. W. C. Chen, D. D. Liu, W. Y. Ma, A. Y. Xie and J. Fang, J. Cryst. Growth, 2002, 236, 413–419 .
16. D. Erdemir, A. Y. Lee and A. S. Myerson, Acc. Chem. Res., 2009, 42, 621–629 .
17. E. L. Paul, H.-H. Tung and M. Midler, Powder Technol., 2005, 150, 133–143 .
18. X. Ni, Innovations Pharm. Technol., 2006, 20, 90–96.
19. NiTech Solutions Ltd, http://www.nitechsolutions.co.uk, Accessed 18th March 2011.
20. X. Ni, H. Jian and A. Fitch, Trans. IChemE, 2003, 81, 842–853 .
21. M. R. Hewgill, M. R. Mackley, A. B. Pandit and S. S. Panu, Chem. Eng. Sci., 1993, 48, 799–809 .
22. M. R. Mackley, G. M. Tweedle and I. D. Wyatt, Chem. Eng. Sci., 1990, 45, 1237–1242 .
23. S. Lawton, G. Steele, L. Shering, L. Zhao, I. Laird and X. Ni, Org. Process Res. Dev., 2009, 13, 1357–1363 .
24. X. Ni, Y. S. De Gelicourt, M. H. I. Baird and N. V. Rama Rao, Can. J. Chem. Eng., 2001, 79, 444–448 .
25. X. Ni and A. Liao, Chem. Eng. J., 2010, 156, 226–233 .
26. C. M. Chew and R. I. Ristic, AIChE J., 2005, 51, 1576–1579 .
27. NiTech Solutions Ltd., Clients› data, unpublished work.
28. C. W. Bunn, in Chemical Crystallography, Oxford Clarendon Press, Oxford, 1945, ch. 2, p. 39.
29. R. M. Pagni and R. N. Compton, Cryst. Growth Des., 2002, 2, 249–253 .
30. C. Mayer and R. Lacmann, Chem. Eng. Technol., 1997, 20, 633–640 .
31. D. K. Kondepudi, K. L. Bullock, J. A. Digits, J. K. Hall and J. M. Miller, J. Am. Chem. Soc., 1993, 115, 10211–10216 .
32. D. K. Kondepudi, R. J. Kaufman and N. Singh, Science, 1990, 250, 975–976 .
33. E. G. Denk and G. D. Botsaris, J. Cryst. Growth, 1972, 13–14, 493–499 .
34. E. G. Denk and G. D. Botsaris, J. Cryst. Growth, 1972, 15, 57–60 .
35. B. Martin, A. Tharrington and X. l. Wu, Phys. Rev. Lett., 1996, 77, 2826–2829 .
36. C. J. Callahan and X.-W. Ni, Cryst. Growth Des., 2012, 12, 2525–2532 .
37. N. A. Mitchell, P. J. Frawley and C. T. Ó'Ciardhá, J. Cryst. Growth, 2011, 321, 91–99 .
38. R. K. Sinnott, in Coulson and Richardson's Chemical Engineering Volume 6 – Chemical Engineering Design, ed. R. K. Sinnott, Butterworth-Heinemann, Oxford, Third edn, 1999, ch. 10, pp. 471–473 .
39. A. Seidell, in Solubilities of inorganic and metal organic compounds, Van Nostrand Company, New York, Third edn, 1940, vol. 1, p. 1250 .
40. S. Nagata, Mixing: Principles and Applications, Kodansha Ltd, Tokyo, 1975
41. M. H. I. Baird and P. Stonestreet, Trans. IChemE, 1995, 73, 503–511 .
42. T. W. Evans, G. Margolis and A. F. Sarofim, AIChE J., 1974, 20, 950–958 .
43. L. G. Bauer, R. W. Rousseau and W. L. McCabe, AIChE J., 1974, 20, 653–659 .
44. The MathWorks Inc., http://www.mathworks.co.uk/help/stats/kruskalwallis.html, Accessed 11th July 2013.
45. M. E. Mackay, M. R. Mackley and Y. Wang, Trans. IChemE, 1991, 69, 506–513 .
46. X. Ni and C. C. Stevenson, J. Chem. Technol. Biotechnol., 1999, 74, 587–593 .
47. X. Ni, J. A. Cosgrove, A. D. Arnott, C. A. Greated and R. H. Cumming, Chem. Eng. Sci., 2000, 55, 3195–3208 .
48. D. Gebauer and H. Colfen, Nano Today, 2011, 6, 564–584 .
49. R. Davey, S. L. M. Schroeder and J. H. ter Horst, Angew. Chem., Int. Ed., 2013, 52, 2166–2179

CHAPTER
8

THE ROLE OF SURFACE HYDROLYSIS OF FERRICYANIDE ANIONS IN CRYSTAL GROWTH OF SNOWFLAKE-SHAPED A-FE_2O_3

Zhong Liu [ab], Chang-Yang Chiang [b], Wu Li [a] and Wuzong Zhou [b]

[a]Key Laboratory of Salt Lake Resources and Chemistry, Qinghai Institute of Salt Lakes, Chinese Academy of Sciences, Xining 810008, China

[b]EaStChem, School of Chemistry, University of St Andrews, St Andrews, KY16 9ST, UK.

ABSTRACT

Selective adsorption and surface hydrolysis of $[Fe(CN)_6]^{3-}$ anions on α-Fe_2O_3 crystals was found to be a crucial process in the formation of a snowflake-like morphology, and the established mechanism is complementary to the classical theories of crystal growth.

INTRODUCTION

Formation of high symmetric polyhedral morphologies of crystals can normally be explained using a kinetic approach of Bravais–Friedel–Donnay–Harker law, *i.e.* appearance of facets of a crystal is attributed to the slow growth rates along these directions. Another theory gives a thermodynamic approach: the equilibrium shape of a free crystal is the shape that minimizes its surface free energy. Both of these classical theories deal with the bonding energy of crystal structures without considering external effects, and therefore, cannot explain the formation mechanisms of many unusual crystal morphologies, *e.g.* polyhedra of polycrystalline particles of calcite and MOFs,[1] complicated snowflake-shaped hematite (α-Fe_2O_3).[2] The anisotropy of the crystal structure of hematite cannot justify the only active crystal growth directions of six equivalent zone axes, parallel to the (*ab*) plane of the hexagonal unit cell.

Hematite, as the most stable iron oxide under ambient conditions, is a good candidate for application in many fields, including catalysis, gas sensors, lithium battery, magnetic materials production, electromagnetic devices and pigments, *etc.*[3] In recent years, it has been found that these applications are greatly affected by the microstructures and morphologies of the crystals.[4] 'Snowflake' α-Fe_2O_3 with hyperbranches has been investigated because of their large surface areas and unique hierarchical structures.[5] However, the formation mechanism of these morphologies is yet to be studied.

Herein, we report our recent re-synthesis of 'snowflake' α-Fe_2O_3 crystals *via* a hydrolysis process of $[Fe(CN)_6]^{3-}$. The growth directions of $\langle 11\bar{2}0\rangle$ instead of commonly reported $\langle 10\bar{1}0\rangle$ are detected from selected area electron diffraction (SAED) patterns and high resolution transmission electron microscopic (HRTEM) images. The growth direction was found to be governed by highly selective surface adsorption followed by on-site hydrolysis of ferricyanide anions. Addition of octylamine in the synthetic solution led to a change in the favourite growth directions from $\langle 11\bar{2}0\rangle$ to $\langle 10\bar{1}0\rangle$ and induced a morphology evolution from 'snowflake' to 'starfish',

Citation: Zhong Liu, Chang-Yang Chiang, Wu Li and Wuzong Zhou. The role of surface hydrolysis of ferricyanide anions in crystal growth of snowflake-shaped α-Fe_2O_3. Chem. Commun., 2015, 51, 9350-9353, DOI: 10.1039/C5CC02011E 'flower' and finally to hexagonal plate, offering further support to the newly established mechanism.

'Snowflake' α-Fe_2O_3 particles were synthesized according to the reported method.[2] 1.52 mmol $K_3Fe(CN)_6$ were dissolved in 40 mL deionized water. The solution was subsequently sealed in a 50 mL autoclave, and maintained at 180 °C for 20 h, followed by cooling down to room temperature gradually. Almost all the particles in the red precipitate have a snowflake-like shape (Fig. S1a, b, ESI†). The XRD pattern of the product shows a single phase of α-Fe_2O_3 with all the diffraction peaks indexed onto the hexagonal unit cell with a = 5.0356 and c = 13.7489 Å (Fig. S2a, ESI†).

A TEM image of a typical 'snowflake' particle is shown in Fig. 1(a), revealing more microstructural details. Although the constructed hyperbranched microstructure of the particle looks quite complicated, the corresponding SAED pattern (Fig. 1b) indicates that the whole particle can be regarded as a single crystal, and the viewing direction perpendicular to the face of the 'snowflake' is along the [0001] zone axis of α-Fe_2O_3. In other words, all the branches and sub-branches in the particle are linked through a crystalline connection. Comparable examples are mesoporous single crystals of metal oxides.[6] It can also be confirmed by HRTEM images that the crystal growth directions are the six $\langle 11\bar{2}0\rangle$ zone axes (Fig. 1d, e). The whole particle is quite flat with a smooth (0001) surface, while the diameter of the 'snowflake' particles is about 10 μm, which is much larger than the particle sizes of other specimens as shown in Fig. S1, ESI.† It again implies that the growth along the $\langle 11\bar{2}0\rangle$ directions is fast. If the diffusion rates and attachment behaviour of the building units, such as Fe^{3+} and OH^-, to various crystal surfaces are similar, such highly selective directions of crystal growth cannot be explained based on the crystallographic anisotropy of α-Fe_2O_3.

Figure 1 (a) TEM image of a typical snowflake-like particle of α-Fe_2O_3. The inset is a photo of a real snowflake for comparison. (b) The corresponding SAED pattern. (c) An enlarged TEM image of a tip. (d) and (e) HRTEM images from the tip areas marked 'd' and 'e' in (c), respectively. The arrows indicate the crystal growth directions. The insets are the corresponding FFT patterns.

We assume, in the initial stage, some nuclei of α-Fe_2O_3 form in the solution of $K_3Fe(CN)_6$ *via* the following hydrolysis process.

$[Fe(CN)_6]^{3-} \leftrightarrow [Fe(H_2O)_x]^{3+} \rightarrow FeOOH/Fe(OH)_3 \rightarrow \alpha\text{-}Fe_2O_3$.

SEM and HRTEM investigations of time-dependant specimens (Fig. S3, ESI†) indicate that the processes of nucleation and growth of spherical crystallites are very slow. Further crystal growth of the branches is much faster. The latter process relies on attachment of precursor ions/molecules on the crystal surface. Since the dissociation of $[Fe(CN)_6]^{3-}$ in water is extremely low with the dissociation constant, $K_d = 1.0 \times 10^{-42}$, there are almost no Fe^{3+} free cations in the solution. On the other hand, hydrolysis of $[Fe(CN)_6]^{3-}$ results in a reduction of pH value from 7.27 to 6.76 (see Table S1, ESI†) and, therefore, the concentration of OH^- anions is also very low. The principal precursor ions in the solution are $[Fe(CN)_6]^{3-}$. These ferricyanide anions contain six lone pairs of electrons on the outer surface, which can bond on the crystal surface if the latter contains some exposed positive charged sites or Lewis acidic sites.

Under the hydrothermal synthetic conditions, the surface of metal oxide crystals is normally terminated with hydroxyl groups. Close examination of the terminal atomic planes on the principal facets of α-Fe_2O_3, such as the $\{10\bar{1}0\}$, $\{11\bar{2}0\}$, and $\{0001\}$ planes, has been previously performed and their electronic structures calculated.[7] The charge of the hydroxyl groups depends on the local configuration, *i.e.* the number of Fe coordination. If a hydroxyl group is coordinated by only one Fe cation, it is negatively charged, designated site I or $OH^{0.5-}$. A doubly coordinated hydroxyl group is neutralized, designated site II or OH^0, while that in a triple coordination is positively charged, designated site III or $OH^{0.5+}$. The $\{0001\}$ surfaces of hematite contain only doubly coordinated hydroxyl groups with a density of 13.7 nm^{-2}, which are neutral and inactive in the surface adsorption of ferricyanide anions. The $\{10\bar{1}0\}$ surfaces have both singly and doubly coordinated hydroxyls (5.8 nm^{-2} and 2.9 nm^{-2}, respectively) and are, therefore, overall negatively charged, which is also inactive for bonding with the ferricyanide anions. The $\{11\bar{2}0\}$ surfaces contain all the three types of co-ordinations (5.0 nm^{-2} for each type of sites). The positively charged $OH^{0.5+}$ in the triply coordinated hydroxyl groups can act as Lewis acidic sites, offering a chance to bond with $[Fe(CN)_6]^{3-}$. This is probably the most important factor of the selective crystal growth along the $\langle 11\bar{2}0 \rangle$ directions.

Fig. 2(a) shows a top view of the $(11\bar{2}0)$ surface. The hydroxyl groups having different Fe coordinations (sites I, II and III) are indicated with the triply coordinated oxygen highlighted by a pink shadow. Fig. 2(b) is a profile view of the $(11\bar{2}0)$ surface *via* rotation of the model in Fig. 2(a) around the *Z* axis for 90°. H atoms on site III are shown. When a $[Fe(CN)_6]^{3-}$ anion is adsorbed on the $(11\bar{2}0)$ surface as shown in the middle of Fig. 2, bonding with $OH^{0.5+}$, partial electron transfer from a CN^- ligand to the crystal surface weakens the Fe–C bond, enhancing the surface hydrolysis of the $[Fe(CN)_6]^{3-}$ anion. All the CN^- ligands would eventually leave Fe on the surface sites followed by dehydrogenation and dehydration, leading to deposition of α-Fe_2O_3 as shown in Fig. 2(c and d). The original triply coordinated oxygen atoms become four coordinated and the singly coordinated sites become triply coordinated, *i.e.* new sites III. The process can go on. Since such a surface hydrolysis only takes place on the $\{11\bar{2}0\}$ surfaces, and the growth on the $\{10\bar{1}0\}$ and $\{0001\}$ surfaces is greatly depressed, the final morphology of the crystals becomes two dimensional thin snowflake-like. When the hydrolysis is not completed, a small amount of CN^- might remain on the crystal surface, as detected by XPS (Fig. S4a, ESI†).

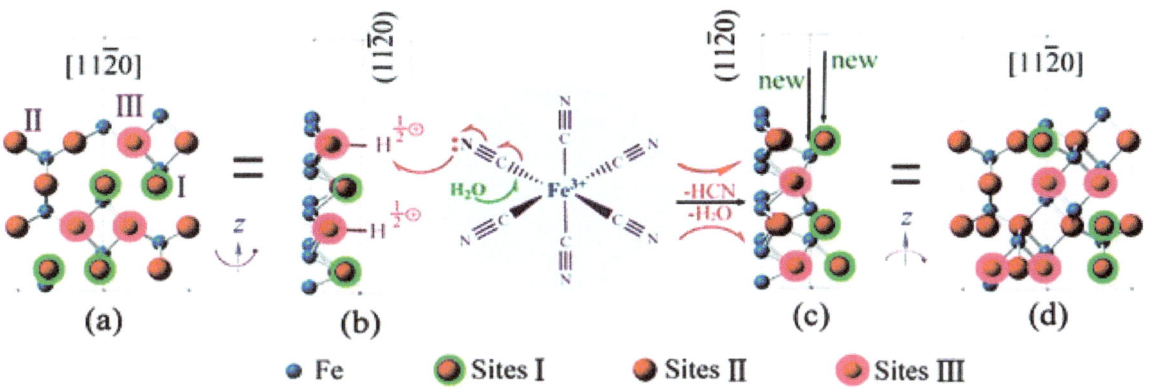

Figure 2 Schematic drawing of the surface adsorption of a $[Fe(CN)_6]^{3-}$ anion to the positively charged triply coordinated hydroxyl group on the $(11\bar{2}0)$ surface and its surface hydrolysis into α-Fe_2O_3. (a) and (b) Top view and profile view with H on site III of the $(11\bar{2}0)$ surface. (c) and (d) Profile view and top view of the surface with a newly deposited layer of α-Fe_2O_3.

When octylamine was added into the solution with a concentration of 0.12 mol L^{-1}, the morphology of product changed from 'snowflake' to 'starfish' with six arms, while the particle size decreases to about 3 μm (Fig. S1c, d, ESI†). Although six primary branches were also developed in each crystal, the details of the hyperbranched structure, as seen in a TEM image inFig. 3(a), is different from 'snowflake' crystals. First, the secondary branches are close to each other, leaving no room for growth of the tertiary branches. Secondly, the thickness of the primary branches along [0001] increases significantly. Therefore, the particles are no longer thin and flat. Finally and most importantly, a careful analysis of SAED and HRTEM revealed that the principal crystal growth directions have been shifted from $\{11\bar{2}0\}$ to $\{10\bar{1}0\}$, as presented in Fig. 3(a,b).

Figure 3 (a) TEM image of a typical 'starfish' particle from the sample prepared with 0.12 mol L^{-1} octylamine. The inset is the corresponding SAED pattern. (b) HRTEM image of a tip area, marked by 'b', of the particle in (a). (c) TEM image of a 'flower' α-Fe_2O_3 particle from the sample prepared with 0.24 mol L^{-1} octylamine. The inset is the corresponding SAED pattern. (d) HRTEM image from a tip area, marked by 'd', of the particle in (c). The insets in (b) and (d) are the FFT pattern from the whole areas. The arrows in (b) and (d) indicate the crystal growth directions of $[10\bar{1}0]$.

With a further increase of the concentration of octylamine to 0.24 mol L^{-1} and 0.36 mol L^{-1}, the hyperbranched structure completely disappears. The centres of the particles grow out, showing a hexagonal flower-like morphology. The diameters of the particles further decrease to 1 μm and below (Fig. S1e–h, ESI†). Fig. 3(c,d) shows TEM and HRTEM images of a typical flower-like particle. Analysis of both the corresponding SAED pattern and *d*-spacings in the HRTEM image confirms that the principal growth directions are along the $\langle 10\bar{1}0 \rangle$ zone axes.

The effects of octylamine can be understood in two aspects. The overall reaction of hydrolysis of $[Fe(CN)_6]^{3-}$ can be written as

$$2[Fe(CN)_6]^{3-} + 3H_2O \leftrightarrow \alpha\text{-}Fe_2O_3 + 6H^+ + 12CN^-.$$

As a weak alkali capping agent, octylamine can enhance the above reaction. Consequently, the concentration of $[Fe(CN)_6]^{3-}$ anions in the solution would be reduced. On the other hand, octylamine (pK_b = 3.35) is a relatively strong alkaline compound than ferricyanide anions (pK_b = 4.86).[8] Therefore, octylamine can directly bond with the $OH^{0.5+}$ on site III of the $\{11\bar{2}0\}$ surfaces, blocking these surfaces. The surface hydrolysis of ferricyanide anions as described in Fig. 2 is greatly suppressed. XPS of flower-like crystals shows a small N 1 s peak at a binding energy of 399.47 eV (Fig. S4a, ESI†), which has a significantly higher intensity than that from snowflake crystals (Fig. S4b, ESI†) and can be attributed to the surface adsorption of octylamine.[9]

Another effect of amine on the surface hydroxyl configuration is from the alkaline property of the solution, as seen in Table S1, ESI.† The $OH^{0.5+}$ groups on the $\{11\bar{2}0\}$ surfaces would be more or less neutralized. When NaOH was added instead of octylamine, no 'snowflake' were produced, and both the particle size and morphology changed with the concentration of NaOH.

The transformation of $[Fe(CN)_6]^{3-}$ to Fe_2O_3 cannot be a direct process. It may be achieved *via* formation of series intermediate ions, *e.g.* $[Fe(H_2O)_x(CN)_{6-x}]^{x+3-}$ ($x = 1$ to 6). The charge of the Fe-containing ions changes from 3− to 3+ step-by-step. When site III is no longer available for adsorption of negatively charged ions, site I with $OH^{0.5-}$ becomes the active site to attract positively charged ions. In other words, Fe-containing cations would go to the $\{11\bar{2}0\}$ and $\{10\bar{1}0\}$ surfaces. Bearing in mind the former is blocked by octylamine, the most active surface for crystal growth would be the latter. This proposed mechanism explains why the crystal growth directions shifted from the $\langle 11\bar{2}0\rangle$ to $\langle 10\bar{1}0\rangle$ zone axes when octylamine was added.

When the concentration of octylamine was increased to 0.48 and 0.60 mol L^{-1}, the pH value of the solution increased further. The charge difference of the hydroxyl groups on all the surfaces became less obvious. The hydrolysis of $[Fe(CN)_6]^{3-}$ only took place in the solution. The crystal growth turned to a normal process, and the morphology of the produced particles gradually changed to hexagonal plates as shown in Fig. S1(i–l), ESI.† The average diameters of the particles reduced to 500 nm to 400 nm, respectively, while their thickness increased significantly. TEM and HRTEM images in Fig. S5, ESI† confirm such a crystal morphology.

In summary, from the formation of snowflake-shaped α-Fe_2O_3, we learnt a new phenomenon of crystal growth, interaction of precursor ions with the crystal surface produces the building units for crystals on selective sites, leading to specific growth directions. Low dimensional and complicated morphologies can form in free crystals with no templates or surface modification with ligands. This mechanism may be used to explain formation of many novel crystal morphologies.

This work was supported by the NSFC (No. 51302280), NSF of Qinghai (2014-ZJ-936Q), and Young Scholar Project of Qinghai Institute of Salt Lakes, Chinese Academy of Sciences. WZ thanks EPSRC for financial support to the Electron Microscopy Laboratory (EP/F019580/1).

REFERENCES

1. A. Ritchie, M. Watson, R. Turnbull, Z. Lu, M. Telfer, J. Gano, K. Self, H. F. Greer and W. Z. Zhou, *CrystEngComm*, 2013, 15, 10266 ; C. M. Zheng, H. F. Greer, C.-Y. Chiang and W. Z. Zhou, *CrystEngComm*, 2014, 16, 1064 .
2. Z. Liu, B. L. Lv, D. Wu, Y. Zhu and Y. H. Sun, *CrystEngComm*, 2012, 14, 4074 .
3. N. D. Carbonare, V. Cristino, S. Berardi, S. Carli, R. Argazzi, S. Caramori, L. Meda, A. Tacca and C. A. Bignozzi, *ChemPhysChem*, 2014, 15, 1164 ; X. L. Gou, G. X. Wang, J. Park, H. Liu and J. Yang, *Nanotechnology*, 2008, 19, 125606 ; Z. Y. Wang, D. Y. Luan, S. Madhavi, C. M. Li and X. W. Lou, *Chem. Commun.*, 2011,47, 8061 ; C. Z. Wu, P. Yin, X. Zhu, C. Z. OuYang and Y. Xie, *J. Phys. Chem. B*, 2006, 110, 17806 ; Y. Kim, J. H. Pee, J. H. Chang, K. Choi, K. J. Kim and D. Y. Jung, *Chem. Lett.*, 2009, 38, 842 .
4. B. L. Lv, Z. Y. Liu, H. Tian, Y. Xu, D. Wu and Y. H. Sun, *Adv. Funct. Mater.*, 2010, 20, 3987 .
5. M. H. Cao, T. F. Liu, S. Gao, G. B. Sun, X. L. Wu, C. W. Hu and Z. L. Wang, *Angew. Chem., Int. Ed.*, 2005, 44, 4197 .
6. W. B. Yue and W. Z. Zhou, *Prog. Nat. Sci.*, 2008, 18, 1329 .
7. R. M. Cornell and U. Schwertman, *The Iron Oxides: Stucture, Properties, Reaction, Occurences and Use*, Wiley VCH GmbH&Co. KGaA, 2nd edn, 2003, p. 222 ; V. Barron and J. Torrent, *J. Colloid Interface Sci.*, 1996, 177, 407 ; C. H. Rochester and S. A. Topham, *J. Chem. Soc., Faraday Trans. 1*, 1979, 75, 1073 .
8. Z. Wan, W. L. Luan and S. T. Tu, *J. Phys. Chem. C*, 2011, 115, 1569 .
9. Z. Liu, B. L. Lv, D. Wu, Y. H. Sun and Y. Xu, *Eur. J. Inorg. Chem.*, 2012, 4076

CHAPTER 9

SOLUTION-CRYSTALLIZATION AND RELATED PHENOMENA IN 9,9-DIALKYL-FLUORENE POLYMERS. I. CRYSTALLINE POLYMER-SOLVENT COMPOUND FORMATION FOR POLY (9,9-DIOCTYLFLUORENE)

Aleksandr Perevedentsev, Paul N. Stavrinou, Donal D. C. Bradley, Paul Smith

Department of Physics and Centre for Plastic Electronics, Imperial College London, London, United Kingdom

ABSTRACT

Polymer-solvent compound formation, occurring via co-crystallization of polymer chains and selected small-molecular species, is demonstrated for the conjugated polymer poly(9,9-dioctylfluorene) (PFO) and a range of organic solvents. The resulting crystallization and gelation processes in PFO solutions are studied by differential scanning calorimetry, with X-ray diffraction providing additional information on the resulting microstructure. It is shown that PFO-solvent compounds comprise an ultra-regular molecular-level arrangement of the semiconducting polymer host and small-molecular solvent guest. Crystals form following adoption of the planar-zigzag β-phase chain conformation, which, due to its geometry, creates periodic cavities that accommodate the ordered inclusion of solvent molecules of matching volume. The findings are formalized in terms of nonequilibrium temperature–composition phase diagrams. The potential applications of these compounds and the new functionalities that they might enable are also discussed.

INTRODUCTION

Poly(9,9-dioctylfluorene) (PFO) is a widely studied conjugated polymer that exhibits a broad range of desirable properties, such as efficient pure-blue photo (PL) and electroluminescence (EL), high charge–carrier mobility and optical gain, good thermal stability and excellent processability from solutions in common organic solvents.[1-3] It has further been shown that solid-state PFO can be controllably fabricated in a range of glassy, liquid- and semi-crystalline microstructures that directly influence the resulting optoelectronic properties.[4-11] These features make PFO well-suited for applications in a variety of optoelectronic devices, such as LEDs,[12] lasers,[3, 13] sensors,[14, 15] and photonic elements,[16-18] while also allowing it to be used as a test-bed for fundamental studies of conjugated polymer photophysics.[3, 7, 19] The 9,9-dioctylfluorene unit is additionally found in a wide variety of copolymer structures.[20-22]

Solution-crystallization of PFO, typically manifested as thermoreversible gelation when solution concentration is sufficiently high for substantial chain overlap,[23] has been demonstrated for a range of organic solvents, such as cyclohexane,[5, 6] α-pinene,[6] dichloroethane,[24] methylcyclohexane (MCH),[25-27] and toluene.[28] In an early study Grell et al.[6] tentatively excluded polymer-solvent compound formation as a contributory factor due to the literature teaching that such process should occur only

Citation: Aleksandr Perevedentsev, Paul N. Stavrinou, Donal D. C. Bradley, Paul Smith. Solution-Crystallization And Related Phenomena In 9,9-Dialkyl-Fluorene Polymers. I. Crystalline Polymer-Solvent Compound Formation For Poly(9,9-Dioctylfluorene). Journal Of Polymer Science, Part B: Polymer Physics 2015, 53, 1481–1491, Doi: 10.1002/Polb.23798 for good solvents;[29]this turns out not to be the case (vide infra). The fundamentally intra-chain nature of the processes responsible for the observed changes in absorption and PL spectra was nevertheless recognized, with the distinct planar-zigzag β-phase chain segment conformation shown to be a common feature of PFO–cyclohexane gels, vapor-annealed or thermally-treated PFO and PFO:polystyrene dilute blend films, and of vapor-annealed PFO quenched nematic glass fibres.[6] Later studies used a combination of light, X-ray and neutron scattering as well as optical spectroscopy to further investigate the photophysical and structural properties of solution-crystallized PFO. In MCH and toluene PFO was found to crystallize into sheet-like structures, usually referred to as "aggregates," with lateral dimensions on the order of 10–100 nm and thickness of a few nm.[25-28, 30] Subsequent agglomeration of these sheets into ribbon-like structures allowed thermoreversible crosslinking and gelation[24, 31] but the sheet-like structures themselves invariably displayed the spectroscopic and crystallographic signatures of β-phase chain segments.[26, 28] Despite the results of these previous studies, solution-crystallization of PFO has been rather poorly understood due to several fundamental ambiguities. First, the number and nature of the phase-transitions that result in the formation of crystalline structures in PFO solutions remains unclear. Second, although both intra- and inter-chain structure formation, mediated by β-phase chain segment and sheet-like domain generation, respectively, has been observed upon solution crystallization, the relationship between these phenomena has not been unequivocally determined.

This confused situation motivated our present study and we report below on the formation and selected properties of solution-crystallized polymer-solvent compounds, comprising PFO and particular organic solvents. Such compounds form by stoichiometric co-crystallization of polymer chains and small-molecular solvents, resulting in a composite molecular structure featuring weak, typically van der Waals, bonding between the two components.**[32, 33]** Depending on the details of their microstructure as well as the scientific context of the study, these compounds have also been termed crystallosolvates or polymer-solvent intercalates, clathrates, and complexes. **[32, 33]** Cellulose represents one well-known example of a natural polymer that forms such compounds; synthetic polymer examples include polyoxyethylene,**[34]** poly(methyl methacrylate),**[29, 35]** poly(vinylidene fluoride),**[36]** as well as isotactic**[37]** and syndiotactic polystyrene.**[38-41]** Despite their unique microstructure, polymer-solvent compounds have, in general, only found relatively limited use as, for example, nanoporous selectively absorbing materials.**[42, 43]** A number of other applications in sensing,**[44, 45]** catalysis,**[46]** and packaging**[47]** have, however, been proposed recently.

In particular, we present the results of a systematic study of the formation and resulting microstructure of PFO-solvent compounds, investigated by a combination of thermal analysis and X-ray diffraction. We clarify the fundamental role played by the β-phase conformation in compound formation. We further describe how these findings augment the established understanding of PFO solution processing and may allow consequent control of β-phase chain segment fraction within resulting films. While so-called bimolecular crystals/intercalates have been previously reported for mixtures of, for instance, thiophene-based conjugated polymers and fullerene derivatives,**[48, 49]** on the basis of this study PFO represents the first conjugated polymer for which polymer-solvent compound formation is clearly demonstrated. We seek, therefore, to also outline new approaches by which the unique molecular-level guest-host crystalline arrangement enabled by compound formation might be exploited to improve the performance of conjugated polymer-based devices.

EXPERIMENTAL

Materials

PFO was synthesized using the Suzuki coupling route by the Sumitomo Chemical Company Ltd. The polymer was subjected to extensive purification prior to shipment and was used as received. The weight-average molecular weight, as determined by polystyrene-equivalent gel-permeation chromatography (GPC), was 2.87 × 10^5 g mol^{-1}, with a polydispersity index = 3.0. Decahydronaphthalene ("decalin") (>98%, mixture of cis and trans isomers, Acros Organics), toluene (99.7%, anhydrous, Sigma-Aldrich), n-dodecane (99%, Acros Organics), 1,2,4-trichlorobenzene ("oTCB") (≥98%, Merck), and n-hexadecane (99%, ABCR) were used as received.

Gel Preparation and Thermal Analysis

Polymer solutions were prepared directly in the standard low-pressure aluminium differential scanning calorimetry (DSC) crucibles. After addition of the required amount of solvent, the crucibles were sealed and carefully weighed before and after measurements to ensure that no solvent loss had occurred. DSC was carried out using a Mettler Toledo DSC 822e instrument that was routinely calibrated using indium standards. As a first step, all mixtures were annealed at temperatures near the boiling point of the solvent for≥20 min to ensure that homogeneous solutions were obtained. Standard 5 °C min^{-1} heating/cooling rates were used, except for preparation of the so-called "slowly crystallized" gel samples, processed for maximal degrees of crystallinity. For these, the solutions were cooled at 1 °C min^{-1} and then annealed at the corresponding crystallization temperature for 45 min. Melting enthalpy ΔH_m of free, that is, crystallizable, solvent in these slowly crystallized gels was determined by integrating the corresponding DSC heating thermograms in the −50–20 °C temperature range, which included both the high- and low-temperature endothermal transitions.

Optical Microscopy

Temperature-dependent optical microscopy was carried out with a Leica DMRX microscope equipped with a Linkam THMS600 hot-stage. Gel samples were sealed between two coverslips, ensuring that no solvent loss occurred.

Critical-Point Drying of Gels

Dried gels for X-ray diffraction measurements were prepared from the slowly-crystallized as-prepared gels by supercritical drying using a CO_2 critical-point dryer (SPI Supplies). This method generally allows for interface-free removal of the solvent, thereby preventing the collapse of the swollen as-prepared gel due to the absence of surface tension and resulting in minimal associated changes to the microstructure of the polymer-rich phase.[50, 51] The sealed DSC crucibles containing the gel samples were opened and immediately flushed with liquid CO_2 at ~15 °C; the samples were then left for 2 h allowing solvent exchange to take place. The temperature was then increased to 37 °C (below the glass transition temperature T_g of neat PFO), allowing supercritical extraction of CO_2. Due to the limited miscibility of CO_2 with the solvents used in this study, the drying process was repeated 3 times to yield maximally solvent-free samples. To ensure data comparability, both as-prepared and dried gel samples were prepared for X-ray diffraction measurements using solutions with identical starting concentrations.

X-Ray Diffraction

Wide-angle X-ray diffraction (WAXD) was performed on an Oxford Instruments XCalibur PX diffractometer using Mo-Kα radiation (0.71 Å wavelength). Gel, polymer and solvent reference samples were individually sealed inside glass capillary tubes (Hilgenberg; 1.5 mm diameter). Sample temperature was controlled using the Cryojet accessory by streaming temperature-stabilized (±0.1 °C accuracy) nitrogen gas over the capillary tube. The measurements were principally carried out at −100 °C, corresponding to a temperature substantially below the expected compound T_g. While it was not possible to directly measure T_g for any of the compounds, we estimated the expected T_g values from the corresponding T_m values using the empirical equation (1) which holds true for the majority of polymers.[52]

$$\frac{T_g}{T_m} \approx 0.5\text{–}0.7 \qquad (1)$$

While the T_g values determined by this method are very approximate at best, we found that further cooling of the gel samples below −100 °C did not lead to any noticeable changes in the diffraction patterns. The samples were equilibrated at each temperature for 15 min prior to measurement and 20 min integration times were used to record the data. The two-dimensional diffraction patterns were radially integrated following correction for background signal.

RESULTS

Thermal Analysis

Figure 1(a) illustrates common manifestations of solution-crystallization of PFO for the specific case of PFO–decalin mixtures. Also shown in Figure 1(b) are the chemical structure of PFO and schematics of its two principal chain conformations. DSC thermograms [cf. Fig. 1(a)] reveal that the solution-crystallization process is thermoreversible. A crystallization exotherm appears at 28 °C in the cooling trace and a melting endotherm at 82 °C in the heating trace. We note further that the cross-polarized micrograph (bottom left) taken at 0 °C on cooling shows strong crystallite scattering but no evidence for the occurrence of macroscopic liquid–liquid (L–L) demixing and that the corresponding micrograph (top right) taken at 125 °C on heating shows clear dissolution of the semicrystalline microstructure resulting in an isotropic solution.

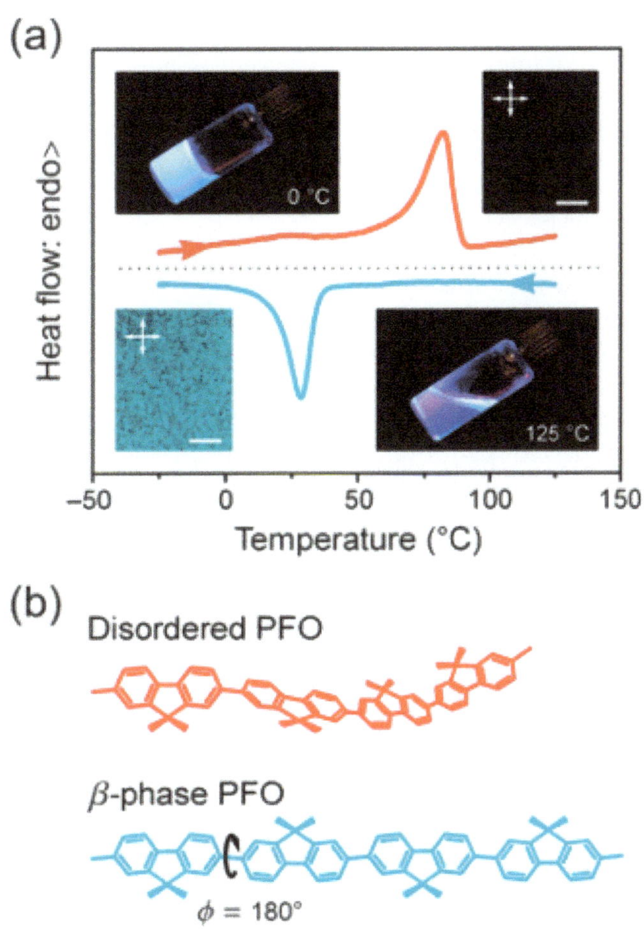

Figure 1. (a) Illustration of the main aspects of the solution-crystallization of PFO, showing, as a specific example, typical DSC thermograms and cross-polarized microscopy images (20 μm scale bars) for PFO–decalin (polymer weight fraction c_p=0.3). The dark isotropic image (top

right) was taken at 125 °C and the bright scattering image (bottom left) at 0 °C. Photographs of vials containing a (more dilute) $c_p = 0.005$ PFO–decalin gel (0 °C, upper left) and solution (125 °C, lower right) under UV illumination are also presented, showing both the thermoreversible solidification and the change in fluorescence color upon gelation. (b) Schematic illustration of the disordered (i.e., wormlike) and β-phase chain conformations of PFO. The octyl (C_8H_{17}) side-chains are omitted for clarity.

The inset photographs of vials containing isotropic (bottom right) and crystallized (top left) PFO–decalin mixtures under UV illumination show additionally that solution-crystallization of PFO leads to gelation, that is, formation of a macroscopically coherent "solid" structure, even at polymer weight fractions in solution, c_p, as low as the 0.005 value used for these photographs. The occurrence of macroscopic gelation at such remarkably low polymer content clearly indicates the chain-extended nature of the crystalline polymer structures that are formed as well as a high degree of crystallinity. By comparison, macroscopic gelation of ultra-high molecular weight polyethylene (UHMWPE) in decalin does not occur for the same polymer fraction,[**53**] despite the fact that the chain contour length is >23 times higher for UHMWPE relative to the PFO used in this study. Hereafter, for the sake of simplicity, all crystallized PFO solutions will be simply referred to as gels.

The photographs also show a distinct change in fluorescence color between the solution and the gel. While in the solution PFO chains typically adopt a disordered "wormlike" conformation with a broad distribution of intermonomer torsion angles, the gels invariably contain a substantial fraction of PFO chain segments in the so-called β-phase conformation; both chain geometries are schematically illustrated in Figure 1(b). The β-phase conformation refers to a planar-zigzag chain geometry, with a well-defined[**10**] inter-monomer torsion angle Φ, which the majority of studies have determined to be Φ = 180°,[**4-7, 54**] and a characteristic alignment of the octyl side-chains along the backbone.[**10, 55**] The red-shifted emission from β-phase chain segments (arising from a more extended conformation that supports greater electronic delocalization) typically dominates PL due to efficient excitation energy transfer from the disordered, that is, higher HOMO-LUMO transition energy, parts of the chain ensemble.[**4, 5, 10, 19**]

In order to investigate the thermal transitions involved in the crystallization and dissolution of PFO, DSC was performed on mixtures with a range of different organic solvents. [N.B. DSC thermograms of neat PFO can be found in Part II of this study.] While all previous reports were for relatively dilute (≤ 50 mg mL^{-1}) PFO solutions,[**30**] in this study the mixtures were prepared (see Experimental) over a much wider concentration range (e.g., 33–3540 mg mL^{-1}for PFO–dodecane). Figure 2(a) presents the temperature–composition diagrams for different PFO-solvent combinations, showing, in each case, the variation of peak dissolution temperature with polymer weight fraction, c_p (g/g). The reported dissolution temperatures correspond to the peak of the endothermal transitions observed in heating DSC thermograms; when two overlapping endotherms were observed then, for the sake of clarity, the endotherm with the highest peak heat flow was selected for determining the dissolution temperature.

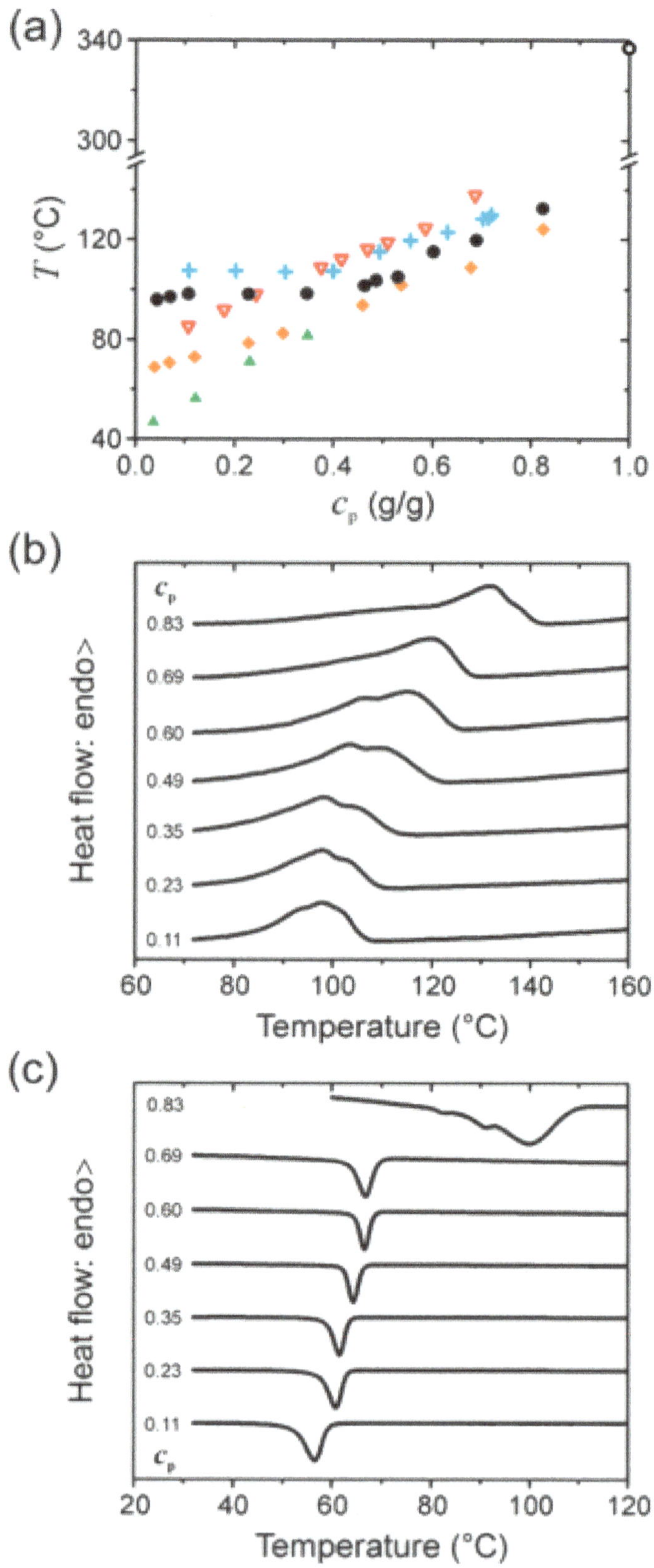

Figure 2. (a) Temperature–composition diagram, representing the variation of peak dissolution temperature T with polymer weight fraction c_p for mixtures of PFO and selected organic solvents. Data is shown for (in order of expected decreasing solvent quality; see Table 1) toluene (▲), decalin (◆), oTCB (▽), hexadecane (+) and dodecane (●). The temperature at which neat PFO undergoes its nematic to isotropic melt transition is also indicated (○, top right corner). For the sake of clarity, dissolution temperature values are reported only for the endotherm with the highest peak heat flow. (b) Heating and (c) cooling DSC thermograms for PFO—dodecane mixtures. The corresponding c_p values are indicated.

As expected,[56] addition of solvents that exhibit non-negligible interaction with the polymer leads to dissolution temperatures that are much lower than the thermal transition to the isotropic melt (~340 °C) observed for neat PFO. The magnitude of this temperature depression, judged from the c_p <0.2 section of the diagram, generally increases with solvent "quality," quantified as the squared difference, $(\Delta\delta)^2$, between the Hildebrand solubility parameters, δ, of PFO and solvent[4, 57] (see Table 1). We note that the Hildebrand solubility parameters are typically derived from the respective cohesive energy densities[57] and therefore quantify the overall attractive forces without explicitly separating them into the dispersive, polar and hydrogen bonding contributions.[58] Nevertheless, since the latter two interactions are likely to be negligible for PFO and the selected solvents, the analysis of their solubilities based on the Hildebrand solubility parameters is deemed to be adequate in the present case.[4]

Table 1: Hildebrand solubility parameters, δ, for PFO[4, 5] and the solvents[57] used in our study

Polymer/Solvent	δ (cal$^{1/2}$ cm$^{-3/2}$)	$(\Delta\delta)^2$ (cal cm^{-3})
$(\Delta\delta)^2$ is the squared difference between the solvent solubility parameter and that of PFO. Lower $(\Delta\delta)^2$ values correspond to lower heats of mixing and, therefore, to higher solvent "quality"; solvents are listed in quality order from good (top) to poor (bottom).		
PFO	9.2	–
Toluene	8.9	0.09
Decalin	8.7	0.25
oTCB	9.9	0.49
Hexadecane	8.0	1.44
Dodecane	7.8	1.96

Also of note are the distinct "kinks," that is, changes in gradient, seen most clearly for dodecane and hexadecane at $c_p \approx 0.45$. As will be shown below, these are due to the occurrence, typically at high polymer weight fractions, of dynamic recrystallization/melting transitions prior to complete dissolution. [N.B. This data will be subsequently revisited (cf. Fig. 6; vide infra), when more rigorous, albeit nonequilibrium, temperature–composition "phase" diagrams will be presented.]

Representative heating and cooling DSC thermograms are shown for PFO–dodecane mixtures in Figure 2(b,c), respectively. Crystallization is observed as a single exothermal peak in the cooling thermograms up to c_p=0.83, at which point solution-crystallization via β-phase chain segment formation then coexists with the standard melt-crystallization of neat PFO into an alternative microstructure termed the "α-phase."[8] This coexistence results from there being insufficient solvent in the mixture to form a uniform, stoichiometric, polymer-solvent compound (vide infra). The process of gel melting/dissolution, as observed in the heating thermograms, is, conversely, somewhat more complicated. For moderately good solvents, that is, toluene, decalin, and 1,2,4-trichlorobenzene (oTCB), gel melting is observed as a single endotherm (see Supporting Information Figs. S1–2). For moderately bad solvents, that is, dodecane (shown here) and hexadecane, two overlapping endotherms are observed in the heating thermograms [cf. Fig. 2(b)]. These were investigated in a control experiment (see Supporting Information Fig. S3), in which a solution was crystallized at a constant cooling rate and re-melted at varied (1–20 °C min^{-1}) heating rates. The relative magnitude of the high-temperature endotherm was observed to decrease at higher heating rates, which indicates that only the low-temperature endotherm corresponds to gel melting, whereas the high-temperature endotherm is due to the process of dynamic polymer recrystallization and simultaneous melting.[38] This assignment is corroborated by the fact that such behavior at low/moderate polymer weight fractions is only observed for the moderately bad solvents, in which the driving force for PFO recrystallization is increased for thermodynamic reasons.

Finally we note that all studied PFO gels exhibited pure β-phase PL, shown in Supporting Information Figure S4, despite the possible coexistence of α-phase crystalline chains for the gels with the highest c_p; this is due to preferential excitation energy transfer to the lower HOMO-LUMO transition energy β-phase chain segments. We also note that the solutions and gels did not exhibit any appreciable green-band (g-band) emission,[59, 60] even after repeated heating-cooling DSC cycles followed by compaction and desiccation of the gel, indicating that thermo-oxidative degradation of PFO was minimal (cf. Supporting Information Fig. S4). More detailed optical spectroscopy results will be presented in Part II of this study, which specifically focusses on the differences in solution-crystallization behavior for PFO and a related polyfluorene copolymer with a modified side-chain structure. However, for the purposes of this study we can take it as implicit from the PL data that solution-crystallized PFO chains adopt the β-phase conformation. This is also confirmed by the close correspondence between X-ray diffraction patterns recorded for PFO gels (vide infra) and the diffraction data reported elsewhere for solution-grown β-phase crystals.[26, 54]

As already discussed in the introduction, polymer-solvent compounds comprise a structure in which the two components co-crystallize, with solvent molecules incorporated into the periodically arranged cavities created by, for instance, a particular helical conformation of the polymer backbone. Depending on solution concentration, there can additionally be so-called "free" solvent molecules that can undergo the usual crystallization/melting transitions expected for the neat solvent. Crystallizing and re-melting these free-solvent molecules allows for the free-solvent melting enthalpy ΔH_m to be determined, thus providing additional information on gel composition and structure.[35, 37, 61] Conversely, the solvent molecules involved in the structure of the compound cannot undergo independent thermal transitions and, therefore, do not contribute to the measured ΔH_m.

A DSC study was carried out between −50 and 20 °C (spanning the free-solvent crystallization and melting temperature range for the selected solvents) for slowly crystallized gels, that is, those possessing maximal degrees of crystallinity (see Experimental). Figure 3(a) shows the extracted variation of free-solvent ΔH_m with gel composition, expressed here as the PFO repeat unit molar fraction x_u, for gels with oTCB, dodecane and hexadecane. Plotting in this way helps to confirm polymer-solvent compound formation[35, 37, 61] and allows the compound stoichiometry to be readily determined.

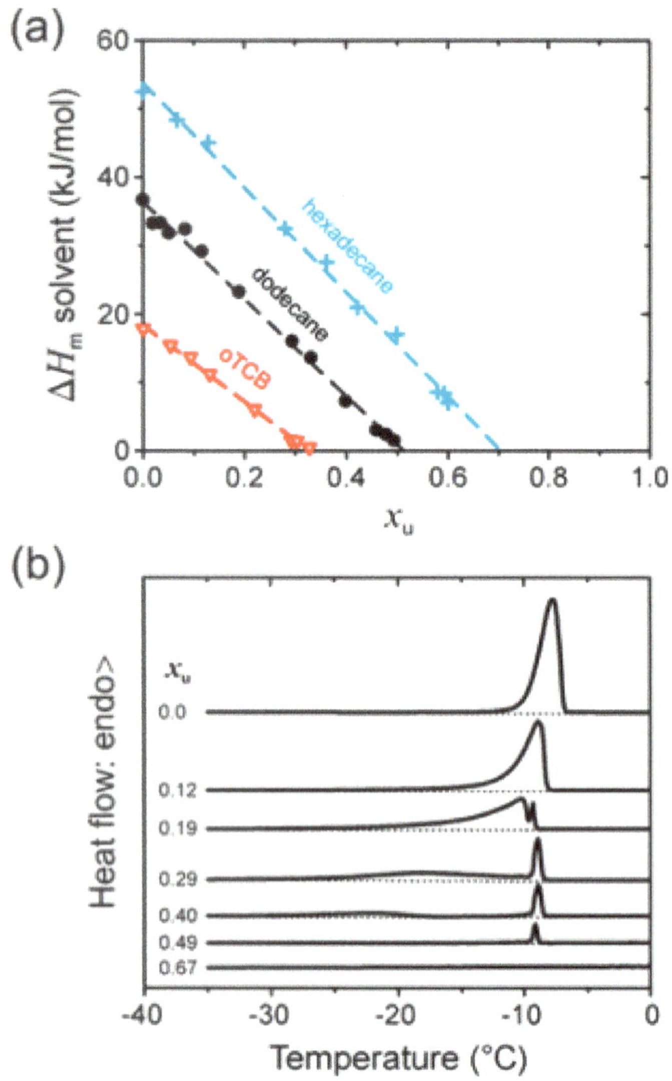

Figure 3. (a) Free-solvent melting enthalpy, ΔH_m, as a function of the PFO repeat unit molar fraction, x_u, for "slowly crystallized" PFO gels with dodecane (●), hexadecane (+) and oTCB (▽). Dashed lines indicate linear fits to the data. (b) Representative DSC heating thermograms (offset for clarity), parametric in x_u, for PFO–dodecane gels; the extrapolated baselines (dotted lines) highlight the presence of additional (weak) endotherms for some of the data.

Gibbs' phase rules dictate that the enthalpy associated with the first-order melting transition of the free-solvent should vary linearly with its concentration. It is indeed observed that increasing the PFO molar fraction, x_u, results in a linear diminution of ΔH_m. Extrapolating the linear fits [dashed lines in Fig. 3(a)] to $\Delta H_m=0$ allows the PFO-solvent compound concentration x_u* and, therewith, stoichiometry (the number of solvent molecules per polymer repeat unit, F8) to be estimated; results are summarized in Table 2. We note that the reported stoichiometries may be somewhat inaccurate since our analysis assumes full transformation of each solution into polymer-solvent compound plus free-solvent and ignores the presence of any residual dissolved polymer fraction for which a higher degree of solvation can be expected. This situation would tend to overestimate the amount of solvent incorporated within the compound. Reassuringly, however, the deduced stoichiometry values are consistent with the results of X-ray diffraction analysis performed on the same samples (vide infra).

Table 2: Compound concentrations, x_u*, expressed in terms of the PFO repeat unit molar fraction, and corresponding stoichiometries determined by (free) solvent melting enthalpy measurements

Solvent	x_u*	Stoichiometry (F8:Solvent)	c_p*
To facilitate data comparison, x_u* values are also given as polymer weight fraction, c_p*.			
Dodecane	0.51±0.01	1:1	0.70
Hexadecane	0.70±0.02	2:1	0.80
oTCB	0.33±0.01	1:2	0.51

The DSC heating thermograms recorded in the free-solvent melting experiments for PFO–dodecane gels are shown in Figure 3(b). These reveal more complex behavior than is evident simply from the ΔH_m versus x_u plots [Fig. 3(a)]. For $x_u=0$ we see a single melting endotherm that peaks at −8 °C as expected for neat dodecane.**[62]** For concentrations in the range $0.19<x_u<0.4$ two endotherms are seen: one especially sharp peak close to −9 °C and a second feature that broadens, weakens and shifts to lower temperatures as x_u increases. These observations point to free-solvent being present in the form of both a subset of very well-defined crystals (sharp endotherm) and a second broader distribution of crystals with a range of decreasing sizes which, in accordance with Gibbs' theory, melt at lower temperature than the quasi-infinite crystals. **[41]** Finally, for $x_u=0.67$ no free-solvent melting is detectable (flat DSC thermogram) as all of the solvent is trapped within polymer-solvent compounds.

X-Ray Diffraction

WAXD was used to determine the crystalline structure of the PFO-solvent compounds. Slowly-crystallized gel samples were prepared with PFO concentrations, x_u, close to but marginally (~0.04) below their respective x_u* values in order to simultaneously minimize the presence of α-phase crystalline PFO and free-solvent. The resulting gel diffraction patterns [cf. Fig. 4(a)] recorded at −100 °C nevertheless still comprise contributions from: (i) PFO-solvent compound as well as any residual amorphous PFO fraction and (ii) crystals of the "free," that is, non-intercalated and crystallizable, solvent. To eliminate the contribution from the latter, we subtracted the normalized diffraction pattern of the neat solvent from that of the as-prepared gel. The reader is directed to Supporting Information Figures S5–7 for further details of WAXD deconvolution and analysis. For comparison, diffraction patterns were also recorded for dried polymer gels obtained by critical-point drying of the as-prepared gels. This approach allows for interface-free removal of the solvent and minimizes the possibility of associated structural changes to the polymer. **[51, 52]** We note that both the as-prepared and dried PFO gels exhibited essentially identical β-phase-dominated PL spectra (not shown) and yielded WAXD patterns with closely corresponding reflection peaks for the polymer (see Supporting Information Fig. S5).

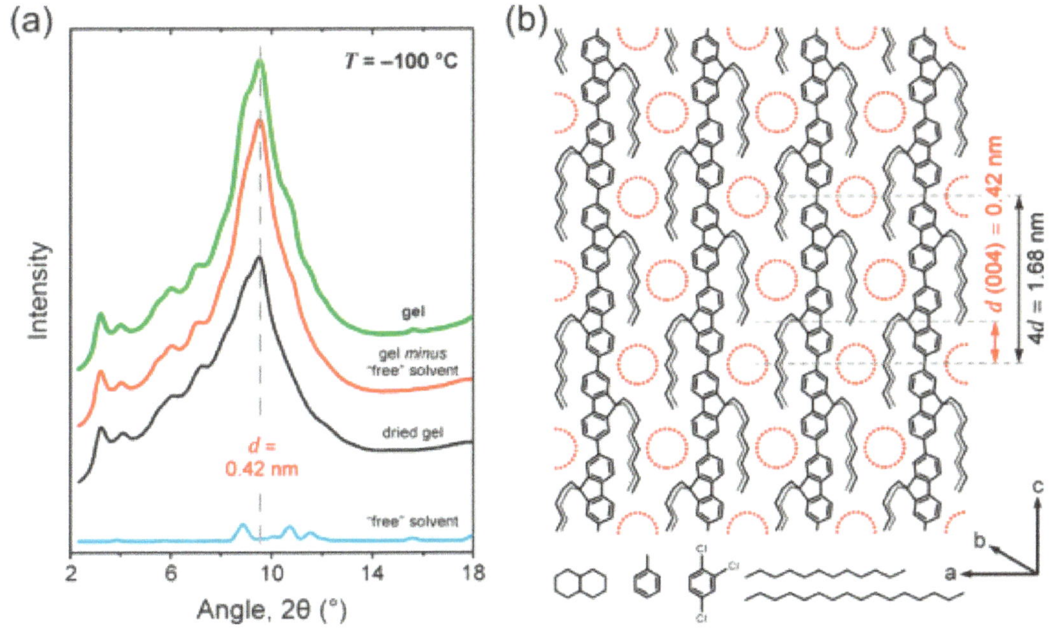

Figure 4: (a) The radially integrated WAXD pattern (green line) recorded for a PFO–hexadecane gel, $x_u \approx 0.6$, at −100 °C (i.e., below the expected glass transition temperature T_g of the compound and T_m of the solvent), together with the corresponding normalized diffraction patterns for the dried gel (black line) and the "free" solvent (blue line). Also shown is the diffraction pattern of the gel following subtraction of the normalized diffraction pattern of the free-solvent ('gelminus "free" solvent'; red line). (b) Schematic illustration of the PFO-solvent compound, showing the β-phase chain segments and the cavities that contain the intercalated solvent (red circles). The (004) reflection assigned to the solvent in the compound (d=0.42 nm) as well as the c-axis periodicity of intercalated solvent molecules (4d=1.68 nm) are also indicated, with the latter being equivalent to the length of two PFO repeat units in the β-phase conformation (see, e.g., refs. 6,10,54). The chemical structures of the solvents used in this study are shown at the bottom (left to right: decalin, toluene, oTCB, dodecane and hexadecane). Note that, while the n-alkanes are shown in their all-trans conformations, the intercalated solvent molecules are likely to be folded inside the cavities.

The corresponding set of diffraction patterns for a $x_u \approx 0.6$ PFO–hexadecane gel is shown in Figure 4(a). The pattern for the gel subtracted with the free-solvent contribution [cf. 'gelminus "free" solvent' data in Fig. 4(a)] features a strong reflection at d=0.42 nm; note that this peak is unique to the gel and does not have a counterpart in the diffraction pattern of the free solvent. Equivalent reflections are seen at the same d-spacing for PFO gels in dodecane and oTCB, thereby demonstrating the generality of the adopted structure for different solvents and emphasizing the dominant role of the solution-crystallized PFO chains in determining this structure. Furthermore, the same reflection peak is also observed in the diffraction pattern of the dried PFO–hexadecane gel, albeit with a reduced intensity in comparison with that of the as-prepared gel [see Fig. 4(a)].

A strong reflection at d=0.42 nm has previously been observed for solution-processed β-phase-rich solid PFO samples[**9, 54, 64**] and was attributed to a c-axis backbone periodicity, corresponding, as it does, to half the length of the PFO repeat unit.[**5**] Specifically, Liu et al.[**54**] used X-ray and selected-area electron diffraction (SAED) to study the structure of solution-grown β-phase crystals of monodisperse PFO oligomers. It was reported that, following solvent removal, the β-phase crystals comprised an orthorhombic unit cell with itsc-axis length (3.36 nm) corresponding to the length of four fully-extended fluorene repeat units. Hence, in that study the peak at d=0.42 nm was indexed as the (008) reflection of the β-phase chain segments within these solution-grown crystals.[**54**]

Since the placement of solvent molecules within the PFO-solvent compound would be dictated by the chain conformation of the polymer, we attribute the peak at d=0.42 nm observed for the as-prepared gels to be in part due to the (004) reflection of the intercalated solvent molecules in the compound. This is corroborated by the fact that the d=0.42 nm reflection (i) is more intense for the as-prepared gel compared with that observed for the dried gel [cf. Fig. 4(a)] and (ii) substantially increases in intensity and sharpens upon cooling the gel samples to −100 °C below the expected T_g of the PFO-solvent compound (see Supporting Information Fig. S7). In this case, 4d corresponds to the c-axis periodicity of the intercalated solvent which, reassuringly, is equivalent to the length of two PFO repeat units in the β-phase conformation.[**6, 10**] We note that the angular resolution of our instrument prevents us from also observing the (001) reflection peak for the intercalated solvent. An additional reason why the observed reflection is unlikely to be related to solvent placement along the a- or b-axis is that, due to the different compound stoichiometries and, hence, solvent volumes per cavity, the d-spacings along these axes would be expected to vary between the three solvents.[**63**] Conversely, due to the rigid chain geometry of the β-phase, the d-spacings along the c-axis would be identical for different solvents and stoichiometries, as indeed was observed experimentally. Additional WAXD data, notably for a different solvent and two contrasting polymer side-chain structures, is presented in Part II of this study.

The proposed structure of the PFO-solvent compound is shown schematically in Figure 4(b). The planar-zigzag chain structure[**4-7**] and the alignment of the octyl side-chains along the backbone[**10, 55**] are unique to the β-phase conformation. The packing distance of adjacent β-phase chain segments along the a-axis is consistent with the interdigitated model proposed previously for β-phase PFO.[**24, 54, 65**] Taken together, these two structural aspects lead to the formation of cavities [red circles in Fig. 4(b)] which allow for the intercalation of small-molecular solvents of appropriate volume, such as those used in this study (cf. chemical structures in the same figure).

As an additional check, we investigated shrinkage of the polymer unit cell upon critical-point drying of the gels in dodecane, hexadecane and oTCB. The (100) reflection of PFO is well-resolved in all three gels and appears at d ≈ 1.26 nm. In the case of a polymer-solvent compound, the amount by which d (100) shrinks following solvent removal is expected to be proportional to the volume of solvent per cavity, V_c, in the as-prepared gel. On the other hand, if PFO does not form a compound then the shrinkage of d (100) (if any) would not be correlated with V_c.

Figure 5 confirms that the decrease in Bragg distance for the (100) reflection of PFO is proportional to V_c, calculated using the previously determined compound stoichiometries and the known molar volumes for the respective solvents. The inset in Figure 5 shows representative WAXD data for the as-prepared and dried $x_u \approx 0.3$ PFO–oTCB gels. The increase in the width of the (100) reflection following drying can be attributed to (i) decreasing crystal size due to the disruptive effect of solvent removal, and (ii) distortion of the crystal lattice during unit cell shrinkage.

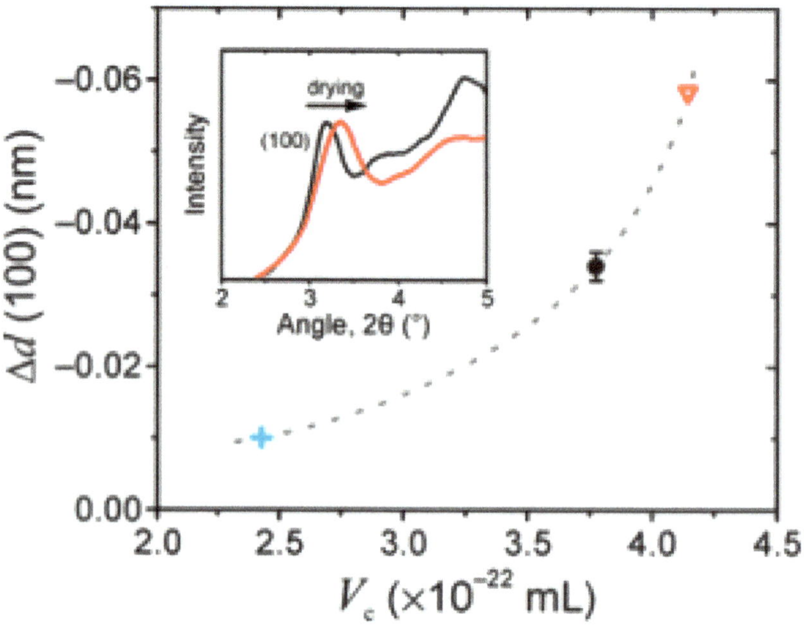

Figure 5: Change in Bragg distance, Δd, of the (100) PFO reflection peak following supercritical drying of PFO gels with dodecane (●), hexadecane (+) and oTCB (▽), plotted as a function of solvent volume per cavity, V_c, of the polymer-solvent compound. The dashed grey line serves as a guide to the eye. Inset shows representative diffraction patterns of $x_u \approx 0.3$ PFO–oTCB gels in the as-prepared state (black line) and following supercritical extraction of the solvent (red line). The arrow indicates the shift of the (100) reflection. All data were recorded at −100 °C and for polymer concentrations marginally (~0.04) below the corresponding compound concentrations x_u*.

The volume of a single cavity in the PFO-solvent compound [cf. Fig. 4(b)] can be estimated from the maximum obtained V_c value (two oTCB molecules per cavity) and the volume of hexadecane which was found to be too large for a single cavity (F8 : solvent stoichiometry = 1:2). This yields an estimated cavity volume of 4.1–4.9 × 10^{-22} mL for the PFO-solvent compound. We note that this value shows good agreement with a back-of-the-envelopecalculation based on the dimensions of the compound structure determined by WAXD, which estimates cavity volume at ~3 × 10^{-22} mL (see Supporting Information Fig. S8).

Phase Behavior

Having confirmed the formation of polymer-solvent compounds with PFO by both thermal analysis and WAXD, we now present selected data from Figure 2(a) in the context of nonequilibrium temperature–composition "phase" diagrams, shown in Figure 6. The temperature–composition phase diagram for PFO–dodecane [cf. Fig. 6(a)] is characteristic of an incongruently-melting compound, whereby the compound C transforms into another phase (in this case a liquid plus, most likely, semicrystalline polymer ($L+S_2$)) at concentrations below x_u* prior to its complete melting.[**33**] Such behavior has been observed for syndiotactic polystyrene (sPS) gels in trans-decalin, for which a compound with stoichiometry 1:1 has been demonstrated.[**41**] Conversely, the temperature–composition phase diagram for PFO–oTCB [cf. Fig. 6(b)] is typical of a singular-melting compound, for which direct melting occurs, with maximum melting temperature found at the compound concentration x_u*.[**33**] This type of compound has been found for sPS gels in tetralin[**41**] and toluene,[**33**] both featuring 1:1 stoichiometry. Typically, incongruently-melting compounds are observed when poorer solvents are used[**39**] which is indeed the case for PFO, considering the less-matched solubility parameter of dodecane compared with oTCB (cf. Table 1). Interestingly, the ~25 °C difference in maximum T_m of the two compounds shown in Figure 6 suggests that the stability of the compound is higher for PFO–oTCB. One possibility is that this is due to a higher volume of solvent per cavity in the compound (4.1 × 10^{-22} mL compared with 3.8 × 10^{-22} mL for dodecane) which results in stronger van der Waals bonding between the polymer and intercalated solvent. An alternative explanation is that the stability of the PFO compound is proportional to the solvent quality; this correlation has been proposed for sPS compounds, although without considering the possible stoichiometric differences.[**39**]

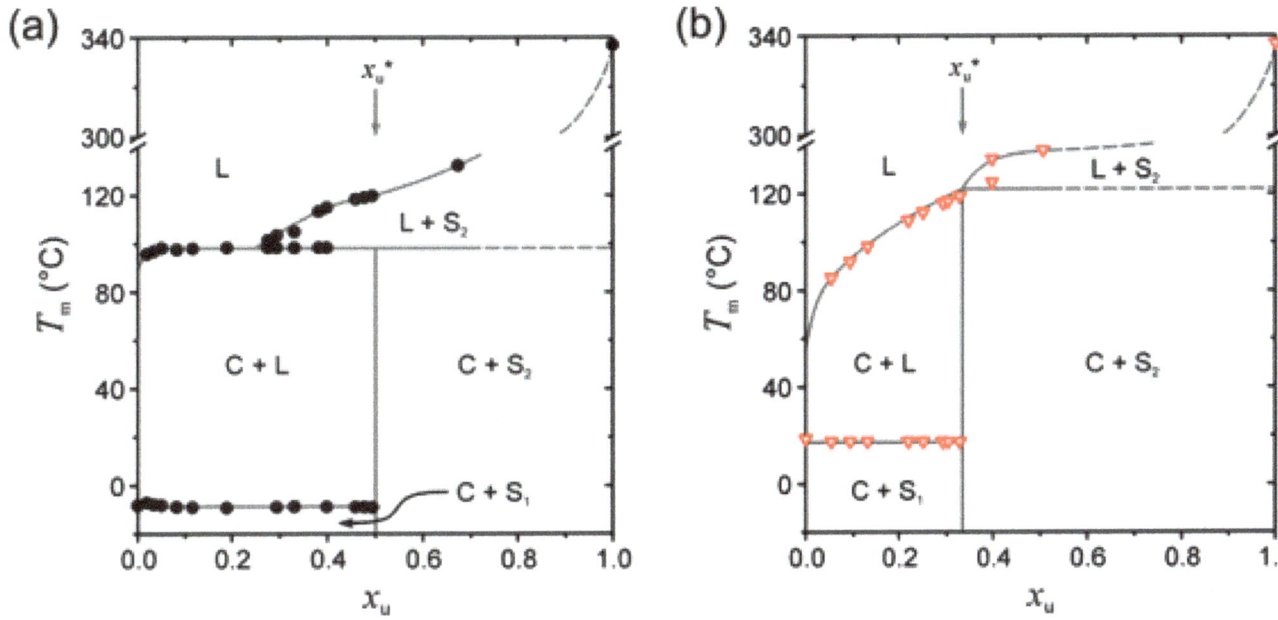

Figure 6: Schematic nonequilibrium temperature–composition phase diagrams for PFO mixtures with (a) dodecane and (b) oTCB. Polymer concentration x_u is expressed as the F8 repeat unit molar fraction. Symbols represent melting temperatures; solid lines are guides to the eye for the observed transitions while dashed lines indicate their probable extensions. The phases are represented as follows: L, liquid solution or melt; S, (semi-)crystalline solid; C, compound; subscripts 1 and 2 refer to the solvent and polymer components, respectively. Compound concentrations x_u* are indicated by the arrows.

On a cautionary note, we should emphasize that the presented phase diagrams can serve only as an approximation to the equilibrium phase behavior. First, although PFO gels were dealt with as two-component systems, strictly speaking the polymer can hardly be treated as a single component given its large polydispersity and incomplete formation of the β-phase conformation. Second, since the gels were prepared by dynamic crystallization at finite cooling rates, their reported melting temperatures do not represent equilibrium values. Nevertheless, as shown in previous reports,**[40, 66]** these nonequilibrium temperature-composition phase diagrams provide an adequate qualitative indication of the phase behavior.

DISCUSSION

Our demonstration of PFO-solvent compound formation invites discussion, as well as further research, in two specific directions:

These observations substantially add to our understanding of the β-phase conformation, and in particular show that, due to its planar-zigzag geometry, this conformation allows for intercalation of solvent into the on-chain cavities, thereby stabilizing the backbone. In this respect, the situation is similar to sPS-solvent compounds that feature chains in the helical, so-called δ-phase, conformation. The δ-phase of sPS can be formed via crystallization from solutions as well as by exposure of amorphous or crystalline sPS films to an appropriate solvent in liquid or vapor form.**[32]** The latter methods, namely immersion in solvent–nonsolvent mixtures and solvent vapor annealing, are also commonly used to introduce the β-phase conformation in solid-state PFO.**[6, 12, 17, 18]** Approaching β-phase chain segment formation via crystallization protocols may stimulate new approaches to fabrication of solid-state PFO samples with β-phase content greater than the ~45% achieved to date.**[11]** Confirmation of the fundamental role played by solvent in formation of the β-phase and first elucidation of the stabilizing effect of polymer-solvent compound formation encourages further studies of solvent and processing temperature optimization. Our preliminary results indicate that the solution-crystallization rate of PFO is reduced in low molar volume solvents (see Supporting Information Fig. S9), adding a new parameter axis to solvent selection. Such considerations can be especially important for fabrication when PFO films maximally free from β-phase chain segments are required.**[3, 18]**

On a separate note, it should be mentioned that the first observations of the β-phase conformation in PFO were made following thermal postprocessing of glassy spin-coated PFO films via in vacuo cooling to and reheating from −196 °C.**[4-6]** Given the seemingly solvent-free formation of the β-phase in that case, as well as other observations of β-phase formation during, for instance, solution-casting on a water surface,**[67, 68]** it would be most interesting to further study the differences between these and more common solvent-based approaches to inducing β-phase in solid-state PFO.

The formation of a polymer-solvent compound is a molecular recognition process, for which the critical solvent properties are: (a) appropriate volume and, to a lesser extent, (b) matching solubility parameter. Hence, molecular PFO compounds might more generally be formed with the organic solvent replaced by a small molecule of desired optoelectronic properties, thereby allowing the fabrication of ultra-regular molecular-level blends comprising a PFO host and a small-molecular guest. A judicious choice of guest molecule might therefore enable improved device performance for PFO compounds in terms of, for example, charge–carrier injection and transport, energy transfer, charge separation, and stability against photo- or thermal-degradation. This is expected to prove an interesting avenue for future research.

CONCLUSIONS

Polymer-solvent compound formation has been demonstrated for PFO and a range of organic solvents, observed as solution-crystallization and accompanied by thermoreversible gelation. The crystalline structures in the gels consist of solvated β-phase crystals, in which solvent molecules are intercalated into the on-chain cavities characteristic of the PFO β-phase chain conformation. The properties (such as volume and solubility parameter) of the small-molecular species were found to play a key role in compound formation, significantly affecting the stoichiometry of the resulting compound, as well as its stability. Comparisons have been drawn between the compound formation for PFO and similar phase behavior in other systems, notably those based on syndiotactic polystyrene. Our findings clarify the nature of the β-phase conformation and suggest new strategies for controllable solution-processing of PFO.

Having shown compound formation with simple molecules like 1,2,4-trichlorobenzene and dodecane, future work will extend this approach to other guest molecules with similar physicochemical properties (e.g., molecular volume) with the aim of engineering new functionalities in solid-state PFO-based blends.

ACKNOWLEDGMENTS

We thank the UK Engineering and Physical Sciences Research Council (grant EP/H000917/1) for financial support. AP expresses his heartfelt gratitude to Prof. Christian Müller (Chalmers University of Technology) for insightful and stimulating discussions on polymer science which were helpful for the preparation of this manuscript. AP also thanks the Polymer Technology group (Department of Materials, ETH Zürich) as well as its Chuchichästli and Geordies' associates for their generosity, creativity and freely shared expertise during his research stay. DDCB is the Lee-Lucas Professor of Experimental Physics.

REFERENCE

1. M. Grell, D. D. C. Bradley, M. Inbasekaran, E. P. Woo, Adv. Mater. 1997, 9, 798–802.
2. 2M. Redecker, D. D. C. Bradley, M. Inbasekaran, E. P. Woo, Appl. Phys. Lett. 1999, 74, 1400–1402.
3. 3B. K. Yap, R. Xia, M. Campoy-Quiles, P. N. Stavrinou, D. D. C. Bradley, Nat. Mater. 2008, 7, 376–380.
4. 4D. D. C. Bradley, M. Grell, X. Long, H. Mellor, A. Grice, M. Inbasekaran, E. P. Woo, SPIE Proc. 1997, 3145,254–259.
5. 5M. Grell, D. D. C. Bradley, X. Long, T. Chamberlain, M. Inbasekaran, E. P. Woo, M. Soliman, Acta Polym.1998, 49, 439–444.
6. 6M. Grell, D. D. C. Bradley, G. Ungar, J. Hill, K. S. Whitehead, Macromolecules 1999, 32, 5810–5817.
7. 7A. J. Cadby, P. A. Lane, H. Mellor, S. J. Martin, M. Grell, C. Giebeler, D. D. C. Bradley, M. Wohlgenannt, C. An, Z. V. Vardeny, Phys. Rev. B 2000, 62, 15604–15609.
8. 8S. H. Chen, A. C. Su, C. H. Su, S. A. Chen, Macromolecules, 2005, 38, 379–385.
9. 9S. H. Chen, A. C. Su, A. A. Chen, J. Phys. Chem. B, 2005, 109, 10067–10072.
10. 10W. Chunwaschirasiri, B. Tanto, D. L. Huber, M. J. Winokur, Phys. Rev. Lett. 2005, 94, 107402.
11. 11J. Peet, E. Brocker, Y. Xu, G. C. Bazan, Adv. Mater. 2008, 20, 1882–1885.
12. 12H.-H. Lu, C.-Y. Liu, C.-H. Chang, S.-A. Chen, Adv. Mater. 2007, 19, 2574–2579.
13. 13Y. Yang, G. A. Turnbull, I. D. W. Samuel, Appl. Phys. Lett. 2008, 92, 163306.
14. 14Y. Yang, G. A. Turnbull, I. D. W. Samuel, Adv. Funct. Mater. 2010, 20, 2093–2097.
15. 15M. Anni, R. Rella, J. Phys. Chem. B 2010, 114, 1559–1561.
16. 16G. Heliotis, D. D. C. Bradley, G. A. Turnbull, I. D. W. Samuel, Appl. Phys. Lett. 2002, 81, 415–417.
17. 17D. Nassyrov, C. Müller, A. Roigé, I. Burgués-Ceballos, J. O. Ossó, D. B. Amabilino, M. Garriga, M. I. Alonso,A. R. Goñi, M. Campoy-Quiles, J. Mater. Chem. 2012, 22, 4519–4526.
18. 18A. Perevedentsev, Y. Sonnefraud, S. Sharma, A. E. G. Cass, S. A. Maier, J.-S. Kim, P. N. Stavrinou, D. D. C. Bradley, Nat. Commun. 2015, 6, 5977.
19. 19A. Perevedentsev, S. Aksel, K. Feldman, P. Smith, P. Stavrinou, D. D. C. Bradley, J. Polym. Sci., Part B: Polym. Phys. 2015, 53, 22–38.
20. 20A.J. Campbell, D. D. C. Bradley, H. Antoniadis, M. Inbasekaran, W. W. Wu, E. P. Woo, Appl. Phys. Lett. 2000,76, 1734–1736.
21. 21R. Xia, G. Heliotis, D. D. C. Bradley, Appl. Phys. Lett. 2003, 82, 3599–3601.
22. 22P. A. Levermore, R. Jin, X. H. Wang, J. C. de Mello, D. D. C. Bradley, Adv. Funct. Mater. 2009, 19, 950–957.
23. 23A. Keller, Faraday Discuss. 1995, 101, 1–49.
24. 24Z.-Q. Lin, N.-E. Shi, Y.-B. Li, D. Qiu, L. Zhang, J.-Y. Lin, J.-F. Zhao, C. Wang, L.-H. Xie, W. Huang, J. Phys. Chem. C 2011, 115, 4418–4424.
25. 25M. Knaapila, V. M. Garamus, F. B. Dias, L. Almásy, F. Galbrecht, A. Charas, J. Morgado, H. D. Burrows, U. Scherf, A. P. Monkman, Macromolecules 2006, 39, 6505–6512.
26. 26M. Knaapila, F. B. Dias, V. M. Garamus, L. Almásy, M. Torkkeli, K. Leppänen, F. Galbrecht, E. Preis, H. D. Burrows, U. Scherf, A. P. Monkman, Macromolecules 2007, 40, 9398–9405.
27. 27C.-Y. Chen, C.-S. Chang, S.-W. Huang, J.-H. Chen, H.-L. Chen, C.-I. Su, S.-A. Chen, Macromolecules 2010, 43,4346–4354.
28. 28J.-H. Chen, C.-S. Chang, Y.-X. Chang, C.-Y. Chen, H.-L. Chen, S.-A. Chen, Macromolecules 2009, 42,1306–1314.
29. 29A. Saiani, J. Spěváček, J.-M. Guenet, Macromolecules 1998, 31, 703–710.
30. 30M. Knaapila, A. P. Monkman, Adv. Mater. 2013, 25, 1090–1108.

31. 31M. Knaapila, D. W. Bright, R. Stepanyan, M. Torkkeli, L. Almásy, R. Schweins, U. Vainio, E. Preis, F. Galbrecht, U. Scherf, A. P. Monkman, Phys. Rev. E 2011, 83, 051803.
32. 32J. M. Guenet, Polymer-Solvent Molecular Compounds; Elsevier: Amsterdam, 2007.
33. 33J. M. Guenet, Macromol. Symp. 2007, 258, 179–185.
34. 34J. J. Point, C. Coutelier, J. Polym. Sci., Part B: Polym. Phys. 1985, 23, 231–239.
35. 35J. Spěváček, A. Saiani, J. M. Guenet, Macromol. Rapid Commun. 1996, 17, 389–395.
36. 36D. Dasgupta, S. Manna, S. Malik, C. Rochas, J. M. Guenet, A. K. Nandi, Macromolecules 2005, 38,5602–5608.
37. 37J. M. Guenet, Macromolecules 1986, 19, 1961–1968.
38. 38F. Deberdt, H. Berghmans, Polymer 1993, 34, 2192–2201.
39. 39T. Roels, F. Deberdt, H. Berghmans, Macromolecules 1994, 27, 6216–6220.
40. 40H. Berghmans, F. Deberdt, Phil. Trans. R. Soc. Lond. A 1994, 348, 117–130.
41. 41S. Malik, C. Rochas, M. Schmutz, J. M. Guenet, Macromolecules 2005, 38, 6024–6030.
42. 42G. Guerra, C. Manfredi, M. Rapacciulo, P. Corradini, G. Mensitieri, M. A. Del Nobile, Italian Patent (CNR),1994.
43. 43G. Guerra, C. Manfredi, P. Musto, S. Tavone, Macromolecules 1998, 31, 1329–1334.
44. 44P. Pilla, A. Cusano, A. Cutolo, M. Giordano, G. Mensitieri, P. Rizzo, L. Sanguigno, V. Venditto, G. Guerra,Sensors 2009, 9, 9816–9857.
45. 45M. Erdogan, Z. Ozbek, R. Capan, Y. Yagci, J. Appl. Polym. Sci. 2012, 123, 2414–2422.
46. 46A. Buonerba, C. Cuomo, S. O. Sanchez, P. Canton, A. Grassi, Chem. Eur. J. 2012, 18, 709–715.
47. 47A. R. Albunia, P. Rizzo, G. Ianniello, C. Rufolo, G. Guerra, J. Polym. Sci., Part B: Polym. Phys. 2014, 52,657–665.
48. 48A. C. Mayer, M. F. Toney, S. R. Scully, J. Rivnay, C. J. Brabec, M. Scharber, M. Koppe, M. Heeney, I. McCulloch, M. D. McGehee, Adv. Funct. Mater. 2009, 19, 1173–1179.
49. 49N. C. Miller, R. Gysel, C. E. Miller, E. Verploegen, Z. Beiley, M. Heeney, I. McCulloch, Z. Bao, M. F. Toney,M. D. McGeehee, J. Polym. Sci., Part B: Polym. Phys. 2011, 49, 499–503.
50. 50C. Daniel, D. Alfano, V. Venditto, S. Cardea, E. Reverchon, D. Larobina, G. Mensitieri, G. Guerra, Adv. Mater. 2005, 17, 1515–1518.
51. 51C. Daniel, A. Avallone, G. Guerra, Macromolecules 2006, 39, 7578–7582.
52. 52G. S. Misra, Introductory Polymer Chemistry; New Age International: New Delhi, 1993.
53. 53P. Smith, P. J. Lemstra, H. C. Booij, J. Polym. Sci., Polym. Phys. Ed. 1981, 19, 877–888.
54. 54C. Liu, Q. Wang, H. Tian, J. Liu, Y. Geng, D. Yan, Macromolecules 2013, 46, 3025–3030.
55. 55C. Volz, M. Arif, S. Guha, J. Chem. Phys. 2007, 126, 064905.
56. 56P. J. Flory, Principles of Polymer Chemistry; Cornell University Press: Ithaca, New York, 1953.
57. 57J. Brandrup, E. H. Immergut, E. A. Grulke, Polymer Handbook, Academic Press, New York, 1993.
58. 58C. M. Hansen, J. Paint Technol. 1967, 39, 505–510.
59. 59E. J. W. List, R. Guentner, P. S. de Freitas, U. Scherf, Adv. Mater. 2002, 14, 374–378.
60. 60M. Sims, D. D. C. Bradley, M. Ariu, M. Koeberg, A. Asimakis, M. Grell, D. G. Lidzey, Adv. Funct. Mater. 2004,14, 765–781.
61. 61M. Klein, J.-M. Guenet, Macromolecules 1989, 22, 3716–3725.
62. 62D. R. Lide, CRC Handbook of Chemistry and Physics, CRC Press, Florida, 2005.
63. 63C. S. J. van Hooy-Corstjens, Degree Thesis; Technische Universiteit Eindhoven: The Netherlands, 2002.
64. 64C. Liu, Q. Wang, H. Tian, J. Liu, Y. Geng, D. Yan, J. Phys. Chem. B 2013, 17, 8880–8886.
65. 65S. Kawana, M. Durrell, J. Lu, J. E. Macdonald, M. Grell, D. D. C. Bradley, P. C. Jukes, R. A. L. Jones, S. L. Bennet, Polymer 2002, 43, 1907–1913.
66. 66J.-M. Guenet, Thermochim. Acta 1996, 284, 67–83.
67. 67J. Hill, S. Y. Heriot, O. Worsfold, T. H. Richardson, A. M. Fox, D. D. C. Bradley, Phys. Rev. B 2004, 69,041303(R).
68. 68A. Dauendorffer, S. Miyajima, S. Nagamatsu, W. Takashima, S. Hayase, K. Kaneto, Appl. Phys. Express2012, 5, 092101.

CHAPTER 10 RAPID VAPOUR DEPOSITION AND IN SITU MELT CRYSTALLIZATION FOR 1 MIN FABRICATION OF 10 μm-THICK CRYSTALLINE SILICON FILMS WITH A LATERAL GRAIN SIZE OF OVER 100 μm

Y. Yamasaki , K. Hasegawa , T. Osawa and S. Noda

Department of Applied Chemistry, Waseda University, 3-4-1 Okubo, Shinjuku-ku, Tokyo 169-8555, Japan

ABSTRACT

We developed a film deposition method which yielded continuous polycrystalline Si films with large lateral grain sizes of over 100 μm and thicknesses of ~10 μm in 1 min on growth substrates other than silicon wafers in a single-step process. The silicon source is heated to ~2000 °C, much higher than the melting point of Si, which enables a high deposition rate. Controlling the temperature of the growth substrate, initially above and later below the melting point of Si, allows the seamless lateral to vertical growth of crystalline silicon grains. Thermally and chemically stable substrates of quartz glass and alumina with a 0.1 μm-thick amorphous carbon layer were effective; liquid silicon wetted well by forming a thin SiC interlayer while substrates stayed stable. Such large-grain polycrystalline silicon films synthesized rapidly in 1 min may be used for low-cost, stable and flexible thin film photovoltaic cells.

INTRODUCTION

Crystalline silicon (Si) wafers and films are supporting our society as key parts of many electronic devices such as semiconductor integrated circuits (ICs), thin-film transistors in flat-panel displays (FPDs) and photovoltaic (PV) cells. The production processes of crystalline Si can be classified into two types; one is wafer production by wire-saw slicing of mono- or polycrystalline ingots made from molten Si,[1] and the other is thin-film deposition on growth substrates by chemical vapour deposition (CVD) of Si precursor gases.[2,3] Monocrystalline Si wafers, which are high-quality, high-performance and expensive, are indispensable for ICs. Polycrystalline Si wafers or films having larger in-plane grain sizes than the thicknesses, which we call large-grain Si hereafter, have been used as fundamental parts of large-area devices such as PV cells and FPDs.[4,5]

The fabrication processes of large-grain Si films on substrates can be divided by the type of growth substrate: monocrystalline Si substrates and other substrates. Monocrystalline Si films can easily grow on monocrystalline Si substrates through homoepitaxy by CVD, physical vapour deposition, liquid-phase epitaxy and other methods. However, Si films need to be transferred to low-cost carrier substrates (e.g. plastic or glass) in a successive process for cost reduction.[3,6,7] The growth substrate and the homoepitaxial film have the same crystal structure, orientation and chemical properties, and therefore it is difficult to separate the film from the growth substrate (i.e. Si wafer). Sacrificial heteroepitaxial interlayers such as $CoSi_2$ have been reported,[8] but the transfer

Citation: Y. Yamasaki , K. Hasegawa , T. Osawa and S. Noda. Rapid vapour deposition and in situ melt crystallization for 1 min fabrication of 10 μm-thick crystalline silicon films with a lateral grain size of over 100 μm. CrystEngComm, 2016, 18, 3404-3410, DOI: 10.1039/C6CE00122J process is complicated. Methods need to be established for easy separation of the films from growth substrates and reuse of growth substrates.

Growth substrates other than monocrystalline Si, such as SiO_2, are attractive in the viewpoint of the separation of films and substrates. On most of such substrates, Si does not grow epitaxially. Si forms either amorphous films on substrates below its crystallization temperature[9] or polycrystalline films on substrates above its crystallization temperature.[10,11] In polycrystalline film growth, each nucleus has a different orientation and generally grows isotopically, comes in contact with other growing nuclei and forms continuous films. Then, each nucleus/grain grows in the out-of-plane direction, yielding polycrystalline films with a lateral grain size smaller than the thickness. To obtain polycrystalline films with large grains (i.e.lateral grain size larger than the film thickness), films are post-treated by laser or lamp annealing[12–14] to cause lateral grain growth through melting and (re)crystallization processes. In particular, the zone melting recrystallization (ZMR) method can enlarge grains in several micrometre-thick films from the micrometre scale to the centimetre scale by sweeping a heater and thus producing several millimetre-wide molten zones at sweep rates of a few centimetres per minute.[15] However, the ZMR process is accompanied with agglomeration of molten Si, the so-called balling-up effect, as a result of surface tension, and therefore Si films have to be covered with a SiO_2 capping layer to prevent agglomeration.[15,16] Among these processes, the excimer laser annealing process has been used for the production of FPDs.[5] However, none of these methods have been practically used for the production of low-cost PV cells. The ribbon Si process, which coats and crystallizes molten Si on heat-resisting substrates, can be used to fabricate large-grain polycrystalline Si films of around 100 μm in thickness,[17–19] but further thinning is difficult because of the surface tension of molten Si.

In this report, we propose and develop a new method called "rapid vapour deposition (RVD) of liquid Si and in situ melt crystallization", which yields large-grain polycrystalline Si films directly on heat-resistant substrates other than monocrystalline Si in a single-step process. RVD is a vacuum deposition process realizing a high deposition rate of over 10 μm min^{-1} by heating a Si source in boats to $T_{boat} \sim 2000$ °C, which is much higher than the melting point of Si, T_m = 1414 °C.[20] During RVD, the substrate temperature T_{sub} is intentionally changed in a controlled manner; T_{sub} is set at $>T_m$ for the first several seconds to deposit Si as a liquid film, and subsequently T_{sub} is decreased to $<T_m$ to nucleate Si crystals and grow them laterally in the liquid film, and finally thickening the film to ~10 μm by continuing the deposition of Si vapours. All these steps are completed seamlessly in one process within 1 min, yielding 10 μm-thick crystalline Si films with a lateral grain size of over 100 μm on a heat-resistant quartz-glass and alumina substrate with a 0.1 μm-thick amorphous carbon (a-C) adhesive layer.

EXPERIMENTAL

Quartz glass (SiO_2; 15 mm square, 0.5 mm in thickness), monocrystalline (100) Si wafer with a 50 nm-thick thermal oxide layer (SiO_2/Si; 20 mm square, 0.65 mm in thickness, Cz-p type, resistivity of 10–20 Ω cm), polycrystalline alumina (Al_2O_3; 15 mm square, 1 mm in thickness) (Nilaco AL-017518, Tokyo, Japan), or sapphire (c-plane, 20 mm square, 0.3 mm in thickness) (Kyocera, Kyoto, Japan) were used as substrates. The substrates were pre-treated by immersing them into a mixed solution of H_2SO_4 (95 wt%) and H_2O_2 (30 wt%) with a volume ratio of 3 : 1 for 5 min, and then rinsing them with purified water. The pre-treated monocrystalline Si substrates were subsequently dipped in HF solution (5 wt%) for 1 min to partially remove the thermal oxide from half of their surface, and were then rinsed with purified water. A 0.1 μm-thick a-C layer was deposited on some substrates by direct current magnetron sputtering under 2.5 Pa Ar. For RVD, (100) Si wafer (3.5 × 30 mm^2, 0.65 mm in thickness, Cz-p type, resistivity of 10–20 Ω cm) was used as the vapour deposition source after pre-treatment by immersion in HF solution (5 wt%) for 1 min and rinsing with purified water.

Fig. 1 shows a digital image and a schematic of the internal structure of the RVD apparatus. The components were set from bottom to top as follows: two tungsten boats for the Si source, a substrate stage made of quartz glass, a $W_{0.95}Re_{0.05}$–$W_{0.74}Re_{0.26}$ thermocouple, a substrate heater made of a graphite sheet and four reflectors made of molybdenum sheets. After setting a substrate on the stage with its surface facing down and the Si source on the boats, the chamber was evacuated to $<3 \times 10^{-4}$ Pa using a turbomolecular pump with an oil rotary pump. The substrate heater was turned on to make the thermocouple reach a target temperature, and then RVD was carried out by resistive heating of the source boats to ~2000 °C in several seconds. Si was deposited for 1 min, and after that, heating of the source boats was turned off to finish the deposition.

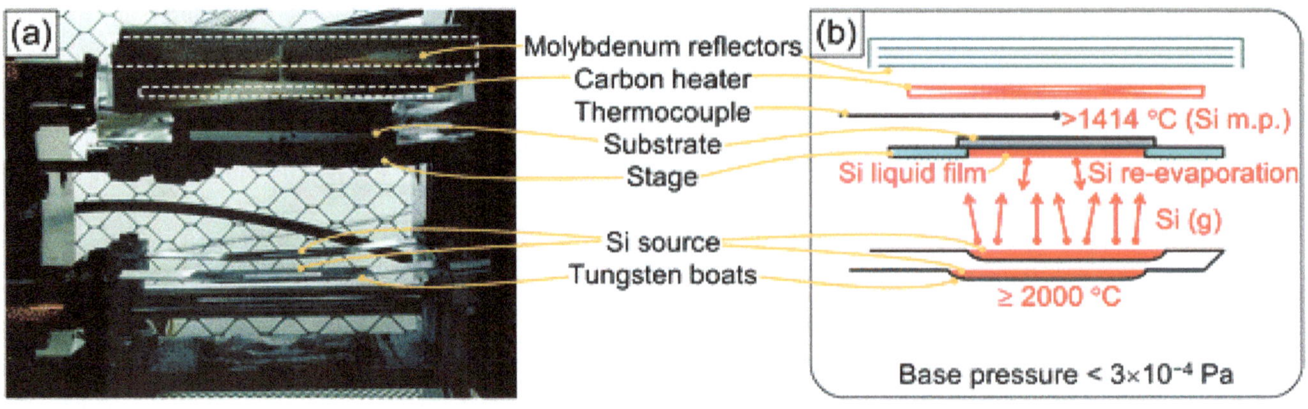

Figure 1 Digital image (a) and schematic (b) of the inner structure of the RVD apparatus.

The structure of the Si films was analysed by scanning electron microscopy (SEM; Hitachi S-4800, Tokyo, Japan) and X-ray diffraction (XRD; Rigaku RINT-TTR III, Tokyo, Japan) with a monochromatised CuKα X-ray source.

RESULTS AND DISCUSSION

The proposed model with typical results of the deposition method and comparison with conventional methods

Fig. 2 shows a schematic comparing the film growth processes by our deposition method and the conventional deposition method. In common methods such as vacuum deposition or CVD, a substrate is kept at a constant temperature with $T_{sub} < T_m$ during deposition (Fig. 2b). On a substrate in which epitaxial growth does not occur (e.g. SiO_2), Si atoms form small nuclei that grow in the out-of-plane direction and form columnar grains. Grains oriented in the fastest growth direction grow preferentially, often resulting in increasing surface roughness with film growth. However, in our method, Si is deposited on a substrate set at $T_{sub} > T_m$, condensing on the substrate to form a liquid film (Fig. 2a). Although Si re-evaporates from the liquid film, the Si source heated at $T_{boat} \approx$ 2000 °C ≫ T_m yields a much higher vapour pressure than the liquid film, realizing a deposition rate which is much larger than the re-evaporation rate and the deposition of the liquid film. Then, we lower the T_{sub} value to $<T_m$ to nucleate Si crystallites from the liquid film, inducing lateral grain growth of planar grains in the liquid film, and thickening the film to ~10 μm by continuing the vapour deposition. Through this growth model, we tried to obtain large-grain polycrystalline Si films in 1 min.

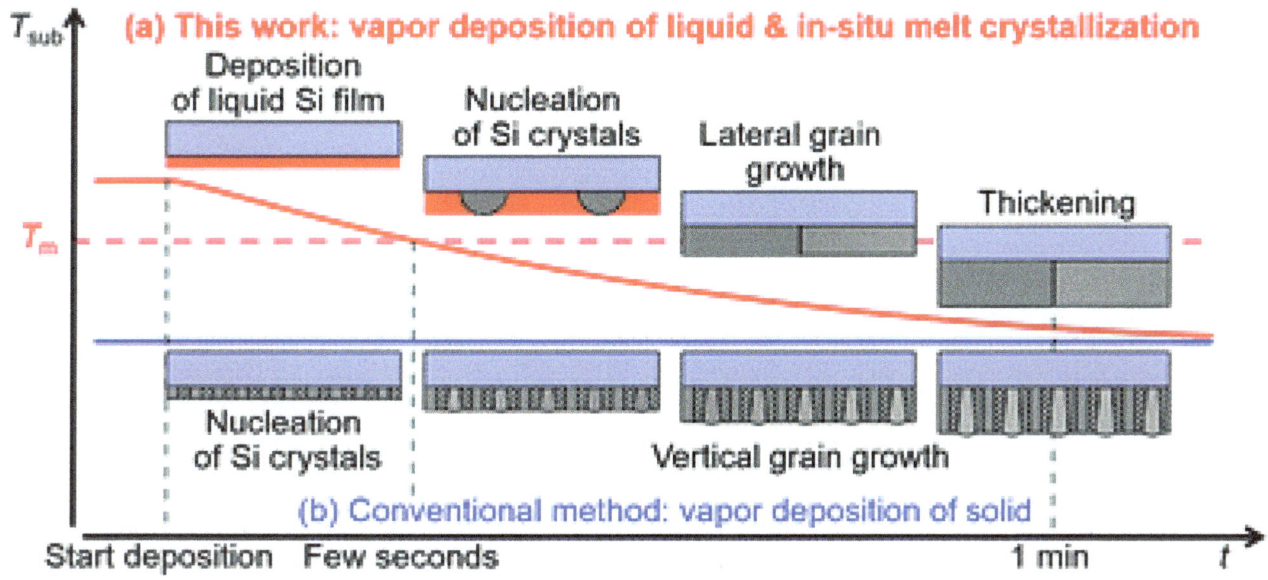

Figure 2 Film growth processes proposed for our method (a) and known for the conventional vapour deposition method (b).

Fig. 3 shows a typical Si film deposited on a 0.1 μm a-C/SiO_2 substrate using our method. Fig. 3a shows the time profile of the temperature of the thermocouple, T_{TC}; the deposition was started at 0 s with $T_{TC} > T_m$, T_{TC} was quickly decreased to $T_{TC} < T_m$ at 5–20 s by decreasing the input power for the substrate heater, T_{TC} was kept almost constant for 20–60 s, and then heating of the boats was turned off at 60 s, resulting in a further decrease in T_{TC} at 60–70 s. The thermocouple was not in contact with the substrate to avoid temperature distribution in the substrate, and therefore, T_{TC} has some deviation from T_{sub} and is used as a reference. We made a reference experiment to measure T_{sub} using a 0.1 μm a-C/SiO_2 substrate with a Pt–$Pt_{0.87}Rh_{0.13}$ thermocouple fixed at the centre of the substrate. We carried out the RVD process with the same time profile of input power for the tungsten boats and the carbon heater but without putting the Si source material in the boats. As shown in the ESI,† Fig. S1, T_{sub} proved to be 100–150 °C lower than T_{TC}, and slightly above the melting point of Si (T_m = 1414 °C) when the RVD process was started. T_{sub} decreased below T_m by decreasing the heating power for the upper carbon heater at 5 s. The digital image of the sample in Fig. 3b shows that Si was deposited on the inner 13 mm square area of the 15 mm square substrate and showed a silver-grey colour with patterns coming from the 100 μm grains. The surface and cross-sectional SEM images of the Si film in Fig. 3c and d show that a Si film with a lateral grain size of over 100 μm and a thickness of 10 μm was actually obtained rapidly in 1 min by our deposition method. The out-of-plane XRD pattern of the Si film in Fig. 3e shows the diffraction peaks from a diamond structure without any preferred orientation. The intensity ratios were different from the powder pattern and changed with the measurement point because only a limited number of the large Si grains (over 100 μm) were detected in each XRD measurement.

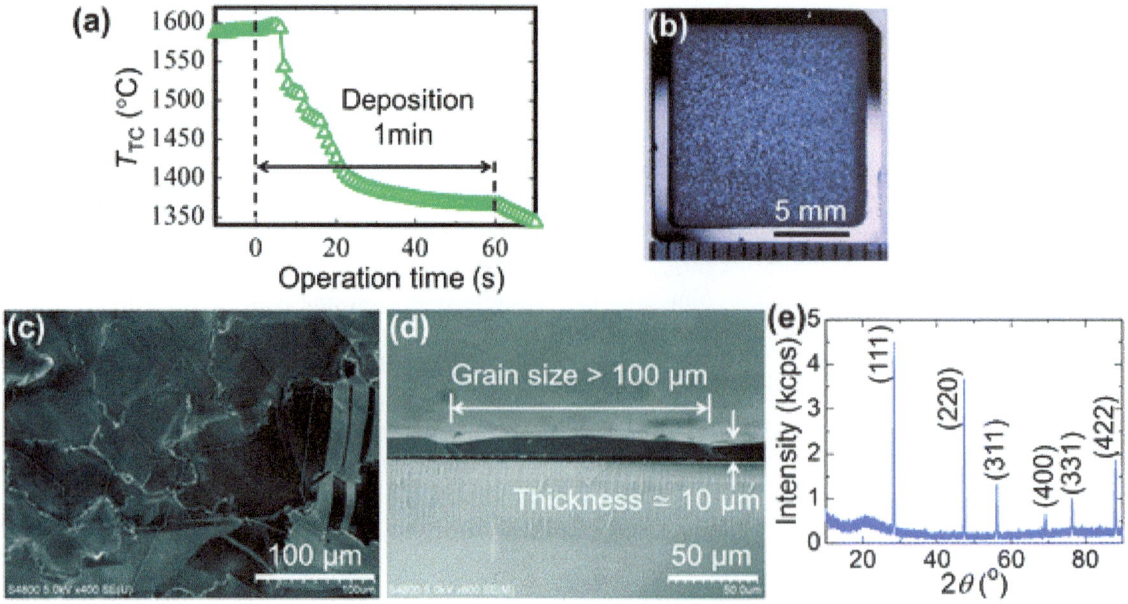

Figure 3 Representative results of the Si film deposited by our method. (a) Time profile of the temperature during deposition. (b) Digital image, (c) surface SEM image, (d) cross-sectional SEM image and (e) ω–2θ-scan XRD spectrum of a Si film deposited on the a-C/SiO_2substrate.

We next show the growth behaviour of Si films at a temperature of T_{TC} ~ 1100 °C < T_m with a fixed substrate heater power. Fig. 4a shows a digital image of the sample deposited on the SiO_2/Si(100) substrate with SiO_2 removed by HF from half of its surface. The Si film deposited on SiO_2 had a cloudy white surface whereas the Si film deposited on Si(100) had a mirror surface. The surface SEM images showed sub-micrometre-sized grains for the Si film on SiO_2 (Fig. 4b) and a flat surface without any texture for the Si film on Si(100) (Fig. 4c). The cross-sectional SEM images showed columnar grains for the Si film on SiO_2 (Fig. 4d) and a flat cross-section without any texture or boundary with the substrate for the Si film on Si(100) (Fig. 4e). The low-magnification cross-sectional SEM image near the boundary between SiO_2/Si and Si showed a clear change in the Si film structure and rapid deposition of 16 μm in 1 min (Fig. 4f). The out-of-plane XRD spectrum of the Si film on SiO_2/Si in Fig. 4g shows a prominent (400) peak coming from the Si(100) substrate with a forbidden (200) peak, which is characteristic of monocrystalline Si.[21] In addition, (111), (220) and (331) peaks are observed, and therefore, the film is polycrystalline Si without any specific orientation in the out-of-plane direction. The Si film on Si(100) showed the same XRD spectra as the Si(100) substrate; the out-of-plane XRD spectrum showed only (400) and (200) peaks (Fig. 4h) and the Φ scan of the (022) diffraction showed a four fold symmetry (Fig. 4i). These results prove that the Si film grew by homoepitaxy on the Si(100) substrate. Our RVD method realized the rapid growth of Si films at 16 μm min^{-1} owing to the high T_{boat} ~ 2000 °C, as a homoepitaxial film on the Si(100) surface, possibly owing to the high mobility of Si adatoms at the high substrate temperature, and as a columnar polycrystalline film on a SiO_2 surface.

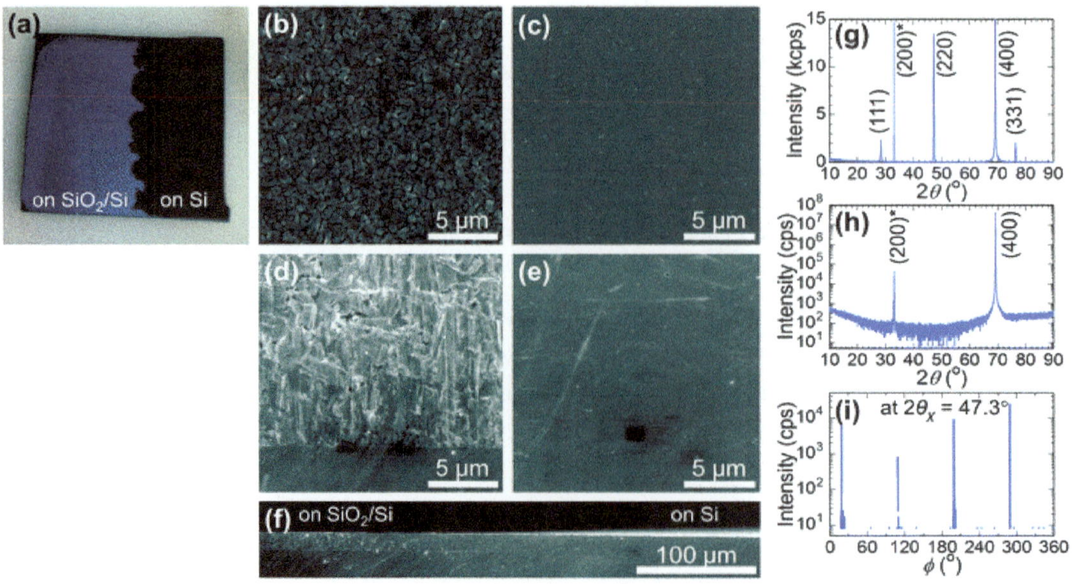

Figure 4 Typical results of Si films deposited on Si(100) substrates at T_{TC} ~ 1100 °C (<T_{mp}). (a) Digital image of the surface of a Si film deposited on a SiO_2/Si substrate with the thermal oxide layer being removed from half of its surface by HF. Surface (b, c)

and cross-sectional (d, e) SEM images of the Si film deposited on SiO_2/Si (b, d) and Si (c, e). (f) Cross-sectional SEM image of the Si film near the boundary between SiO_2/Si and Si regions. (g–i) XRD spectra: ω–2θ scan (g) of a Si film deposited on a SiO_2/Si(100) substrate and ω–2θ scan (h) and Φ scan (i) of a Si film deposited on Si(100). The symbol * denotes the forbidden (200) diffraction characteristic of monocrystalline Si.[21]

As mentioned above, compared with the conventional vapour deposition method at a constant substrate temperature below the melting point, our method can yield Si films with a significantly different structure with planar large grains by changing the substrate temperature, starting from T_{TC} (~T_{sub}) > T_m and decreasing to T_{TC} (~T_{sub}) < T_m during deposition. The conditions for large-grain polycrystalline Si films are discussed in detail below.

Effect of cooling rate on the grain sizes of the Si films deposited on a-C/SiO_2 substrates

Si films were deposited on a-C/SiO_2 substrates with different cooling rates. Deposition was started with T_{TC} > T_m and continued at the fixed substrate heater power for 5 s, and then the heater power was decreased at three different rates (Fig. 5a). The digital images of the surface (Fig. 5b–d) showed shinier surfaces with coarser patterns for the samples deposited with slower cooling rates. The SEM images of the surface (Fig. 5e–g) showed polycrystalline films with lateral grain sizes changing with the cooling rates, from ~50 μm by fast cooling to >100 μm by slow cooling. The grain size of the deposited Si films changed with the cooling rate, similar to the ordinal crystal growth from molten Si.

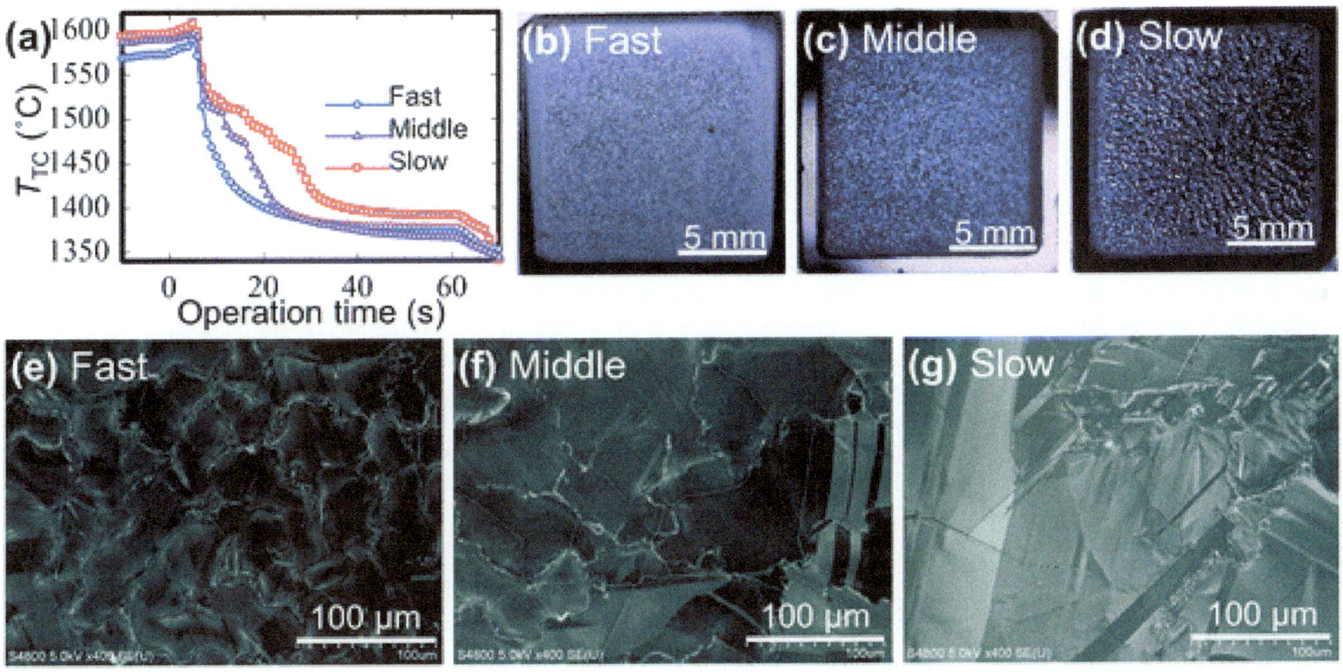

Figure 5 Si films deposited on a-C/SiO_2 substrates at various cooling rates. (a) Time profiles of T_{TC} for fast, middle and slow cooling. (b–d) Digital images and (e–g) SEM images of the surface of the Si films formed with different cooling rates.

Si film growth on SiO_2 substrates without the a-C layer

Fig. 6 shows a Si film deposited on the SiO_2 substrate without the surface a-C layer. In the digital image (Fig. 6a), the part surrounded by red broken lines is the deposited area; however, a continuous Si film remained only on the grey-coloured central region of a 10 mm circle. A spherical particle with a diameter of ~1 mm was found at the centre, which should have solidified from a Si droplet repelled from the SiO_2 substrate. The SEM images of the surface and cross-section of the Si film at the central region showed a polycrystalline film with grain size as small as several micrometres (Fig. 6b and c). In the digital image (Fig. 6a), transparent SiO_2 was exposed at the outer region. The surface and cross-sectional SEM images showed small spherical Si particles of ~10 μm in diameter. Liquid Si should easily repel from and be discontinuous on the substrates with poor wettability.

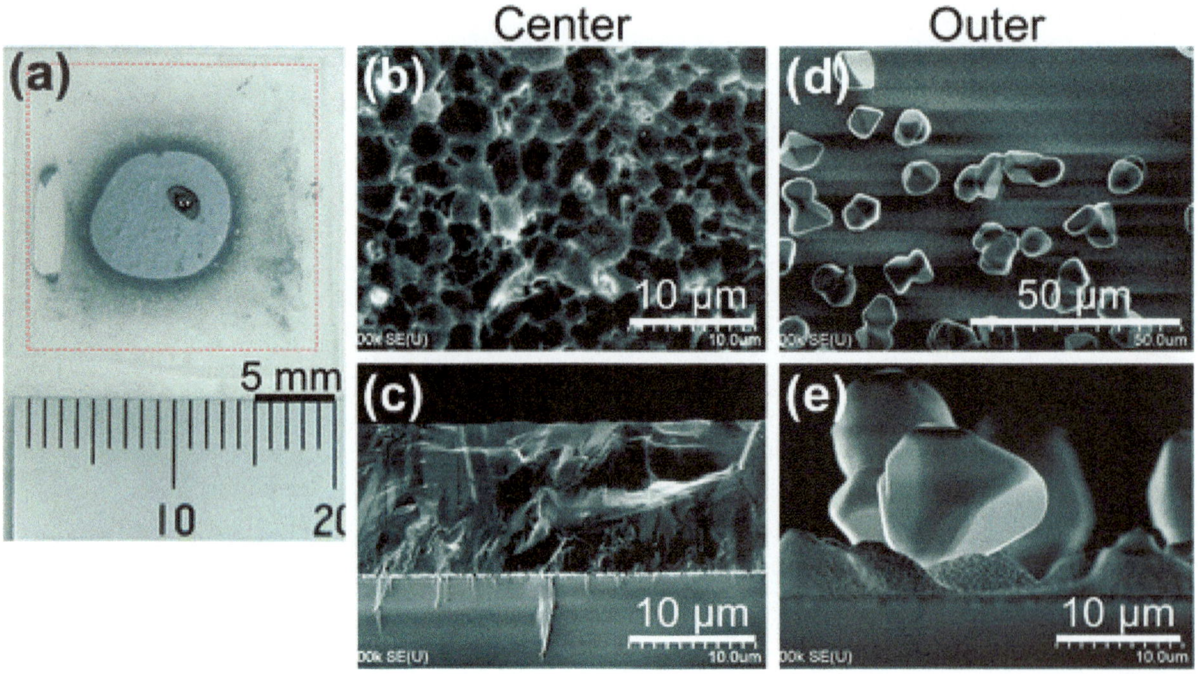

Figure 6 Si film deposited on a SiO_2 substrate without the surface a-C layer. (a) Digital image of the surface. The square in red broken lines shows the Si deposition area. Surface (b, d) and cross-sectional (c, e) SEM images of the Si film at the central (b, c) and outer (d, e) regions.

Conversely, the SiO_2 substrates with a 0.1 μm-thick a-C surface layer yielded large-grain Si films uniformly over the deposited area as shown in Fig. 3 and 5. Next, we examined the very early stage of Si deposition on a-C/SiO_2 to check the behaviour of Si on the a-C layer. We deposited Si for only 5 s, quickly cooled down the substrate, and then analysed the sample by SEM and XRD, as shown in Fig. 7. There are many droplet-like Si grains with smaller heights than the lateral dimensions and a much smaller wetting angle of ~50° (Fig. 7c) than the SiO_2substrate without an a-C layer (Fig. 6e). The thin Si layer might have de-wetted during the cooling down process in this experiment of 5 s deposition. During actual deposition for 1 min, however, Si was depositing rapidly on the Si layer, and the thickening Si layer formed a continuous film without de-wetting on the substrate. Fig. 7d is the grazing-incidence XRD spectrum of this sample, which clearly shows the formation of β-SiC. Contact angles were previously reported for liquid Si on vitreous carbon (a-C) to be 40–50° at 1426 °C (ref. 22) and 36° at 1430 °C,[23] which agree well with our observation (~50°, Fig. 7c). A 0–100 nm a-C layer was previously applied to a microcrystalline alumina substrate to improve the wettability of Ni-63 at% Si alloy, where the formation of SiC was the key for the improved wettability.[24] Thus, we conclude that a-C reacted with liquid Si and formed a β-SiC layer, on which liquid Si wetted well.

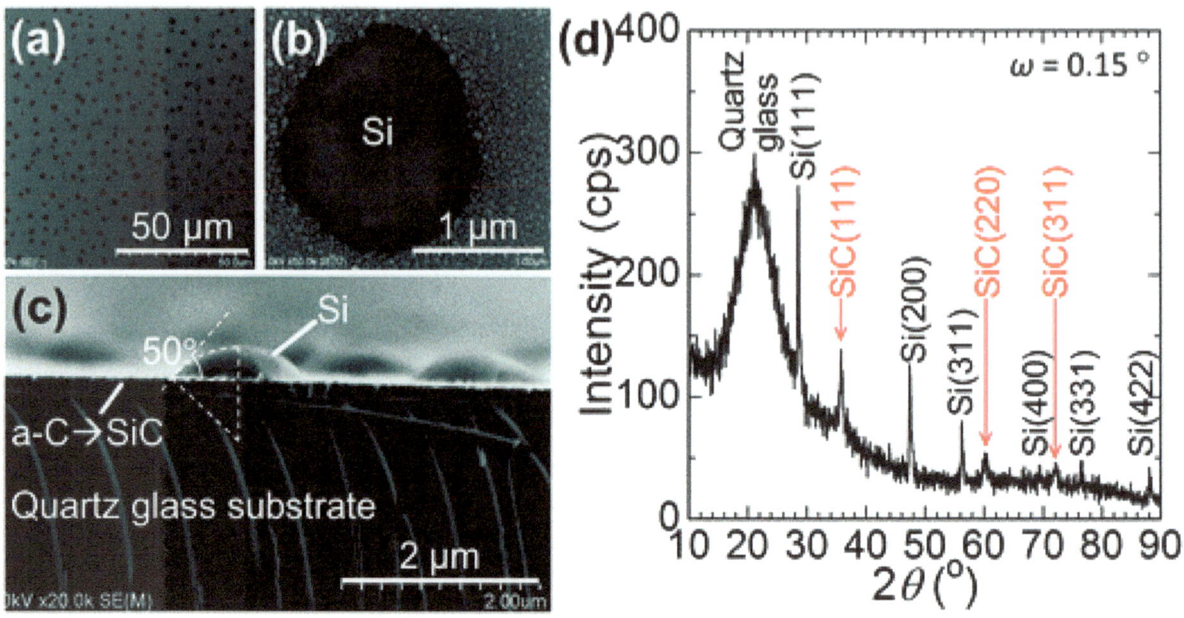

Figure 7 SEM images of a Si sample deposited on a-C/SiO_2 for 5 s and cooled down immediately. Surface SEM images at low (a) and high (b) magnifications and a cross-sectional SEM image (c). Many hemispherical Si islands of a lateral size of approximately 1–2 μm were observed. (d) Grazing-incidence XRD spectrum taken with an incidence angle of $\omega = 0.15°$. Three main peaks for β-SiC (ref. 25) were observed in addition to those for Si.

As discussed above, Si formed continuous films with large grains on a-C because of the good wettability of Si with a-C. In contrast, Si had poor wettability with SiO_2, yielding the repelled Si droplets and the discontinuous film of the solidified Si particles on the SiO_2 substrate without an a-C layer. Because liquid Si reacts with an a-C layer to form SiC, it is important to have an a-C layer thin (0.1 μm) enough not to consume much Si by this reaction.

Effects of substrate materials: comparison of SiO_2 and Al_2O_3 substrates

Residual stress between films and substrates can cause breaking and/or delamination of the films. In our method, a mismatch of the linear thermal expansion between the Si films and the substrates causes thermal stress during the cooling process. The linear thermal expansion of Si, fused SiO_2, polycrystalline Al_2O_3, and a-axis Al_2O_3 at the temperature range of 20–1500 °C are estimated using previously reported equations[26] and summarized in Fig. S2.† It clearly shows that fused SiO_2 has the smallest value, Si has the medium value and Al_2O_3 has the largest value. Fig. 8 shows the Si films formed on a-C/SiO_2 and a-C/Al_2O_3 substrates. Linear cracks can be seen on a-C/SiO_2 at intervals of several tens of micrometres whereas no cracks were observed on a-C/Al_2O_3. Tensile stress worked on the Si film on a-C/SiO_2 during cooling, resulting in crack formation. It should be important to choose appropriate substrates with thermal linear expansion similar to or somewhat larger than Si to obtain crack-free continuous Si films.

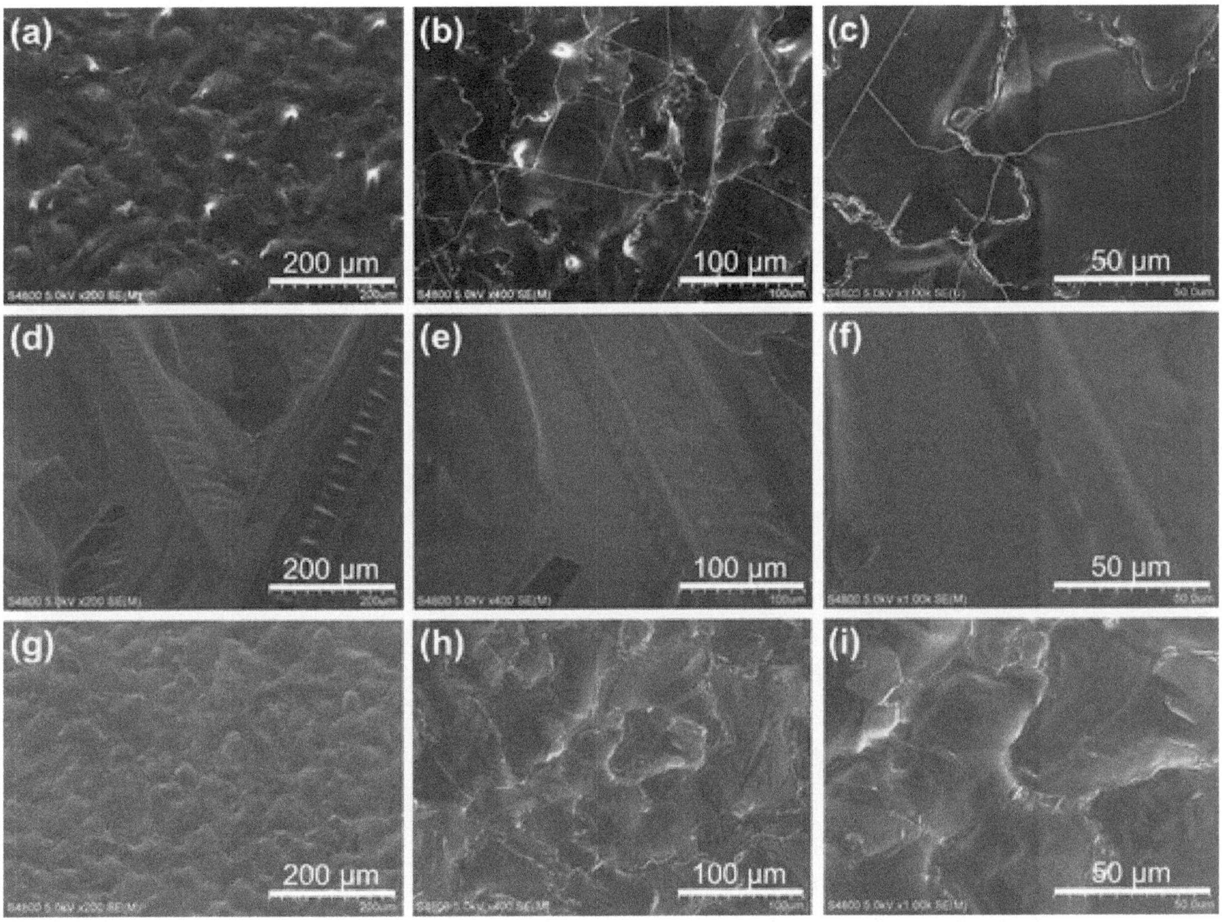

Figure 8 Surface SEM images of the Si films deposited on SiO_2 (quartz glass) (a–c), Al_2O_3(polycrystalline alumina) (d–f) and Al_2O_3 (c-plane sapphire) (g–i) substrates having a 0.1 μm-thick a-C layer. Cracks were observed only in the Si film on the SiO_2 substrate (a–c).

CONCLUSIONS

We proposed a film deposition method, in which Si is rapidly deposited on a substrate heated above the melting point of Si and the substrate is subsequently cooled down below the melting point while maintaining rapid deposition. The method yielded continuous polycrystalline Si films with large lateral grain sizes of over 100 μm and thicknesses of ~10 μm in 1 min in a single process on SiO_2 and Al_2O_3 substrates with a 0.1 μm-thick a-C layer. To realize such films without using monocrystalline Si substrates, the choice of substrates is important. The substrates need to be thermally stable at temperatures above the Si melting point, chemically stable enough against liquid Si to be reusable, and in addition, need to have good wettability with liquid Si and to apply small/no tensile stress to the resulting Si films. An Al_2O_3 substrate with an a-C layer was a good combination, yielding crack-free continuous Si films with large grains owing to the good wettability of liquid Si with a-C via the reaction forming SiC and the larger thermal linear expansion of Al_2O_3 than Si (and thus compressive stress instead of tensile stress working on the resulting Si films).

To practically produce Si films for real applications such as PV cells, how to avoid contamination to Si films at a high temperature and how to scale-up the process are the crucial issues. The maximum substrate temperature was slightly above the melting point of Si (Fig. S1†). This shows that our process has a similar temperature to but much shorter time than the ingot production process. Al is an effective dopant in making a back surface field in bulk Si PV cells[27] and thus the Al_2O_3 substrate would be promising in avoiding the negative effects of contamination. For scale-up, we are planning a continuous process carrying growth substrates over a Si melt bath. It may be possible to directly convert Si melt in a crucible to Si films on Al_2O_3 with a takt time of ~1 min, instead of making and slicing Si ingots into thick wafers in many hours/days. There is a long way to reach such goal, but this work should be the very important first step. As the next step, we are now working on a method to transfer large-grain Si films from a growth substrate to device substrates (glass and/or plastic) and reuse the growth substrate to achieve practical low-cost production of such Si films.

ACKNOWLEDGEMENTS

The authors thank Mr. Takahiro Goto at Materials Characterization Central Laboratory, Waseda University, for XRD analysis of the Si films. This work is financially supported in part by a Grant-in-Aid for Challenging Exploratory Research (15K14216) from the Japan Society for the Promotion of Science, Japan.

REFERENCES

1. W. Koch, A. L. Endrös, D. Franke, C. Häßler, J. P. Kalejs and H. J. Möller, in Handbook of Photovoltaic Science and Engineering, ed. A. Luque and S. Hegedus, John Wiley & Sons, Ltd, Hoboken, 2003, ch. 6, pp. 218–254.
2. K. R. Catchpole, M. J. McCann, K. J. Weber and A. W. Blakers, Sol. Energy Mater. Sol. Cells, 2001, **68**, 173.
3. M. J. McCann, K. R. Catchpole, K. J. Weber and A. W. Blakers, Sol. Energy Mater. Sol. Cells, 2001, **68**, 135.
4. T. Saga, NPG Asia Mater., 2010, **2**, 96.
5. S. Uchikoga, Low-Temperature Polycrystalline Silicon Thin-Film Transistor Technologies for System-on-Glass Displays, MRS Bull., 2002, **27**, 881 .
6. R. B. Rendel, Jpn. J. Appl. Phys., 2001, **44**, 4431 .
7. C. H. Lee, D. R. Kim and X. Zheng, ACS Nano, 2014, **8**, 8746 .
8. Y. Tsuji, M. Mizukami and S. Noda, Thin Solid Films, 2008, **516**, 3989 .
9. L. Qi, Z. Hu, W. Li, X. Qin, G. Du, W. Han and W. Shi, Mater. Sci. Semicond. Process., 2012, **15**, 412 .
10. Y. Kajikawa, S. Noda and H. Komiyama, Solid State Phenom., 2003, **93**, 411 .
11. Y. Kajikawa, T. Tsuchiya, S. Noda and H. Komiyama, Chem. Vap. Deposition, 2004, **10**, 128
12. T. Sameshima, S. Usui and M. Sekiya, IEEE Electron Device Lett., 1986, **7**, 276 .
13. K. Sera, F. Okumura, H. Uchida, S. Itoh, S. Kaneko and K. Hotta, IEEE Trans. Electron Devices, 1989, **36**, 2868 .
14. H. Morikawa, Y. Kawama, Y. Matsuno, S. Hamamoto, K. Imada, T. Ishihara, K. Kojima and T. Ogama, Sol. Energy Mater. Sol. Cells, 2001, **65**, 261.
15. T. Kieliba, PhD Thesis, University of Konstanz, 2006 .
16. T. Kunz, M. T. Hessmann, R. Auer, A. Bochmann, S. Christiansen and C. J. Brabec, J. Cryst. Growth, 2012, **357**, 20.
17. H. Lange and I. A. Schwirtlich, J. Cryst. Growth, 1990, **104**, 108 .
18. G. Hahn, S. Seren, M. Kaes, A. Schönecker, J. P. Kalejs, C. Dubé, A. Grenko and C. Belouet, in Proc. 4th WCPEC, Waikoloa, 2006, pp. 972–975.
19. B. Heilbronn, F. De Moro, E. Jolivet, E. Tupin, B. Chau, R. Varrot, B. Drevet, S. Bailly, D. Rey, H. Lignier, Y. Xi, T. Riberi-Béridot, N. Mangelinck-Noël, G. Reinhart and G. Regula,Cryst. Res. Technol., 2014, **114**, 101.
20. J. Lee and S. Noda, RSC Adv., 2015, **5**, 2938 .
21. P. Zaumseil, J. Appl. Crystallogr., 2015, **48**, 528 .
22. T. J. Whalen and A. T. Anderson, J. Am. Ceram. Soc., 1975, **58**, 396 .
23. O. Dezellus, S. Jacques, F. Hodaj and N. Eustathopoulos, J. Mater. Sci., 2005, **40**, 2307 .
24. A. Koltsov, A. Crisci, F. Hodaj and N. Eustathopoulos, J. Mater. Sci., 2010, **45**, 2062 .
25. R. J. Iwanowski, K. Fronc, W. Paszkowicz and M. Heinonen, J. Alloys Compd., 1999, **286**, 143 .
26. Y. S. Touloukian, R. K. Kirby, R. E. Taylor and T. Y. R. Lee, Thermophysical Properties of Matter-The TPRC Data Series. Volume 13. Thermal Expansion-Nonmetallic Solids, IFI Plenum, New York, 1977, p. 154, 176 and 358.
27. S. Narasimha, A. Rohatgi and A. W. Weeber, IEEE Trans. Electron Devices, 1999, **46**, 1363

CHAPTER 11

ANNEALING-INDUCED PERIODIC PATTERNS IN SOLUTION GROWN POLYMER SINGLE CRYSTALS

Bin Zhang [abc], Jingbo Chen [c], Hui Zhang [a], Moritz C. Baier [d], Stefan Mecking [d], Renate Reiter [af], Rolf Mülhaupt [ef] and Günter Reiter [af]

[a]Institute of Physics, University of Freiburg, 79104 Freiburg, Germany.

[b]Hermann Staudinger Graduate School, University of Freiburg, 79104 Freiburg, Germany

[c]School of Materials Science & Engineering, Zhengzhou University, Zhengzhou 450002, People's Republic of China

[d]Chair of Chemical Materials Science, Department of Chemistry, University of Konstanz, 78464 Konstanz, Germany

[e]Institute of Macromolecular Chemistry, University of Freiburg, 79104 Freiburg, Germany

[f]Freiburg Materials Research Centre, University of Freiburg, 79104 Freiburg, Germany

ABSTRACT

Applying a slow annealing procedure, we have transformed geometrically simple, faceted polymer single crystals into periodically branched crystals. Interestingly, the period of the branches increased logarithmically with annealing time and depended on crystallization temperature in a similar fashion as the thickness of the lamellar crystal. We tentatively relate the periodic pattern to meta-stable states, differing in crystalline order and thus melting temperature. The degree of meta-stability and its variance depend on lamellar thickness but can also change with the degree of molecular order, causing differences in melting behaviour. Our results propose that periodic variations in thermal stability within a polymer single crystal can be made visible by annealing.

INTRODUCTION

Polymer crystals usually grow under conditions far away from thermodynamic equilibrium,[1–5] resulting in lamellar crystals of highly folded chain-like molecules.[6,7] These crystals may exhibit many different, often complex, morphologies.[8–11] However, in particular in solutions, for many decades polymer crystals with a simple faceted morphology have also been frequently observed.[12–15] At the corresponding comparatively fast growth rates,[16–18] polymers cannot equilibrate and meta-stable states of folded polymers are the rule. Consequently, although of simple shape, faceted polymer crystals should consist of polymers differing in degree of molecular order.[19–21] According to the Gibbs–Thomson relation,[22] the melting temperature depends on the size of the crystal or the size of the domain of molecules of equal degree of order. Smaller or less ordered regions melt at lower temperatures than larger or better ordered ones.[23–27] Thus, when annealing faceted crystals at temperatures below the nominal melting temperature,[28,29] one may anticipate observing morphological changes which may provide information on the spatial distribution of less ordered regions within faceted crystals and, possibly, on how the faceted crystals were grown.

Annealing-induced morphological changes in polymer single crystals at temperatures well below the nominal melting point have been observed previously.[6,30–33] Local molecular reorganization processes were identified as the central mechanism which allowed to partially remove chain folds with increasing annealing time or annealing temperature.[31,34,35] As a consequence, the mean lamellar thickness increased, leading to enhanced thermal stability.[36,37] Pioneering microscopy works[38–42] on individual single crystals have demonstrated that, as a mere consequence of mass conservation, such annealing induced local thickening[43] is accompanied by the formation of a "Swiss cheese" morphology or "saw-tooth" patterns.[44–47] Such

Citation: Bin Zhang, Jingbo Chen, Hui Zhang, Moritz C. Baier, Stefan Mecking, Renate Reiter, Rolf Mülhaupt and Günter Reiter. Annealing-induced periodic patterns in solution grown polymer single crystals. RSC Adv., 2015, 5, 12974-12980, DOI: 10.1039/C4RA10563J spatially varying changes in morphology induced by annealing reflect heterogeneities, i.e., differences in molecular order within these crystals. However, so far annealing did not reveal any evident spatial correlations of these heterogeneities. No relation to the initial growth conditions and to the conditions of annealing could be established.

One reason for not observing correlated patterns induced by annealing may be attributed to the typically small differences in degree of order within a single crystal. Thus, to uncover heterogeneities,[48,49] differences in thermal stability have to be amplified first. Here, we have chosen a lengthy series of annealing steps which started at low temperatures and slowly moved to higher temperatures. Annealing has to remove sequentially, and with high accuracy, crystalline parts which have an only slightly lower thermal stability than more perfect ones. Upon proper annealing, less stable regions will melt. Simultaneously, the remaining regions will become even more stable, e.g., thicker. Consequently, the finally surviving regions reflect, to a large extent, the more stable crystalline parts formed already during crystal growth. By applying suitable annealing sequences, we attempted to unveil the less stable parts of a polymer crystal and aim to answer the following question: are heterogeneities within facetted lamellar single crystals distributed homogeneously or are they arranged according to some pattern?

EXPERIMENTAL

As a model system for answering these general questions we have chosen an archetypical crystallisable polymer: polyethylene.[50,51] The here reported phenomenon was first revealed and studied in detail for a partially deuterated sample of low polydispersity index (PDI). To demonstrate the generality of the phenomenon we verified that other PE-samples, including commercial products with high PDI, also exhibited analogous morphological changes induced by appropriate annealing procedures. However, the changes were more accentuated and more uniform for polymers with a narrow molecular weight distribution.

As shown in Table 1, we studied four high molecular weight polyethylene samples, including two homopolymers of protonated polyethylene (PE), a deuterated PE (dPE) and a diblock copolymer of hydrogenated and deuterated polyethylene (hPE-b-dPE) of different molecular weights.[52] For all these polymers, single crystals were prepared by a self-seeding method,[53] from dilute solutions at appropriate crystallization temperatures (T_c). The polymers were dissolved in p-xylene at rather low concentrations of 0.001 wt% or 0.0001 wt% by heating the solution (ca. 1 ml in a closed glass vial) first to 130 °C for 30 min, i.e., well above the nominal dissolution temperature of ca. 100 °C. This so homogenized polymer solution was then rapidly crystallized and aged for 2 hours at 70 °C in a thermostated bath filled with silicone oil. The solution was subsequently heated to the respective self-seeding temperature T_s, where it was kept for 10 minutes. For growing single crystals, the samples were subsequently quickly transferred to another thermostated bath at a pre-set crystallization temperature T_c. Drops of such solutions containing suspended PE single crystals were then deposited onto a silicon wafer, an often used substrate.[47,54–57] After deposition, the solvent was allowed to evaporate completely in vacuum at 25 °C for 12 hours.

Table 1 Characteristics of PE-b-dPE, dPE and PE samples and preparation conditions of solution grown single crystals

Code	Material	M_w [g mol^{-1}]	PDI	T_s (self-seeding temperature), [°C]	Crystallization concentration, [wt%]	T_m [b], [°C]
a 29 mol% ethylene-d4 (determined by IR spectroscopy). b The temperature control program for DSC was as follows: (a) heat the as-received polymer bulk sample from room temperature to 200 °C at a rate of 10 °C min^{-1}. (b) Keep the polymer melt by holding the temperature at 200 °C for 5 min. (c) Cool at a rate of 10 °C min^{-1} down to −50 °C. (d) Heat the polymer sample from room temperature to 200 °C at a rate of 10 °C min^{-1}.						
PE1	PE-b-deuterated PE[a]	3.8×10^6	1.23	105	0.0001	136.1
PE2	Deuterated PE	5.76×10^5	1.08	100	0.001	132.5
PE3	PE	3.5×10^6	7.5	106	0.0001	135.8
PE4	PE	7.5×10^5	7.5	100	0.001	135.4

AFM Measurements and annealing of single crystals

Characterization of the morphology and thermal behaviour of the single crystals was performed by atomic force microscopy (AFM, JPK Instruments, Germany). The microscope was equipped with a high-temperature heating stage accessory, controlling the temperature of the sample, allowing to perform in situ experiments at elevated temperatures. Fig. 1a illustrates a typical temperature protocol used in this study for the in situ AFM experiments. By conducting in situ AFM measurements during annealing of the samples, it was possible to observe, in addition to changes in morphology at the chosen annealing temperature, also changes in the phase-signal of the tapping-mode AFM. This phase-signal reflects viscoelastic properties of the sample and thus allowed to distinguish between crystalline and molten regions. However, the relatively high annealing temperatures caused stickiness between the AFM-tip and the molten polymers and made measurements more complex. In addition, due to the required scanning velocities, it was often difficult to obtain high-resolution in situ images with a low noise level.

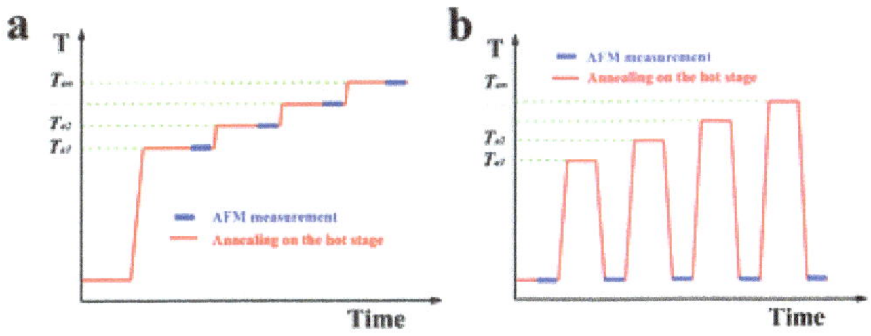

Figure 1 Experimental protocol employed for the annealing experiments. (a) In situ, (b) ex situ.

Thus, some high-resolution AFM images of the morphology evolution after annealing at increasing temperature (T_a) and for prolonged times were performed ex situ at room temperature. For ex situ experiments, the single crystals were annealed first on a Linkam THMS 600 hot stage (Linkam Scientific Instruments, Tadworth, UK) under nitrogen atmosphere at a desired temperature for a chosen time (mostly 20 minutes). The sample was subsequently cooled to room temperature and investigated by AFM. The sample was further annealed on the Linkam hot stage at a slightly higher temperature and re-examined by AFM. This procedure was repeated several times. However, at room temperature the molten polymers re-crystallized and the AFM phase-signal was not able to distinguish them from the crystalline regions existing at elevated T_a. Thus, after a temperature protocol as the one shown in Fig. 1b including a quench to room temperature, AFM measurements could only visualize the morphology obtained after the annealing process but could not indicate the amount of molten polymers and their distribution within the single crystal.

RESULTS AND DISCUSSION

In Fig. 2, we show a sequence of in situ AFM height images of a truncated lozenge-shaped polyethylene single crystal of PE1, taken after different annealing steps. At each temperature, the sample was kept for 20 min, followed by a temperature increase of a maximum of a few degrees (see also Fig. 1a). Up to an annealing temperature (T_a) of ca. 122 °C, the crystal morphology almost did not change. Only the crystal periphery became periodically modulated. Notches, bounded by a thickened rim, appeared along the edges of the crystal. These localized changes reflect modulations of stability favoured by the higher mobility of the polymer chains located at the edges of the crystals, i.e., the most unstable parts of crystal. There, reorganization of thermodynamic meta-stable states towards more stable states occurred first.

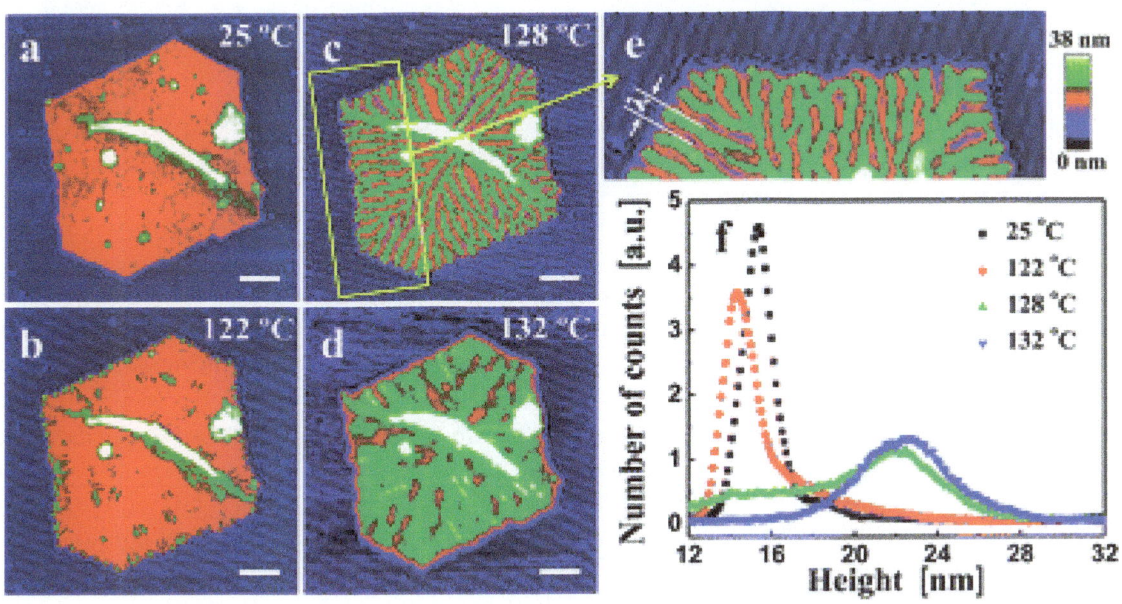

Figure. 2 Morphological changes in a single crystal of polyethylene, induced by annealing at increasing temperatures. (a) PE1 single crystal obtained by isothermal growth [(crystallization temperature T_c, crystallization time t_c) = (83.5 °C, 60 min)] (see Experimental section). Height images were taken in situ during annealing at temperatures (T_a) increasing stepwise from 25 °C to 138 °C with a constant annealing time (t_a) = 20 min per step. T_a = (b) 122 °C, (c) + (e) 128 °C, (d) 132 °C. (f) Height histogram of the PE1 single crystal after annealing at various T_a. The size of the scale bar is 500 nm. λ is the periodic distance between branches.

Upon annealing at T_a > 122 °C, thickened parts became more prominent and grew from the periphery of the crystal in the direction towards the interior of the crystal. The resulting branches progressively carved regularly spaced valleys into the single crystal. Eventually, the branches reached the centre of the crystal, i.e. the initial nucleation site. When heating slightly above 130 °C, valleys were refilled by molten polymers. However, the memory of the branched patterns still existed as the branches re-appeared upon cooling (see also Fig. 3). Only after heating above ca. 135 °C, all memory of a crystalline pattern was lost. It is worth noting that annealing the crystal by jumping right away to temperatures above ca. 130 °C caused the formation of "Swiss-cheese"-like patterns rather than branches. Thus, slowly removing less perfect crystalline regions by providing enough time for re-organisation within the crystal seems to be required to generate periodic patterns formed within a single crystal.

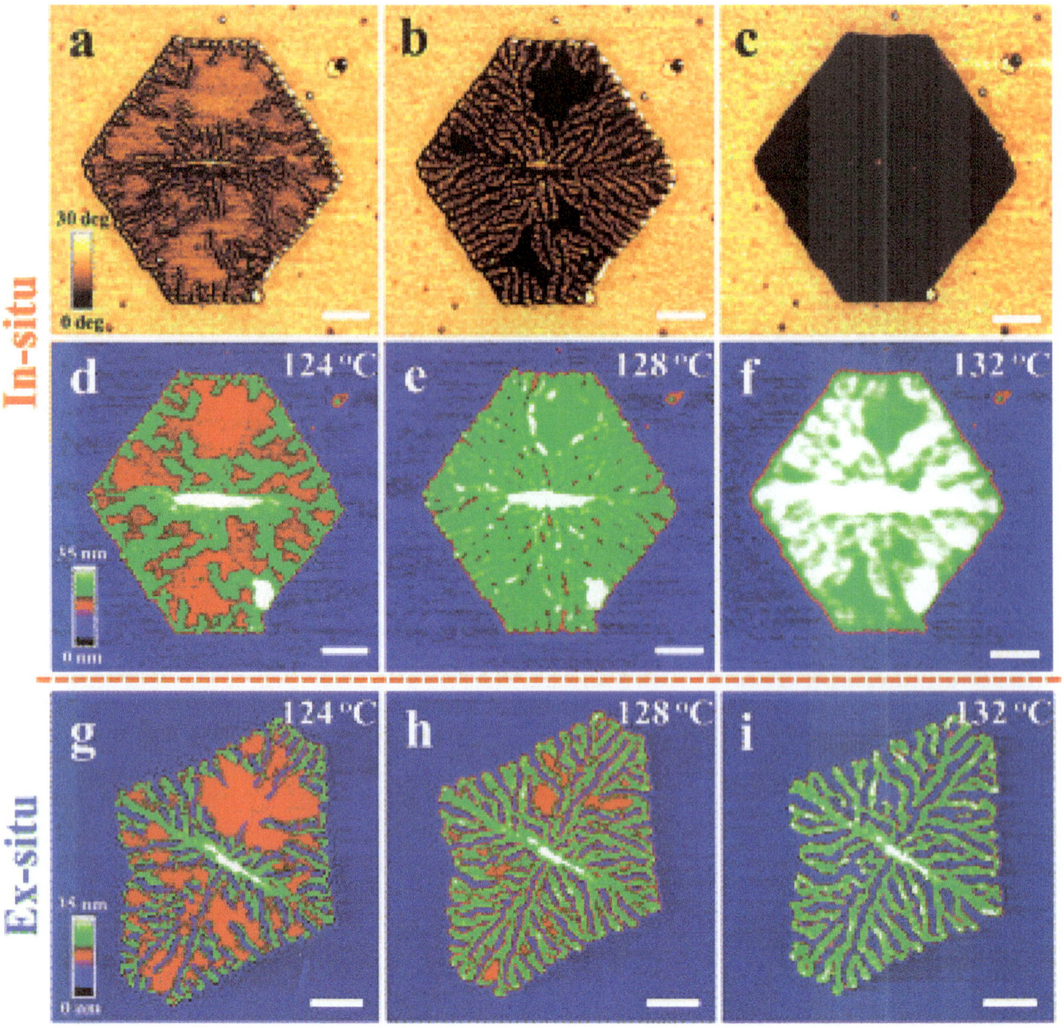

Figure 3 (a)–(f) In situ annealing. A PE1 single crystal obtained by isothermal growth [(T_c, t_c) = (83.5 °C, 60 min)]. Phase images (a)–(c) and corresponding height images (d)–(f) were takenin situ during annealing at various temperatures T_a, increasing stepwise from 25 °C to 132 °C. (g)–(i) Re-crystallization of branched morphology. A different PE1 single crystal obtained by isothermal growth [(T_c, t_c) = (82 °C, 20 min)]. Height images were taken ex situat room temperature after annealing at various temperatures, increasing stepwise from 25 °C to 132 °C for a constant annealing time t_a of 20 min at each step. The size of the scale bar is always 500 nm.

In addition to identifying local morphological changes, we also determined the mean height of the crystal, recorded during annealing. These mean values allowed a direct comparison with results frequently obtained via averaging techniques like X-ray scattering.[34,47] As shown in the histogram of Fig. 2f, taken from the recorded AFM images, a sequence of stages can be identified, consistent with previous observations.[21] For the early stages (low T_a and/or short annealing times (t_a)), the mean height of the crystal (its lamellar thickness (l_d)) did not change much. Nonetheless, even at such early stages, the morphology already started to change at the crystal periphery. At progressively higher T_a, or after prolonged annealing, a stage of local melting and significant reorganization followed (Fig. 2c and d), causing a shoulder to appear at higher thickness values and evolving into a peak in the height histogram. With increasing T_a, this new peak, which is typically attributed to lamellar thickening,[29,31,58,59] slowly shifted to higher values of l_d. In combination with the direct space observation by AFM, we can conclude that the lamellar thickening process was accompanied by the formation of a regular pattern within the single crystal. For T_a > 130 °C, the height distribution became rather broad, related to the simultaneous melting of large parts of the crystal and thickening of the remaining parts, the branches.

Using the phase-signal of tapping mode AFM, it is possible to visualize differences in viscoelastic properties within an annealed single crystal. Under the imaging conditions used here, the contrast visible in the phase images reflects crystalline (light colours) and molten (dark colours) areas. As can be seen in Fig. 3a, the AFM phase images indicated that in the course of the appearance of the branches, they carved regularly spaced valleys into the single crystal which were partially filled by molten polymers as indicated by the change in the phase signal (the dark colour of the phase signal is indicating molten polymers).

The short waiting time of 10 min at T_a = 124 °C did not allow for the formation of branched structures within the whole crystal. After raising the temperature to 128 °C, only the thickened parts, i.e. the parts forming the branched morphology, exhibited a solid-like response in the AFM phase-signal. At even higher temperatures, such as 132 °C (Fig. 3c), the branched structure seemed to disappear. Under the chosen imaging conditions, it was not visible anymore in the AFM phase-signal. However, when increasing the tapping force,[60,61] it became clear that branches still existed but were "hidden" under a layer of molten polymers. In addition, upon quenching to room temperature, the branched structure became clearly visible (compare Fig. 3f and i).

Upon cooling, all molten polymers re-attached to the remaining branches, making them visible again in the AFM topography images. A series of high resolution AFM high images of PE1 single crystals were taken at room temperature after annealing at various T_a (see Fig. 3g–i). When annealing at temperatures lower than 128 °C, the morphological evolution of the annealing-induced branched pattern was obviously very similar to what was demonstrated in the above in situ AFM images (compare Fig. 2b and c and 3b and e). Crack-like structures could be seen both in the central region and at the crystal edges after annealing at 128 °C. As shown in the AFM height image (Fig. 3h), the cracks formed a branched pattern. In addition, the lamellar thickness within such single crystals increased upon annealing.

Upon quenching the sample to room temperature, re-crystallisation of polymers from the partially molten state in between the branches contributed to the morphological changes. This became particularly visible for annealing temperatures above 128 °C. It can be deduced by comparison of Fig. 3d–f with Fig. 3g–i that re-crystallisation of the molten polymers was guided by the crystalline branched pattern.

Besides varying T_a, we also varied t_a while keeping T_a constant. For example, at T_a = 125 °C, the morphological evolution of one solution-grown faceted crystal is shown in Fig. 4a. Already after t_a = 2 min, numerous notches formed along the crystal edges. With increasing t_a, these notches penetrated further towards the centre of the crystal. Elongated branches formed in each sector of the initial crystal, all of them pointing preferentially towards the centre (i.e. the nucleation site) of the crystal. As shown in the height histogram of Fig. 4b, a second broad peak emerged at the expense of the initial peak. The comparison of Fig. 2 and 4 reveals a kind of time–temperature superposition. Interestingly, the value of the periodic spacing (λ) of the branches increased with t_a, reflected by a decrease in number of branches per unit length. This evolution of $\lambda(t_a)$ followed the same trend as the mean crystal thickness (l_d) (Fig. 4c), consistent with previous scattering experiments[34,36,47] which averaged over a large number of crystalline domains. Both, λ and l_d increased approximately with the logarithm of t_a.

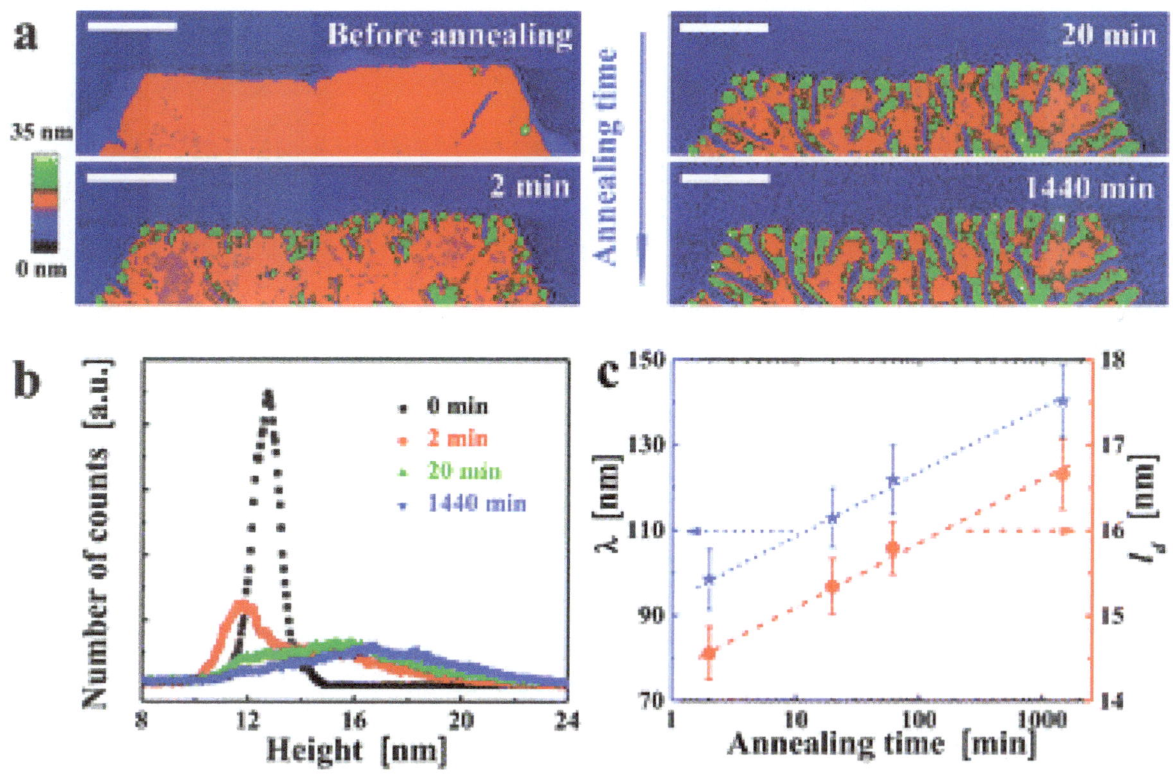

Figure 4 Temporal evolution of morphological changes induced by annealing. (a) Ex situ height images of a PE1 single crystal (see Experimental section), obtained by isothermal growth at (T_c, t_c) = (82 °C, 20 min), taken after annealing at 125 °C for various annealing times t_a, indicated in the figure. Size of the scale bar is 500 nm. (b) Height histograms obtained from the AFM height images in (a), after background subtraction. (c) Evolution of period of the branches (λ, stars) as a function of annealing time for the measurements shown in (a), compared with the mean thickness (l_d, circles) of the lamellar crystal.

In order to demonstrate the generality of our approach, we have repeated analogous experiments for a variety of PE molecules of high molecular weight, including commercial products (see Experimental section). For all samples, we grew single crystals having the shape of a truncated lozenge. Applying similar annealing procedures, we could generate analogously branched patterns for all studied samples. As shown in Fig. 5a for a constant annealing time of 20 minutes, λ and the length of the branches increased with T_a. Keeping T_a constant (Fig. 5b) and varying t_a, a similar increase of λ was observed.

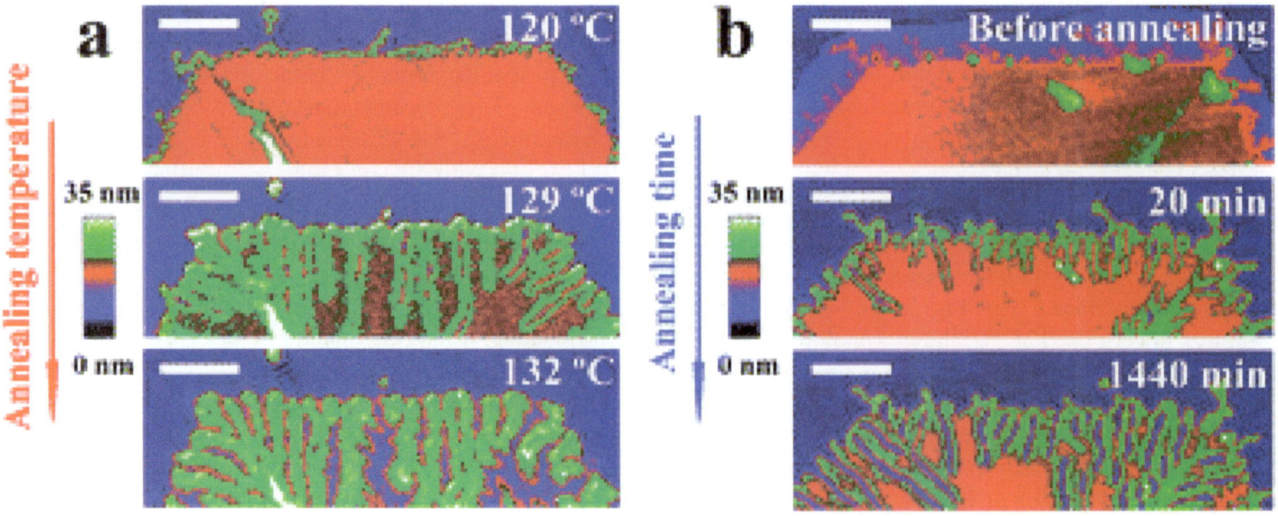

Figure 5 Confirming the generality of the annealing induced patterns. Series of AFM height images of a single crystal of PE3 recorded after annealing performed at (a) various T_a with t_a= 20 min and (b) T_a = 127 °C for various t_a. PE3 single crystals were obtained by isothermal growth at (T_c, t_c) = (85 °C, 90 min). The size of the scale bar is 500 nm.

The initial lamellar thickness (l_{d0}) depends on the growth rate, which, in turn, can be controlled by crystallization temperature T_c.[35,62] In addition, the thickness (l_d) of a crystalline lamella can vary during growth in the regions behind the growth front or during post-growth annealing stages. In Fig. 6, we demonstrate that λ also depends on T_c. This suggests that a relation between λ and l_{d0} (or l_d) exists.

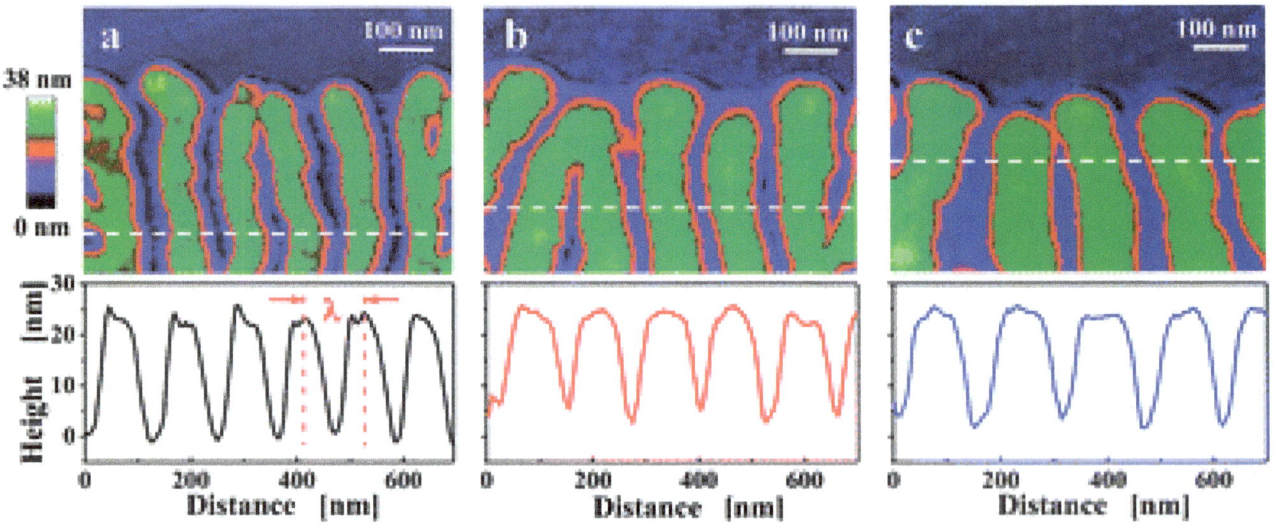

Figure 6 Dependence of the width of the branches on crystallization temperature of the initial crystal. Tapping mode AFM height images of representative regions close to the edge of truncated-lozenge crystals. Measurements were performed at room temperature after annealing of a PE1 single crystal at 132 °C for 20 min. Crystals were grown isothermally at T_c for time t_c: (T_c, t_c) = (a) (82 °C, 20 min), (b) (83.5 °C, 60 min), (c) (85 °C, 120 min).

As has been shown previously,[34,58] lamellar thickening induced by annealing is a complex process involving (1) a molecular diffusion process of the detached/molten molecules to more stable, thicker crystalline regions, (2) attachment and incorporation of molecules into these regions, followed by (3) further improvement of stability by removing even more folds and disappearance of less stable regions. Consequently, as shown in Fig. 2–6, annealing temperature and time have a crucial influence on the evolution of morphology and lamellar thickness (l_d) during/after growth. The rate of such annealing-induced changes depended on the initial degree of order and the morphology of the single crystal, mainly determined by the crystallization temperature.

To shed some light on the origin of annealing-induced branched patterns, we searched for relations between the conditions under which the faceted polymer single crystals were grown and the periodic patterns observed after annealing. For various annealing procedures and crystallization conditions, we analysed a series of single crystals and compared the evolution of λwith the accompanying changes in the parameters l_{d0} (see Fig. 7a) or l_d (see Fig. 7a). The results are summarized in Fig. 7, from which we can draw the following conclusions. λ was found to be much smaller than the size of the faceted crystals and only about ten times larger than l_d. As shown in Fig. 4, for constant T_a, λ increased approximately with log(t_a). Because l_d also increased with log(t_a), a clear relation between λ and l_d was found. Single crystals prepared at increasing T_c showed an increase in l_{d0}. As shown in Fig. 7b for annealing times increasing with T_c, λ also increased with T_c, giving rise to an approximately linear relation between λ and l_{d0}. A more complex behaviour was found when keeping t_a constant (20 minutes) and varyingT_a (from 115 to 130 °C). At low T_a, λ increased rapidly with l_d. But at high T_a, λ increased only slowly with l_d. At low annealing temperatures (from ca. 115 to 120 °C), changes in morphology could only be observed for crystals grown at low T_c. At such conditions, λ was found to be small, i.e., the distances between branches were short, allowing for relatively efficient transport of polymer chains between branches. Thus, a coarsening of λ was observed even for short annealing times. For thicker lamellar crystals, higher temperatures and longer times were required to allow for the annealing-induced changes in morphology to become visible.

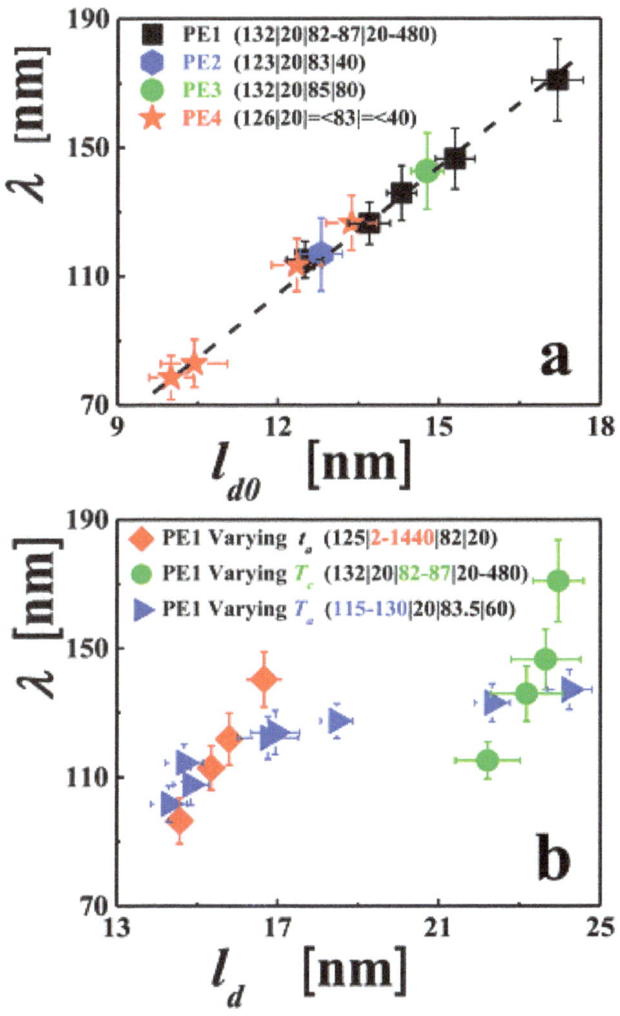

Figure 7 Correlating the annealing-induced patterns, characterized by the length-scale (λ) of the rather periodic branches, with parameters characterizing growth and annealing processes. (a) The generality of the phenomenon is demonstrated through results from four high molecular weight polyethylene samples as a function of the initial crystallization conditions characterized via the lamellar thickness (l_{d0}) of the not yet annealed crystals, measured directly after growth. For a fixed annealing time (t_a) of 20 min, the annealing temperature (T_a) was adjusted to result in well-developed branched patterns. The crystallization temperature (T_c) was varied to allow for crystals having varying values of l_{d0}. The crystallization time (t_c) was varied in order to have crystals of similar lateral size. (b) Comparison of the influence of various annealing procedures, characterized via the annealing-induced changes of lamellar thickness (l_d) and the changes of λ. All crystals were similar in lateral size, obtained by adjusting t_c. Only one parameter (either t_a, T_c or T_a, respectively) was varied for each of the three data sets. Annealing and crystallization conditions are represented as: (T_a|t_a|T_c|t_c).

CONCLUSIONS

Using solution grown polymer single crystals of polyethylene, one of the simplest and most widely studied polymers, we have been able to change the morphology of faceted crystals by annealing. Under the here applied low-temperature and long-lasting annealing conditions, we were able to unveil a rather regularly branched pattern within a polymer single crystal which we tentatively relate to variations in the degree of thermal stability. Upon annealing, the less stable regions, characterized by a lower melting temperature, were redistributed between the more stable regions. The branches, which emerged after annealing, all had similar width and were preferentially oriented towards the centre, the initial nucleation site. Like the thickness of the lamellar crystal, the observed period of the branches depended on the crystallization conditions of the starting crystals but also on the conditions of subsequent annealing. Although within experimental error no regions of different lamellar thickness could be identified within as-grown single crystals, we speculate that such regions of differing degrees of thermal stability existed. These observed annealing-induced morphological changes from a faceted to a branched pattern within polymer single crystals may provide an instructive approach for studying differences in organization and thermal stability created within polymer single crystals in the course of growth.

ACKNOWLEDGEMENTS

The authors would also like to thank Gert Strobl, Wenbing Hu, Liangbin Li, Jun Xu, and Murugappan Muthukumar for fruitful discussions. We also would like to acknowledge financial support from the Sino-German Center for Research Promotion and the German Science Foundation. The authors are grateful to the National Science Foundation of China (no. 11172272 and 11372284) and to the China Scholarship Council (CSC) for a grant to BZ and HZ. M.C.B thanks the Fonds der Chemischen Industrie for a Chemiefonds scholarship. The authors thank Lars Bolk for GPC and DSC measurements and Chemtura for the kind donation of MAO solution.

REFERENCES

1. T. Haxhimali, A. Karma, F. Gonzales and M. Rappaz, Nat. Mater., 2006, 5, 660–664 .
2. G. R. Strobl, The physics of polymers, Springer, 1997 .
3. E. Ben-Jacob and P. Garik, Nature, 1990, 343, 523–530 .
4. R. Trivedi, S. Liu and S. Williams, Nat. Mater., 2002, 1, 157–159 .

5. G. Reiter and G. R. Strobl, Progress in understanding of polymer crystallization, Springer, 2007 .
6. X. Jiang, X. Liu, Q. Liao, X. Wang, D.-D. Yan, H. Huo, L. Li and J.-J. Zhou, Soft Matter, 2014, 10, 3238–3244 .
7. A. Keller, Polymer, 1962, 3, 393–421 .
8. G. Tegze, G. I. Tóth and L. Gránásy, Phys. Rev. Lett., 2011, 106, 195502 .
9. L. Gránásy, T. Pusztai, T. Börzsönyi, J. A. Warren and J. F. Douglas, Nat. Mater., 2004, 3, 645–650 .
10. P. Ball, Nature, 2011, 480, 455 .
11. X. Liu, Y. Zhang, D. K. Goswami, J. S. Okasinski, K. Salaita, P. Sun, M. J. Bedzyk and C. A. Mirkin, Science, 2005, 307, 1763–1766 .
12. G. Reiter, G. telein and J.-U. Sommer, Phys. Rev. Lett., 2001, 86, 5918 .
13. K. Rahimi, I. Botiz, N. Stingelin, N. Kayunkid, M. Sommer, F. P. V. Koch, H. Nguyen, O. Coulembier, P. Dubois and M. Brinkmann, Angew. Chem., Int. Ed., 2012, 51, 11131–11135 .
14. M. Voigt, S. Dorsfeld, A. Volz and M. Sokolowski, Phys. Rev. Lett., 2003, 91, 026103 .
15. S. Z. D. Cheng, Nature, 2007, 448, 1006–1007 .
16. S. Organ and A. Keller, J. Polym. Sci., Part B: Polym. Phys., 1986, 24, 2319–2335 .
17. B. Wunderlich and P. Sullivan, J. Polym. Sci., 1962, 61, 195–221 .
18. P. H. Geil and D. H. Reneker, J. Polym. Sci., 1961, 51, 569–582 .
19. K. Kuwabara, H. Kaji, M. Tsuji and F. Horii, Macromolecules, 2000, 33, 7093–7100 .
20. Y. Jiang, X.-G. Jin, C. C. Han, L. Li, Y. Wang and C.-M. Chan, Langmuir, 2003, 19, 8010–8018 .
21. Y. Kikkawa, H. Abe, T. Iwata, Y. Inoue and Y. Doi, Biomacromolecules, 2001, 2, 940–945 .
22. B. Wunderlich and G. Czornyj, Macromolecules, 1977, 10, 906–913 .
23. N. Dubreuil, S. Hocquet, M. Dosière and D. A. Ivanov, Macromolecules, 2004, 37, 1–5 .
24. J. Loos and M. Tian, Polymer, 2006, 47, 5574–5581 .
25. M. Fujita and Y. Doi, Biomacromolecules, 2003, 4, 1301–1307 .
26. T. Liu, J. Petermann, C. He, Z. Liu and T.-S. Chung, Macromolecules, 2001, 34, 4305–4307 .
27. T. Liu, W. C. Tjiu and J. Petermann, J. Cryst. Growth, 2002, 243, 218–223 .
28. J. Xu, Y. Ma, W. Hu, M. Rehahn and G. Reiter, Nat. Mater., 2009, 8, 348–353 .
29. N. Basu, A. Osichow, S. Mecking and G. Reiter, Eur. Phys. J. E: Soft Matter Biol. Phys., 2012, 35, 1–12 .
30. J. Nakamura, M. Tsuji, A. Nakayama and A. Kawaguchi, Macromolecules, 2008, 41, 1358–1363 .
31. M. Wang, H. Gao, L. Zha, E.-Q. Chen and W. Hu, Macromolecules, 2012, 46, 164–171 .
32. F. Zhang, G. G. Baralia, B. Nysten and A. M. Jonas, Macromolecules, 2011, 44, 7752–7757 .
33. B. Zhang, J. Chen, M. C. Baier, S. Mecking, R. Reiter, R. Mülhaupt and G. Reiter, Macromol. Rapid Commun., 2015, 36, 181–189 .
34. Y.-X. Liu, J.-F. Li, D.-S. Zhu, E.-Q. Chen and H.-D. Zhang, Macromolecules, 2009, 42, 2886–2890 .
35. G. Reiter, Chem. Soc. Rev., 2014, 43, 2055–2065 .
36. C. N. Rochette, S. Rosenfeldt, K. Henzler, F. Polzer, M. Ballauff, Q. Tong, S. Mecking, M. Drechsler, T. Narayanan and L. Harnau, Macromolecules, 2011, 44, 4845–4851 .
37. B. Zhang, J. Chen, J. Cui, H. Zhang, F. Ji, G. Zheng, B. Heck, G. N. Reiter and C. Shen, Macromolecules, 2012, 45, 8933–8937 .
38. W. O. Statton and P. H. Geil, J. Appl. Polym. Sci., 1960, 3, 357–361 .
39. P. Dreyfus and A. Keller, J. Polym. Sci., Part B: Polym. Phys., 1970, 8, 253–258 .
40. P. H. Geil, Polymer Single Crystals, Interscience Publishers, New York, 1963 .
41. A. Peterlin, J. Polym. Sci., Part B: Polym. Phys., 1963, 1, 279–284 .
42. R. J. Roe, C. Gieniewski and R. G. Vadimsky, J. Polym. Sci., Polym. Phys. Ed., 1973, 11, 1653–1670 .
43. H. E. Bair, R. Salovey and T. W. Huseby, Polymer, 1967, 8, 9–20 .
44. R. I. Gearba, N. Dubreuil, D. V. Anokhin, Y. K. Godovsky, J.-J. Ruan, A. Thierry, B. Lotz and D. A. Ivanov, Macromolecules, 2006, 39, 978–987 .
45. X. Wang, J. Zhou, L. Li and C. M. Chan, Macromol. Rapid Commun., 2007, 28, 2001–2006 .
46. N. Sanz, J. Hobbs and M. Miles, Langmuir, 2004, 20, 5989–5997 .
47. S. Hocquet, M. Dosiere, A. Thierry, B. Lotz, M. H. J. Koch, N. Dubreuil and D. A. Ivanov, Macromolecules, 2003, 36, 8376–8384 .
48. S. Rastogi, D. R. Lippits, G. W. Peters, R. Graf, Y. Yao and H. W. Spiess, Nat. Mater., 2005, 4, 635–641 .
49. H. Li, J. Liu, D. Wang and S. Yan, Colloid Polym. Sci., 2003, 281, 973–979 .
50. U. W. Gedde and A. Mattozzi, in Long Term Properties of Polyolefins, Springer, 2004, pp. 29–74 .
51. D. C. Bassett and A. Keller, Philos. Mag., 1962, 7, 1553–1584 .
52. S.-M. Yu and S. Mecking, J. Am. Chem. Soc., 2008, 130, 13204–13205 .
53. D. Blundell, A. Keller and A. Kovacs, J. Polym. Sci., Part B: Polym. Phys., 1966, 4, 481–486 .
54. Y. Ryousho, S. Sasaki, T. Nagamura, A. Takahara and T. Kajiyama, Macromolecules, 2004, 37, 5115–5117 .
55. M. Tian and J. Loos, J. Polym. Sci., Part B: Polym. Phys., 2001, 39, 763–770 .
56. J. Nakamura and A. Kawaguchi, Macromolecules, 2004, 37, 3725–3734 .
57. Y. Wang, M. Rafailovich, J. Sokolov, D. Gersappe, T. Araki, Y. Zou, A. D. L. Kilcoyne, H. Ade, G. Marom and A. Lustiger, Phys. Rev. Lett., 2006, 96, 028303 .

58. S. J. Organ, J. K. Hobbs and M. J. Miles, Macromolecules, 2004, 37, 4562–4572 .

59. M. I. Abo el Maaty and D. C. Bassett, Polymer, 2005, 46, 8682–8688 .

60. A. Knoll, R. Magerle and G. Krausch, Macromolecules, 2001, 34, 4159–4165 .

61. G. Bar, Y. Thomann, R. Brandsch, H.-J. Cantow and M.-H. Whangbo, Langmuir, 1997, 13, 3807–3812 .

62. J. D. Hoffman and R. L. Miller, Polymer, 1997, 38, 3151–3212

CHAPTER 12 RAPID THERMAL ANNEALING AND CRYSTALLIZATION MECHANISMS STUDY OF SILICON NANOCRYSTAL IN SILICON CARBIDE MATRIX

Zhenyu Wan, Shujuan Huang, Martin A Green and Gavin Conibeer

ARC Photovoltaics Centre of Excellence, University of New South Wales (UNSW)

ABSTRACT

In this paper, a positive effect of rapid thermal annealing (RTA) technique has been researched and compared with conventional furnace annealing for Si nanocrystalline in silicon carbide (SiC) matrix system. Amorphous Si-rich SiC layer has been deposited by co-sputtering in different Si concentrations (50 to approximately 80 v%). Si nanocrystals (Si-NC) containing different grain sizes have been fabricated within the SiC matrix under two different annealing conditions: furnace annealing and RTA both at 1,100°C. HRTEM image clearly reveals both Si and SiC-NC formed in the films. Much better "degree of crystallization" of Si-NC can be achieved in RTA than furnace annealing from the research of GIXRD and Raman analysis, especially in high-Si-concentration situation. Differences from the two annealing procedures and the crystallization mechanism have been discussed based on the experimental results.

INTRODUCTION

Shockly and Queisser [1] have calculated the upper theoretical efficiency limitation for on p-n junction silicon solar cell as 30%. In order to further obtain a higher efficiency, multi-junction solar cells with different materials have been designed and fabricated [2]. However, to create different band gap solar cell layers, expensive and perhaps toxic materials have to be involved and this is assumed to be the main obstacle for the wide use of multi-junction solar cell. As a result, in recent years, the theory of "all silicon multi-junction solar cell" has been developed [3, 4], and silicon nanocrystals (Si-NCs) in various dielectric materials study have gained researchers' interests in all silicon multi-junction solar cell applications [5]. Due to quantum size effect, three-dimensional quantum-confined silicon dots have been proven to be able to tune the bandgap in a wide range by controlling the dot size. The bandgap of each cell layer can be adjusted by the wavelength of different light spectrum and all silicon multi-junction solar cells with high efficiency can be well expected.

Many research efforts have been allocated in looking for a better dielectric material as a matrix to embed the Si-NC. Comparing the band gap with different materials such as silicon dioxide (approximately 8.9 eV) and silicon nitride (approximately 4.3 eV), the band gap of silicon carbide (approximately 2.4 eV) is the lowest [5]. The small SiC bandgap increases the electron tunnelling probability. Increased carrier transportation performance and greater current can be expected from these multi-junction solar cells. Kurokawa et al. and M. Künle et al. [6, 7] have reported the fabrication of good quality Si-NC in SiC matrix film by plasma-enhanced

Citation: Zhenyu WanEmail author, Shujuan Huang, Martin A Green and Gavin Conibeer. Rapid thermal annealing and crystallization mechanisms study of silicon nanocrystal in silicon carbide matrix. Nanoscale Research Letters 2011, 6:129, DOI: 10.1186/1556-276X-6-129

chemical vapor deposition (PECVD) system. However, the main disadvantages of PECVD deposition are extremely time consuming in superlattice structure and in toxic, explosive, and expensive gases involved, such as silane (SiH_4), monomethylsilane (MMS), methane (CH_4), and hydrogen (H_2) etc. In our group, Si-NCs in a SiC matrix deposited by a sputtering process have been intensively investigated in order to overcome the disadvantages listed above.

In our previous research, Si-NCs are fabricated by post-deposition annealing of Si-rich SiC (SRC) layer in a nitrogen furnace for a long time (more than 1 h) [8, 9]. Both Si and SiC NC have been clearly observed in x-ray diffraction (XRD) and transmission electron microscopy (TEM) measurements when annealing temperature rise above 900°C. After annealing, SiC-NCs in beta phase (β-SiC) as well as amorphous Si are found surrounding the Si-NC. Rapid thermal annealing (RTA) has been considered as a primary annealing technique in semiconductor industry because of the low energy cost and better crystallization result [10, 11] In nanocrystalline system, better crystallization has also been reported in RTA because heating of the structure is caused by light directly absorbed in the layers [12]. In this paper, we compare two annealing techniques: conventional furnace annealing and RTA upon Si and SiC nanocrystalline system, and subsequently research the differences of structural characterization. By investigating the crystallization differences, we try to explain the crystallization mechanism of Si and SiC-NC.

Experimental Details

The SRC films are deposited by magnetron co-sputtering a Si and a SiC target at room temperature using a multi-target sputtering machine (AJA International, ATC-2200, North Scituate, MA, USA). Radio frequency (RF, 13.56 MHz) power supplies are connected to the targets. The Si concentration in the SRC films is controlled by adjusting the RF supply power connected to the Si target. The base pressure of the main chamber of deposition was 8.0×10^{-7} Torr and the working pressure is 2.0×10^{-3} Torr. Table 1 includes the sample details reported in this paper.

Table 1: Sample names and deposition conditions

Sample name	Silicon-rich concentration (volume percentage v%)	Sample structure/thickness (nm)
SRC80	80	Single layer/approximately 600
SRC70	70	Single layer/approximately 600
SRC60	60	Single layer/approximately 600
SRC50	50	Single layer/approximately 600
SiC	0	Single layer/approximately 600

After deposition, either furnace or RTA annealing is carried out for the purpose of Si precipitation from the matrix. The furnace annealing is processed in nitrogen (N_2) ambient at 1,100°C for 1 h with 40 min ramping-up time from 500°C to 1,100°C. The RTA annealing is also processed in N_2 ambient at 1,100°C, but with a very short ramping time of 45 s in the same temperature range and much shorter annealing time of 2 min. A detailed temperature ramping profile is listed in Table 2.

Table 2: Temperature ramping profile for conventional furnace annealing and RTA

	Room temperature, approximately 500°C	500°C to approximately 900°C	900°C to approximately 1,100°C	1,100°C
Conventional furnace annealing	N/A	25 min	15 min	60 min
RTA	15 min	30 s	15 s	2 min

The structural properties including the nanocrystal size, shape, and phase separation are studied using TEM (Phillips CM200) at 200 kV. The crystalline properties are evaluated by grazing incidence XRD using a Philips's X'Pert Pro material research diffraction system at a voltage of 45 kV and a current of 40 mA, using Cu Kα radiation (λ = 1.5418 Å). The glancing angle of the incident x-ray beam is optimised by omega scan and set between 0.2° and 0.4° The nanocrystal size is estimated using the Scherrer equation. Additional structural properties such as phase separation and crystallinity are studied by Raman spectroscopy (Renishaw, RM2000) in backscattering configuration. The power of the Ar ion laser (514 nm) was reduced below 8 mW to avoid local crystallization by laser beam.

RESULTS AND DISCUSSION

TEM study

Figures 1 and 2 show the plan view TEM images of the sample SRC50 after RTA and furnace annealing. The volume percentage of Si over SiC is 50 v% from RF sputter rates of Si and SiC are calibrated by crystal thickness monitor. Both images clearly reveal the formation of NC. The NC which is circled by solid lines with a fringe spacing 3.1 Å corresponds to Si (111) lattice plane; and the dash-line which is circled with a fringe spacing of 2.5 Å corresponds to the lattice plane of β-SiC (111) [8]. The nanocrystal size and shape are similar in both annealing conditions, with Si size 6-7 nm and SiC size 2-3.5 nm.

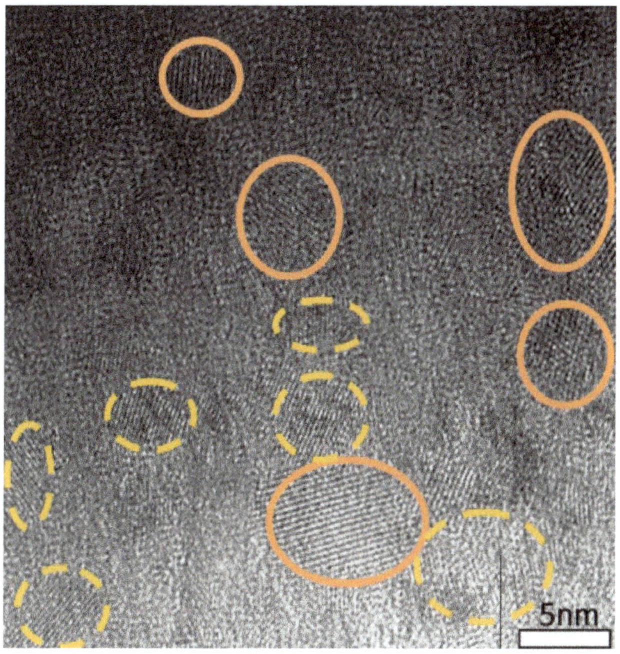

Figure 1: HRTEM plan view of image of SRC50 sample annealed by RTA.

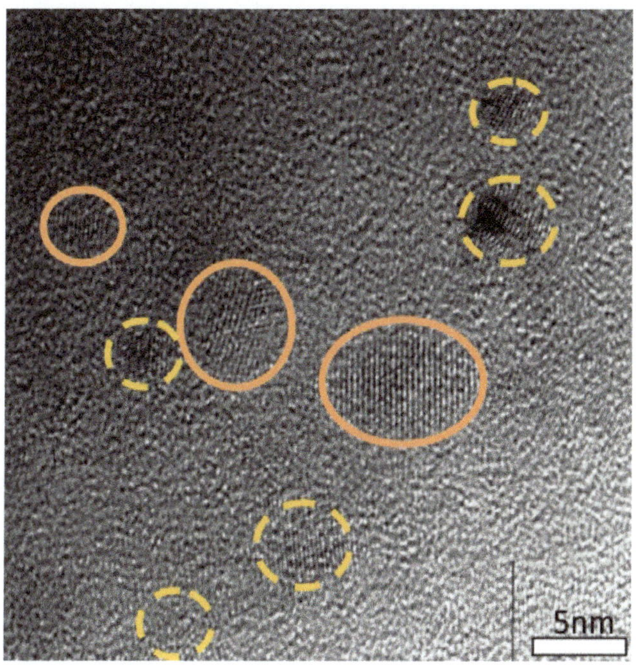

Figure 2: Cross-section TEM image of SRC50 sample annealed by furnace.

X-ray Diffraction Investigation

The crystalline properties of samples annealed by RTA and furnace are studied by XRD. Figure 3 shows a wide scan XRD curve of the sample SRC60 annealed by furnace. The Bragg peaks can be assigned to cubic Si nanocrystal as well as β-SiC nanocrystal, as shown by the indexes in the graph. This suggests the formation of both Si and β-SiC-NC which is consistent to TEM results.

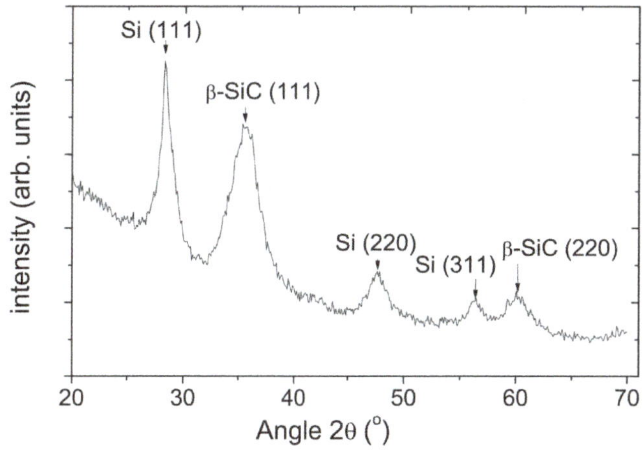

Figure 3: Wide scan XRD curve of the sample SRC60 annealed by furnace.

Figure 4 compares the XRD spectra of the samples with different Si concentrations after 1,100 C annealing. All the annealed samples show clear Bragg peaks from Si and β-SiC crystallization. In addition, the intensity of Si Bragg peak increases while the SiC peak decreases with the increasing of Si concentration. This phenomenon can be explained by more amorphous silicon (a-Si) is involved in precipitation and crystallization, as a result, higher crystallization volume of crystallized-Si can be achieved. This reason can also be used to explain SiC peaks: when Si concentration increase, SiC concentration decreases, and the volume of SiC crystallinity decreases due to less available a-SiC.

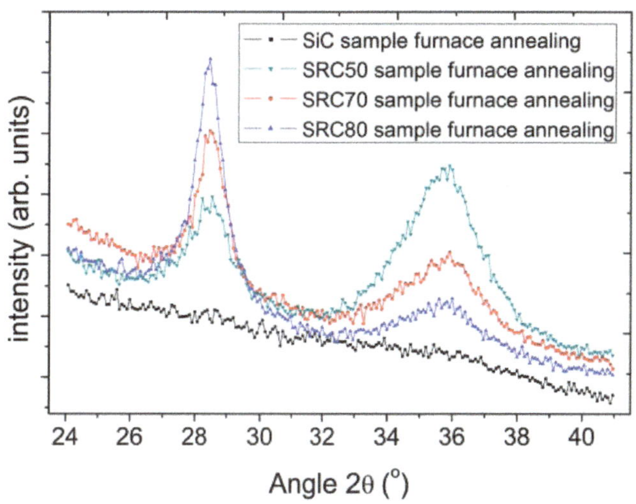

Figure 4: XRD curves of the samples with different Si concentrations after furnace annealing.

It should be noted that there is no Bragg peak of β-SiC phase detected from a sputtered stoichiometric SiC film, indicating that SiC film does not crystallize under 1,100°C annealing condition itself due to insufficient kinetic energy [13]. That both Si and SiC-NC appear in silicon-rich carbide samples could be due to the Si inducement. Some researchers reported sputtered Si starts to crystallize at 900°C [14]. Si and SiC-NC could be observed after annealing at 900°C in our previous research [8, 9]. From these results, we propose that at annealing temperatures of 900°C, the formation of Si-NC [8], act as nuclei for SiC nanocrystal growth. As a result, both Si and SiC diffraction peaks could be observed in silicon-rich carbide samples while no SiC peak observed in sputtered stoichiometric SiC film.

The full width at half maximum (FWHM) of each XRD peak were carefully measured, and the nanocrystal size was calculated by Scherr formula,

$$G=k\lambda/\Delta(2\theta)\cos\theta \qquad (1)$$

where λ is the wavelength of the X-rays, θ is the Bragg diffraction angle at the peak position in degrees, $\Delta(2\theta)$ is the FWHM in radian, and k is a correction factor. The value of k is usually chosen to be 0.9 for Si films. Nanocrystal sizes from RTA and furnace annealing samples are calculated by this formula and are indicated and compared in Figure 5.

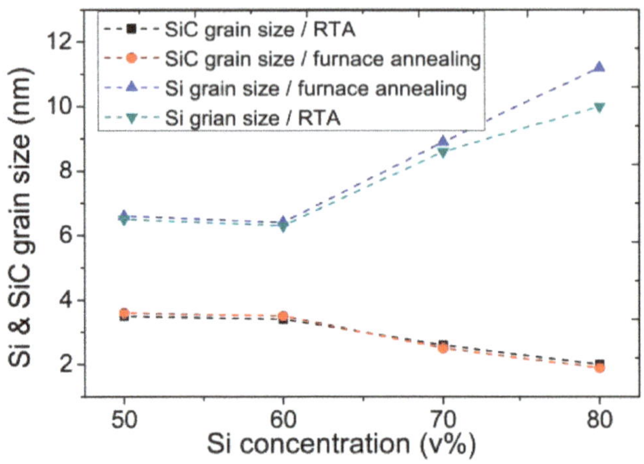

Figure 5: Si and SiC grain size from RTA and furnace annealing in different Si concentration.

In both RTA and furnace annealing samples, we can see that when Si concentration increases, Si grain size which is calculated from formula (1) also tends to increase. But the change is not significant until the Si concentration reaches 60 v% and grain size in furnace annealing samples tends to increase faster in high Si concentration (>70 v%). The same trend can also be observed in SiC-NC, the grain size of SiC crystal start to decrease when Si concentration falls below 60 v%.

The degree of Si crystallization can be estimated by the relative intensity of XRD peaks [15]. Figures 6 and 7compare the RTA and the furnace annealing samples in different concentration. The relative intensity of two Si peaks (at 28.4°) is almost the same under low Si concentration at 50 v% (Figure 6). The intensity difference changes significantly when Si concentration increased to 80 v% (Figure 7). However, the difference of SiC peak intensity barely changes in both Si concentrations.

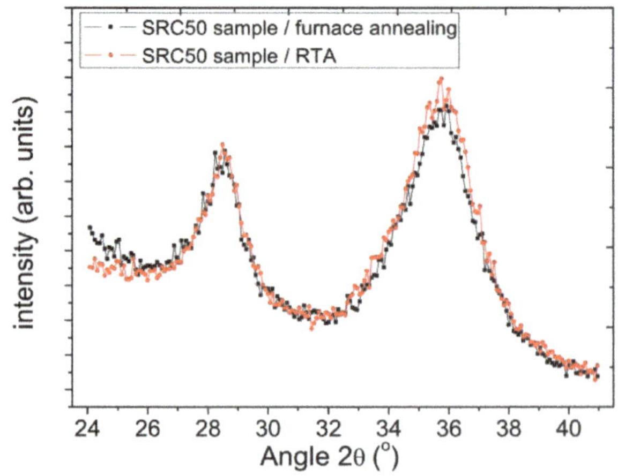

Figure 6: XRD curve comparison of SRC50 sample by RTA and furnace annealing.

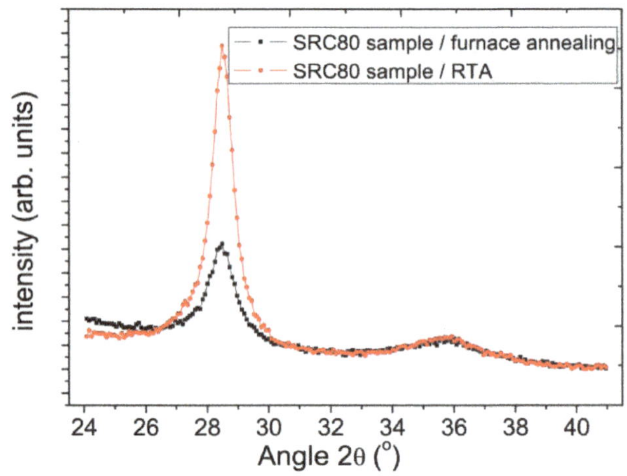

Figure 7: XRD curve comparison of SRC80 sample by RTA and furnace annealing.

We then further measure the intensity of Si peak from XRD result carefully as shown in Figure 8. Under low Si concentration range (50 and 60 v%), Si peak intensity of samples annealed by either RTA or furnace are almost the same. The intensity of RTA samples increased dramatically to two to three times higher compared to the furnace annealing samples when Si concentration increased above 60 v%.

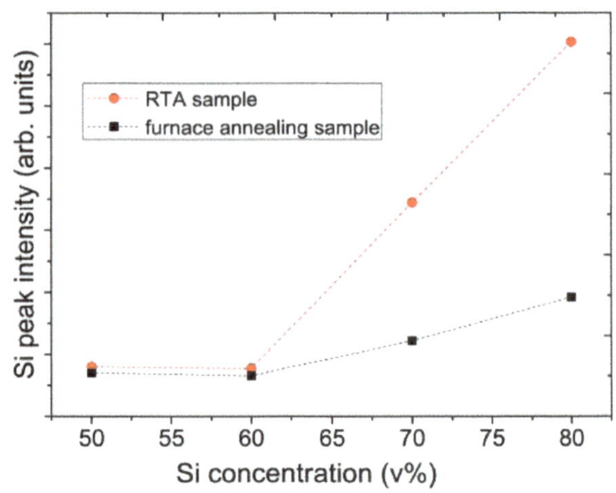

Figure 8: Si peak intensity of different Si concentration by RTA and furnace annealing.

RAMAN INVESTIGATION

Figure 9 shows Raman spectrum of furnace annealed SRC60 sample. As we can see, the peak within the range of 400 to 600 cm^{-1} can be de-convoluted to two main components: the peak centred at approximately 511 cm^{-1} corresponds to Si nanocrystal phase and the peak centred at approximately 480 cm^{-1} corresponds to the amorphous Si phase [6]. The hump at 400 cm^{-1} may be assigned as partial breakdown of Raman selection rules [16]. Meanwhile, two small SiC peaks are also observed at approximately 800 and 940 cm^{-1} attributed to the TO and LO of cubic and hexagonal SiC poly types [17, 18].

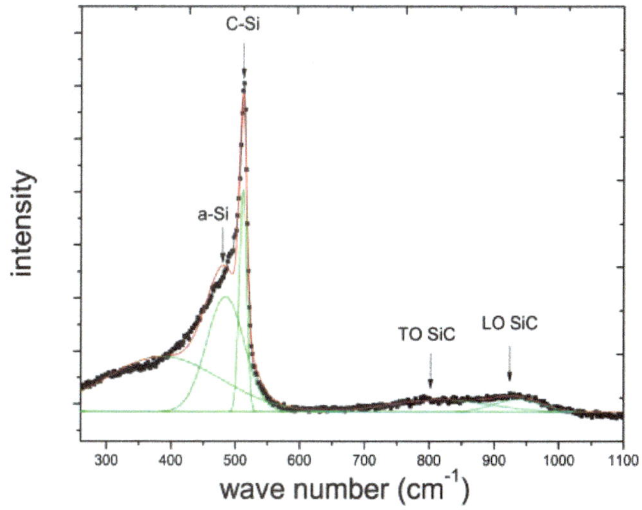

Figure 9: Raman spectrum of SRC60 after furnace annealing.

The degree of crystallization of Si nanocrystal could also be evaluated by calculating the intensity ratio of the crystalline Si peak and amorphous Si peak: I_{C-Si}/I_{a-Si}[6]. Figure 10 shows the relation of Si peak intensity ratio and silicon concentration in the SRC layers. The results indicate, for both RTA and furnace annealing conditions, when Si concentration increases, higher degree of silicon crystallization and less residual amorphous Si tend to be observed. Meanwhile, the samples from RTA show higher degree of Si crystallization in the matrix, comparing to the furnace annealing, especially in high Si concentration level.

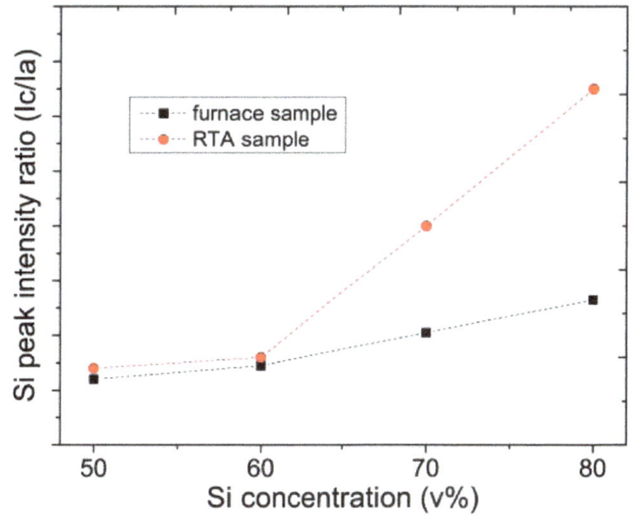

Figure 10: Calculated Si peak intensity ratios (I_{C-Si}/I_{a-Si}) in different Si concentration.

Discussion of structural difference and crystallization mechanism

RTA is considered as a positive annealing method in Si/SiC nanocrystalline system compared with furnace annealing. For the purpose of quantitative investigation, we calculate the degree of crystallization in all Si concentration range by comparing the RTA and furnace value ratio ($D_{RTA}/D_{furance}$) from the result of both XRD Si peak intensity (Figure 8) and Raman peak intensity ratio (Figure 10).

As shown in Table 3, from XRD analysis, the ratio remains at 1 when Si concentration is low (50-60 v%). The value comes to 2.4 under 70 v% Si concentrations and 2.8 under 80 v% Si concentrations. From Raman analysis, we can see the ratio stays also around 1 when in low Si concentration range (50-60 v%), and 2.2 in 70 v% Si concentration and 2.6 in 80 v% Si concentration.

Table 3: Degree of crystallization from RTA and furnace annealing in all Si concentration

	Si concentration (50 to approximately 60 v%)	**Si Concentration (70 v%)**	**Si Concentration (80 v%)**
++Degree of crystallization: $D_{RTA}/D_{furnace}$ (from XRD)	1	2.4	2.8
Degree of crystallization: $D_{RTA}/D_{furnace}$ (from Raman)	1	2.2	2.6

The Si degree of crystallization ratio behaves in a similar overall increase trend from both XRD and Raman results. It's further confirmed that better Si nanocrystal crystallization could be obtained from RTA since more Si-NC are formed and less amorphous Si remained, especially under high Si concentration.

There are two possible crystal mechanisms to explain the main structural difference coming from RTA and furnace annealing procedure as we discussed above:

Si-NC have not reached nucleation equilibrium in RTA

In classical theory of nucleation [19], free energy related to the formation of nanocrystal with radius r in an amorphous matrix can be described as:

$$\Delta G_{total}=4/3\pi r^3\Delta G_{phase}+4\pi r^2\gamma \quad (2)$$

Here, ΔG_{total} is the difference in free energy between the nanocrystal phase and the matrix phase, and γ is the interface energy, ΔG_{phase} is the difference in free energy between the nanocrystal phase and the matrix phase. For negative ΔG_{phase}, the critical nanocrystal size

$$r^*=-2\gamma\Delta G phase \quad (3)$$

When r < r*, because of the decrease of the total free energy, NC tend to reduce in size and vanish in equilibrium. On the other hand, when r > r*, the NC must grow in size to reduce the total free energy until they reach equilibrium.

In our situation, obtaining reliable γ is extremely difficult, but J. K. Bording›s group predicted the r* theoretically to be about 2 nm [20] for crystals and this value matches well with all our measured average SiC-NC size value in Figure5. Basing on this theory, we may conclude, especially in high Si concentration, Si-NC may have not reached the equilibrium before the annealing temperature (1,100°C) drops in RTA. So, Si-NC whose grain size less is than 2 nm may have not completely vanished, thus more Si-NCs would be observed. The grain size increase trend in Figure 5can further prove this point, we can see in high Si concentration region (70-80 v%) the Si grain size in RTA is smaller than furnace. This means Si-NCs in RTA could still grow up compare with samples of same Si concentration in furnace, which indicates Si-NC have not reached the equilibrium in RTA.

Less SiC-NC pre-existed during ramping-up period before Si nanocrystal grow fast at high temperature

This explanation relies on the crystallization sequence. For both annealing techniques, the peak annealing temperatures (1,100°C) are the same, however the duration of temperature raise (from 500-1,100°C) is different. For the RTA system, it takes 45 s to increase but 40 min are needed to ramp up in furnace annealing situation. We believe the time period of temperature ramping up is crucial to Si crystallization process. From the result of Si degree of crystallization, much larger quantity of Si-NC are observed in RTA, which means Si-NC can be crystallized better in short ramping time situation. It may be because of the existence of SiC-NC before Si nanocrystal fast grows. As discussed earlier, Si nanocrystal start to form around 900°C, meanwhile, SiC-NC are induced to crystallize. Short ramping-up time in RTA may lead to less SiC nanocrystal before 1,100°C. As soon as the temperature rise up to Si fast crystallization point at 1,100°C, more Si-NC could be formed in RTA due to the decrease in SiC-NC.

CONCLUSION

Si-rich SiC (SRC) layers with various Si concentrations were prepared by co-sputtering Si and SiC targets. Furnace annealing and RTA techniques were compared by studying the precipitation and crystallization of Si and SiC-NC with varying Si/SiC ratio after annealing.

Si and SiC-NC were observed by TEM in both furnace and RTA annealed at 1,100°C. SiC-NC are believed to be induced by Si nuclei from XRD spectra analysis. Meanwhile, when silicon concentration raised from 50 to 80 v%, increased size of Si nanocrystal (from 6 nm to 10 to approximately 12 nm) are observed but SiC nanocrystal size remains same (2 to approximately 4 nm).

Compared with furnace annealing, RTA samples reveal a better degree of crystallization on Si nanocrystal and less amorphous Si residual. More Si-NCs are detected by XRD and Raman analysis for this approach. This could possibly be explained by Si-NC not reaching nucleation equilibrium in the RTA or that less SiC-NC are present during the ramping-up period which increases Si-NC crystallization at high temperatures.

ACKNOWLEDGEMENTS

The authors thank other members of the Third Generation Group at the ARC Photovoltaics Centre of Excellence for their contributions to this project. This work was supported by the Australian Research Council ARC via its Centres of Excellence scheme.

REFERENCES

1. Shockley W, Queisser HJ: Detailed balance limit of efficiency of p-n junction solar cells. Journal of Applied Physics 1961, 32(3):510–519. 10.1063/1.1736034
2. King RR, Law DC, Edmondson KM, Fetzer CM, Kinsey GS, Yoon H, Sherif RA, Karam NH: 40% efficient metamorphic GaInP/GaInAs/Ge multijunction solar cells. Applied Physics Letters 2007, 90(18):183516. 10.1063/1.2734507
3. Conibeer G, Green M, Cho EC, König D, Cho YH, Fangsuwannarak T, Scardera G, Pink E, Huang Y, Puzzer T, Huang S, Song D, Flynn C, Park S, Hao X, Mansfield D: Silicon quantum dot nanostructures for tandem photovoltaic cells. Thin Solid Films 2008, 516(20):6748–6756. 10.1016/j.tsf.2007.12.096
4. Conibeer G, Green M, Corkish R, Cho Y, Cho EC, Jiang CW, Fangsuwannarak T, Pink E, Huang Y, Puzzer T, Trupke T, Richards B, Shalav A, Lin KL: Silicon nanostructures for third generation photovoltaic solar cells. Thin Solid Films 2006, 511–512: 654–662. 10.1016/j.tsf.2005.12.119
5. Jiang C, Green MA: Silicon quantum dot superlattices: Modeling of energy bands, densities of states, and mobilities for silicon tandem solar cell applications. Journal of Applied Physics 2006, 99(11):114902. 10.1063/1.2203394
6. Kurokawa Y, Miyajima S, Yamada A, Konagai M: Preparation of nanocrystalline silicon in amorphous silicon carbide matrix.Japanese Journal of Applied Physics Part 2: Letters 2006, 45: 37–41. 10.1143/JJAP.45.L1064
7. Künle M, Hartel A, Löper P, Janz S, Eibl O: Preparation Of Si-Quantumdots In Sic: Single Layer Vs Multi Layer Approach. In24th European Photovoltaic Solar Energy Conference. Hamburg, Germany; 2009.
8. Song D, Cho EC, Conibeer G, Huang Y, Flynn C, Green MA: Structural characterization of annealed Si1-x Cx/SiC multilayers targeting formation of Si nanocrystals in a SiC matrix. Journal of Applied Physics 2008, 103(8):83544. 10.1063/1.2909913
9. Song D, Cho EC, Cho YH, Conibeer G, Huang Y, Huang S, Green MA: Evolution of Si (and SiC) nanocrystal

precipitation in SiC matrix. Thin Solid Films 2008, 516(12):3824–3830. 10.1016/j.tsf.2007.06.150

10. Wang Y, Liao X, Ma Z, Yue G, Diao H, He J, Kong G, Zhao Y, Li Z, Yun F: Solid-phase crystallization and dopant activation of amorphous silicon films by pulsed rapid thermal annealing. Applied Surface Science 1998, 135(1–4):205–208. 10.1016/S0169-4332(98)00230-X
11. Szekeres A, Gartner M, Vasiliu F, Marinov M, Beshkov G: Crystallization of a-Si:H films by rapid thermal annealing. Journal of Non-Crystalline Solids 1998, 227–230(Part 2):954–957. 10.1016/S0022-3093(98)00263-4
12. Arguirov T, Mchedlidze T, Kittler M, Rolver R, Berghoff B, Forst M, Spangenberg B: Residual stress in Si nanocrystals embedded in a SiO[sub 2] matrix. Applied Physics Letters 2006, 89(5):053111. 10.1063/1.2260825
13. Schmidt H, Fotsing ER, Borchardt G, Chassagnon R, Chevalier S, Bruns M: Crystallization kinetics of amorphous SiC films: Influence of substrate. Applied Surface Science 2005, 252(5):1460–1470. 10.1016/j.apsusc.2005.02.116
14. Rüther R, Livingstone J, Dytlewski N: Large-grain polycrystalline silicon thin films obtained by low-temperature stepwise annealing of hydrogenated amorphous silicon. Thin Solid Films 1997, 310(1–2):67–74.
15. Carvalho AP, Brotas de Carvalho M, Pires J: Degree of crystallinity of dealuminated offretites determined by X-ray diffraction and by a new method based on nitrogen adsorption. Zeolites 1997, 19(5–6):382–386. 10.1016/S0144-2449(97)00101-2
16. Zi J, Büscher H, Falter C, Ludwig W, Zhang K, Xie X: Raman shifts in Si nanocrystals. Applied Physics Letters 1996, 69(2):200–202. 10.1063/1.117371
17. Kuenle M, Janz S, Eibl O, Berthold C, Presser V, Nickel KG: Thermal annealing of SiC thin films with varying stoichiometry.Materials Science and Engineering: B 2009, 159–160: 355–360. 10.1016/j.mseb.2008.10.056
18. Nakashima S, Harima H: Raman investigation of SiC polytypes. Physica Status Solidi (A) Applied Research 1997, 162(1):39–64. 10.1002/1521-396X(199707)162:1<39::AID-PSSA39>3.0.CO;2-L
19. Riabinina D, Durand C, Margot J, Chaker M, Botton GA, Rosei F: Nucleation and growth of Si nanocrystals in an amorphous Si O2 matrix. Physical Review B 2006, 74(7):075334. 10.1103/PhysRevB.74.075334
20. Bording JK, Taftø J: Molecular-dynamics simulation of growth of nanocrystals in an amorphous matrix. Physical Review B2000, 62(12):8098. 10.1103/PhysRevB.62.8098

CHAPTER 13

CRYSTALLIZATION KINETICS OF MELT-BLENDED LOW-DENSITY PHTHALIC ANHYDRIDE WITH A LOW-TG TIN-PHOSPHATE

Laila M. Al-Harbi

Chemistry Department, Faculty of Science, King Abdulaziz University, Jeddah, Saudi Arabia

ABSTRACT

Polyamide prepared by the reaction of ethylene diamine with different ratios between 1,6-glyconic acid and phthalic anhydride. The differential scanning calorimetry (DSC) is used to study the nonisothermal and isothermal crystallization of low density polyamides in phosphate glass (Pglass)-polymer hybrid blends. The % crystallinity decreased as the Pglass volume increased. The half-life ($t_{1/2}$) for crystallization decreased as the propagation rate constant rose for both the polymer matrices with increasing Pglass concentrations. The results show that the Pglass is improved nucleating agent for formation of the Pglass-polymer hybrids up to 35% Pglass and the energy necessary to break decreased. The stability of the reaction of ethylene diamine with phthalic anhydride and glyconic acid was qualitatively investigated using molecular orbital calculations.

INTRODUCTION

Polyamides form a class of industrially useful materials because of their excellent properties [1] . Glass is a popular fiber that is blended with commercial thermoplastic such as polyethylene and poly propylene [2] -[4] . Addition of fiber to polymers increases their properties such as increase their temperature resistance, and crystallinity [5] .

Recently, the scientists studied the different theoretical variables that have effect on the structure of the polycondenseation polymer, and they concluded that a difference in the reactivity of two functional groups of nonsymmetrical monomer is necessary to produce condensation polymer with an ordered structure [6] , and feedspeed of asymmetric monomer to nonsymmetrical monomer in polymerization is important factor that should be considered [7] . In this work, some polyamides were synthesised by the reaction of different ratios of ethylene diamine with 1,6-glyconic acid and phthalic anhydride. The crystallization behaviour of the material depends on kinetic and properties of hybrids.

EXPERIMENTAL

Materials and Instruments

Phthalic anhydride, 1,6-gliconic acid and ethylene diamine, all chemical used in this study are of BDH quality. Differential thermal analysis (DTA) of the samples was carried out at heating rate of 10°C/min in nitrogen as an atmosphere using a Shimadzu XD-30 Thermal Analyzer, within the temperature range 25°C - 800°C. Scanning electron micrographs (SEM) of

Citation: Al-Harbi, L. (2014) Crystallization Kinetics of Melt-Blended Low-Density Phthalic Anhydride with a Low-Tg Tin-Phosphate. Open Journal of Polymer Chemistry, 4, 38-46. doi: 10.4236/ojpchem.2014.43005.

the samples were obtained using a Hitachi S-2460N VP-SEM (Hitachi, Ibaraki, Japan).

Preparation of Polyamide

The polyamide was prepared by the melt condensation of ethylene diamine and phthalic acid with bis 1,6-glyconic acid as recommended [8] . The reaction was conducted in a slow stream of dry deoxygenated nitrogen gas to facilitate the removal of the water produced. When the reaction was completed, the polyamide was dissolved in DMF and then precipitated by adding a large amount of distilled water and stirring. The precipitate was then collected by filtration and dried in a vacuum oven at 40°C according to [9] .

Densities were measured by means of a pycnometer using hexane at 30°C. Viscosity measurements were performed with an Ostwald viscometer using a 1% solution in dioxane and DMF for polyamide resins and polyamide complexes, respectively.

The low-Tg tin-phosphate glass (Pglass), with a molar composition of 0.45 SnF_2 + 0.25 SnO + 0.30 P_2O_5, an average density of 3.65 g·cm^{-3} and a Tg of 132.4°C (determined using DSC), the polyamide was synthesized with the procedures reported in references [10] [11] .

RESULTS AND DISCUSSION

A differential scanning calorimeter was used to determine the crystallization behaviour of the samples. For nonisothermal tests, 10 mg of sample is heated at 50°C to 5 minute to eliminate any pre-existing crystals. While for isothermal crystallisation analyses, the samples were quenched to a particular crystallisation temperature (Tc) and held for an amount of time (t_c) that depended on the temperature. The information on the crystallisation kinetics was extracted from the raw experimental data. Finally, the samples were reheated above the polymer melting temperature to obtain the data to calculate the equilibrium melting temperature (Tom) as described in [12] Table 1 summarises the thermal history described above for Pglass-glyconic amide (Pglass-GA) 0.5, Pglass-GA 0.7 and Pglass-GA 1.0 hybrids (Pglass-GH). A scan of an indium metal standard at 10°C×min^{-1} was used to check the accuracy of the experiments.

Crystallization exothermic analysis for isothermal crystallization experiments cannot be performed because the matrices of the hybrids of 40% Pglass in Pglass-glyconic amide 0.5, Pglass-GA 0.7 and Pglass-GA 1.0 these values could not be discernible in crystallizationexotherms. In the nonisothermal experiment the enthalpy of fusion ΔHfo was determined by "pyris software"

$$\% \text{ Crystallinity} = \left(\Delta H_f / Q_{\text{polymer}} \Delta H_f^{\circ}\right) \times 100 \quad (1)$$

where ΔH_f is the enthalpy of polymerization and $Q_{polymer}$ is the volume fraction of the polymer (Table 1).

Non Isothermal Crystallization Behavior

The Pglass-polymer hybrids were concentrated with 0.1, 10, 20 and 40% of Pglass-polymer. It is clear from Figure 1 that the Pglass-polymer effect the crystallization of polymer GAL and it is slightly higher than that of pure polymer for $Q_{polymer}$ 25% Pglass-polymer, which is due to the small degree of interaction between the components as well as that Pglass-polymer droplets may by inducing the formation of larger polymer crystal matric.

Table 1 Non-isothermal crystallization from DSC measurements on Pglass-GAL hybrids.

(%) Pglass Width	T_m (°C)	ΔH_f(J/g)	(%) Crystallinity	Onset (°C)	(°C)
		Pglass-GAL			
0	106.60	97.13	34.44	53.15	67.21
1	108.10	93.43	33.47	53.48	68.32
10	106.64	42.02	25.01	55.85	64.24
20	107.62	36.65	22.32	58.56	63.23
25	107.54	32.05	20.51	61.75	58.62
40	106.08	22.68	16.87	67.54	46.24
		Pglass-GAH			
0	166.67	87.74	44.65	105.97	88.87
1	167.84	89.52	46.32	95.42	83.23
10	166.54	75.64	41.22	91.35	76.43
20	165.62	66.76	38.53	88.54	70.52
30	162.58	48.72	28.43	90.54	57.95

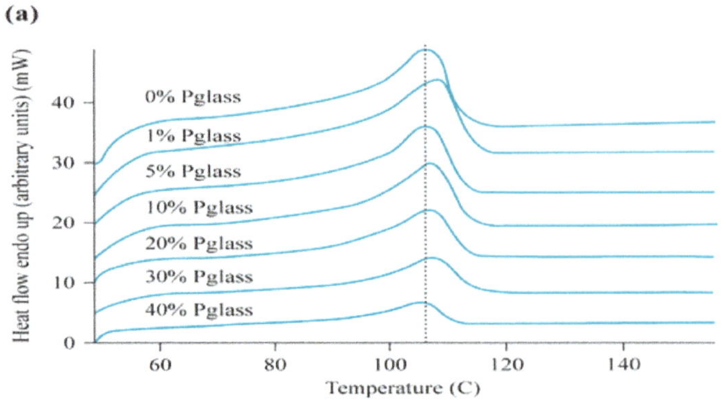

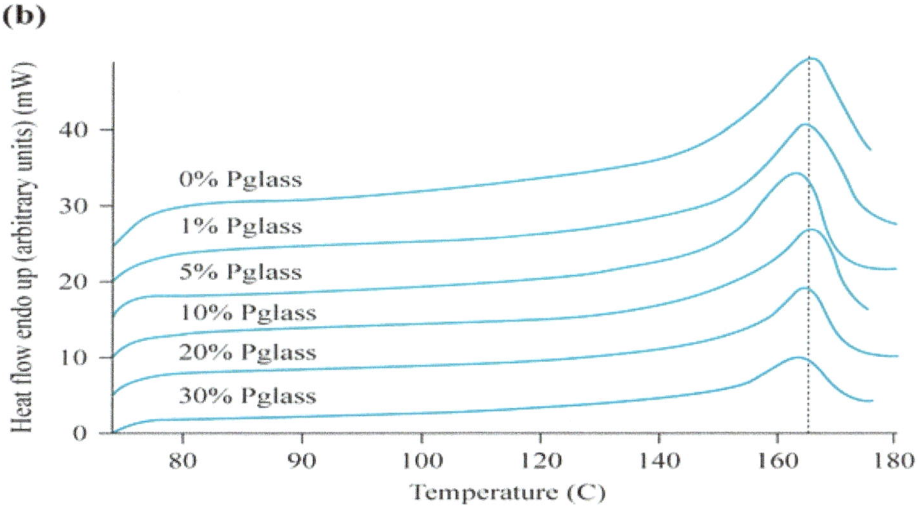

Figure 1: DSC non-isothermal thermograms of (a) Pglass-GAL and (b) PglassGAH hybrids.

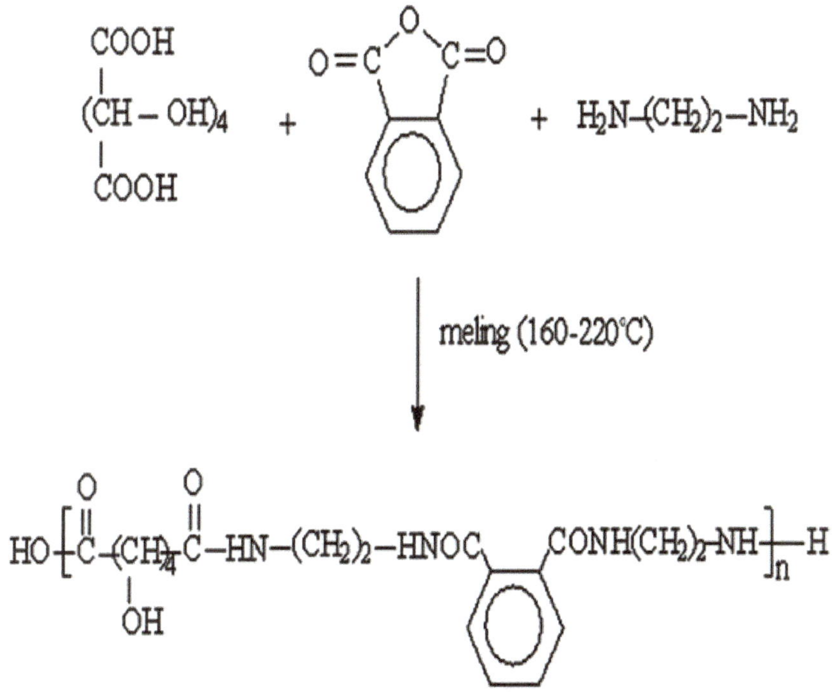

Scheme 1: The reaction mechanism of molten technique.

The addition of Pglass to matrix decrease both the heat of fusion and % crystallization because it inhibits crystallite formation beyond a particular concentration. Other authors have observed that glass fibres actually increase the crystallinity of some polymer matrices. They attributed this increase in the crystallinity to the formation of some polymer crystallites along the surface of the fiber [2] [3] . The slight increase of the melting temperature due to the presence of a large crystal results in a higher temperature necessary to melt the crystal. The width of the melt endotherm decreases by addition of Pglass which is explained as narrowing of the crystallite size distribution [13] . Comparing the thermal behavior of pure GAH and Pglass hybrid (Figure 1), there is slight decrease in melting temperature as more Pglass which is not significant (no more than 2.6) attributed to the molten Pglass promotes the growth of slightly smaller crystallites in the hybrids compared with GAH.

Figure 1 show melting point at 150°C due to the presence of β-spherulite which disappears after addition of Pglass, followed by main melting pointed at 167°C crystallinity (α and β spherulite) decrease (50% for Pglass-GAL when Q Pglass = 21%, Q Pglass = 25% Pglass-GAH) when Pglass content increased which may imply that liquid Pglass is less of a hindrance to GAH chain mobility than the solid Pglass phase is to GAL.

The glass-transition temperature of Pglass is higher than the melting temperature of the GAL and lower than that of the GAH. Hence, the GAL crystallites are produced in the presence of a solid Pglass phase, and the GAH crystallites are generated while the Pglass is in the molten state (Scheme 1).

Isothermal Crystallisation Behaviour

To describe the kinetics of the crystallization behaviour, Avrami's equation was employed to show the correct relationship between the degree of crystallisation X and t [14] .

$$X_c(t,T) = 1 - \exp\left[-(kt)n\right] \quad (2)$$

where the k and n parameters are constants: k (units of s^{-1}) is defined as the propagation rate constant of the crystal, and n (dimensionless) is a number that depends on the nucleation, geometry and control of the growth process.

X is defined as the timeand temperature-dependent ratio of the crystallized mass to the original amorphous polymer mass. Avrami treated intermediate heterogeneous nucleation cases in which the rate of nucleation decreases exponentially with time [14] . An analysis of the relative crystallinity at each temperature was performed through the division of the area of the crystallization peak (at various times) by the value of the theoretical heat of melting over the entire thermogram area, represented by the following equation:

$$X_C = \frac{\int_0^t \left(\frac{dH}{dt}\right) dt}{\int_0^m \left(\frac{dH}{dt}\right) dt} \quad (3)$$

Because this equation yields a mass fraction of crystallinity, one must convert it to a volume fraction for consistency in the use of Equation (2):

$$Xv_c = Xm_c P_a / P_c - Xm_c\left(P_c - P_a\right) \quad (4)$$

where P_a and P_c represent the amorphous and crystalline densities, respectively, and the subscripts v_c and m_c are the volume and mass fractions of crystallinity, respectively. Through application of the Avrami equation to the partially integrated heat flow changes during isothermal crystallization, the parameters k and n could be determined. The rearrangement of Equation (2) by applying a double logarithm to both sides of the equation yields a linear form with variables Ln[-ln(1-X)] and Lnt, slope n and intercept nLnk. Once the k and n parameters are known, the half-time ($t_{1/2}$), which is the time required to obtain a relative crystallinity of 50% (X = 0.5), can be found. Applying this value to Equation (2) gives the following equation:

$$t_{1/2} = (\ln 2)1/n/k \quad (5)$$

For the isothermal kinetics studies, hybrids with Pglass concentrations of 0, 1, 10, 25 and 40 vol% were tested in the temperature range of 85°C to 105°C (Pglass-GAL hybrids). In experiments performed at temperatures below 85°C, the crystallization occurred too quickly to measure it accurately, and above 100°C, the crystallisation was too slow. However, stable thermograms were obtained for the following temperatures: 92°C, 98°C and 105°C. Figure 2(a) shows the Pglass-GAL thermograms at 92°C.

The Pglass affects the crystallization kinetics of GAL crystals; the minimum position of the heat flow of the thermograms is observed to shift to the left as more Pglass is added. The time needed for the crystallization peak to develop is a strong function of temperature such that at higher temperatures, more crystallization time is required. The higher degree of thermal energy that the material possesses at higher melt crystallization temperatures inhibits the crystal growth. Once sufficient energy has dissipated during the quenching and subsequent cooling, the crystal growth increases.Table 2 shows the results of the isothermal crystallization analysis of the Pglass-GAL hybrids and lists the Avrami parameters and their dependence on the Pglass content. The Avrami exponent n of the pure GAL remained within the range of 1.64 and 2.38 for all crystallization and Pglass concentrations, suggesting instantaneous, heterogeneous, two-directional, and diffusion-controlled crystallite growth in a disc-like growth geometry. The addition of Pglass appears to reduce the value of the exponent slightly at crystallization temperatures of 92°C and 98°C and to increase it at T_c = 102°C, but overall, the exponent does not appear to change very much with Pglass concentration.

The crystallization rate constant k shows an increase in value with Pglass concentration, with more pronounced trends at higher crystallization temperatures.

Further evidence of a hastened crystallization of the GAL phase is found in the $t_{1/2}$ data [Table 2]. The halftime behaves inversely to the rate constant in that it decreases with Pglass content in a more pronounced manner at higher crystallization temperatures. Ultimately, the Pglass causes faster crystal growth as more is added to the matrix, but it does not permit the formation of new GAL crystals.

Figures 3(a) and Figures 3(b) show optical micrographs of the morphology of pure GAL and a 1% PglassGAL hybrid, respectively. There is a profound decrease in the spherulitic structure of the Pglass-GAL hybrid, which is attributed to the addition of only 1% Pglass to the GAL matrix in Figure 3(b).

The Pglass-GAH hybrids with Pglass concentrations of 0, 1, 10, 25 and 40 vol% were tested in the temperature range of 125°C to 140°C. Figure 2(b) shows an example of the Pglass-GAH thermograms at 135°C. This graph alone clearly shows that Pglass has a more profound effect on the crystallization kinetics of GAH than GAL. According to the calculated Avrami parameters in Table 2, the rate constant k and half-time of crystallization $t_{1/2}$ change with Pglass content, as was observed with the Pglass-GAL hybrids.

The key difference between the two hybrid systems studied is that this effect is suppressed as the crystallization temperature is increased for the Pglass-GAH hybrid. The main observation of the data presented above is the significant decrease in the half-time of crystallization with Pglass concentration, as noted by the shift in the minimum position of the isothermal exotherms. The marked difference in crystal growth rate between the Pglass-polymer hybrids may be attributed to the phase of the Pglass. This very large interface can be expected to provide sites for numerous covalent bonds or other compatibilisation between phases. The exponent appears to fall within the range of 3.0 - 4.0 at all three temperatures and Pglass compositions according to Table 2, signifying homogenous, three-dimensional, interface-controlled growth. Figure 4(a) and Figure 4(b) show the spherulite growth for pure GAH and for the 1% Pglass-GAH hybrid, respectively.Figure 4(b) reveals that the crystals grow at and around the Pglass interface. An enlargement of one of the Pglass droplets with GAH crystals at the interface is shown in Figure 4(c), confirming the presence of spherical crystallites.

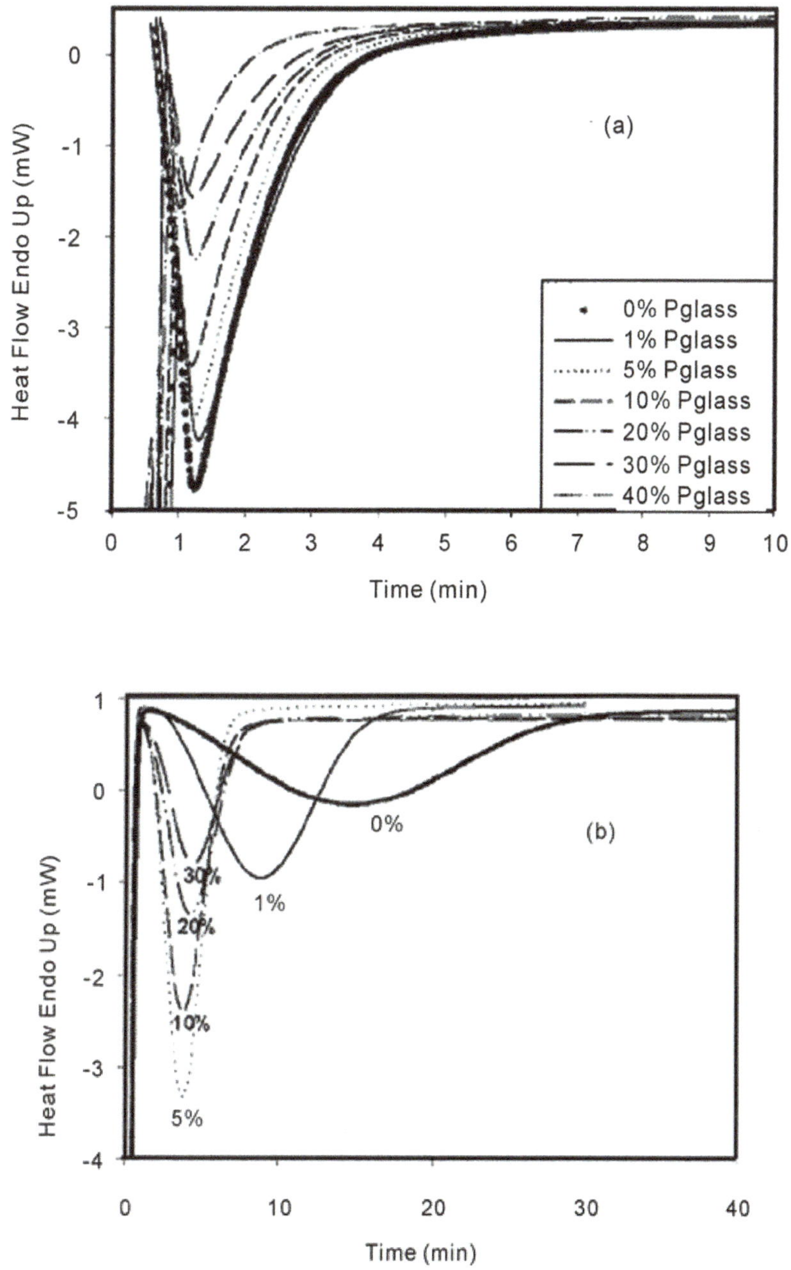

Figure 2: DSC isothermal thermograms of (a) Pglass-GAL hybrids at T_c.

Table 2: Isothermal crystallization kinetic parameters from DSC measurements on Pglass-GAL and Pglass-GAH hybrids.

Pglass (%)	$T_c = 92°C$			$T_c = 98°C$			$T_c = 102°C$		
	n	K (s^{-1})	$t_{1/2}$ (min)	n	K (s^{-1})	$t_{1/2}$ (min)	n	K (s^{-1})	$t_{1/2}$ (min)
Pglass-GAL									
0	2.32	0.435	1.92	1.90	0.175	4.35	1.62	0.072	9.56
1	3.29	0.431	1.90	1.88	0.181	4.29	1.58	0.069	10.45
10	3.35	0.448	1.85	1.79	0.177	3.86	1.65	0.102	8.63
25	2.24	0.418	1.96	1.84	0.220	3.58	2.06	0.105	7.54
40	2.16	0.456	1.86	1.44	0.193	3.84	2.126	0.142	6.65
Pglass-GAH									
0	2.42	0.425	1.82	1.80	0.165	4.25	1.64	0.082	9.56
1	2.24	0.418	1.96	1.84	0.220	3.58	2.06	0.105	7.54
10	2.32	0.435	1.92	1.90	0.175	4.35	1.62	0.072	10.56
25	3.29	0.431	1.90	1.88	0.181	4.29	1.58	0.069	10.45
40	3.35	0.448	1.85	1.79	0.177	3.86	1.65	0.102	8.63

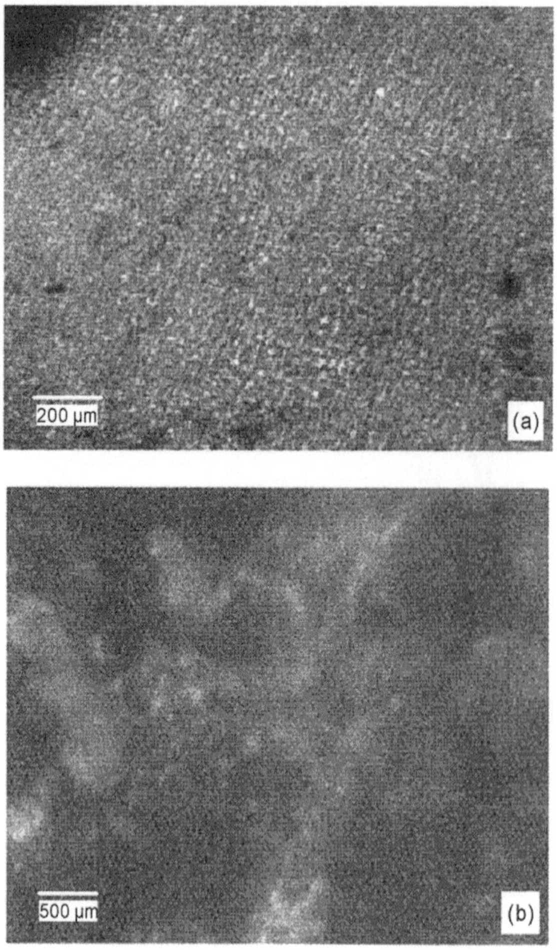

Figure 3: Optical micrographs showing crystallization of (a) pure GAL and (b) 1% Pglass in GAL= 92°C and Pglass-GAH at T_c = 102°C.

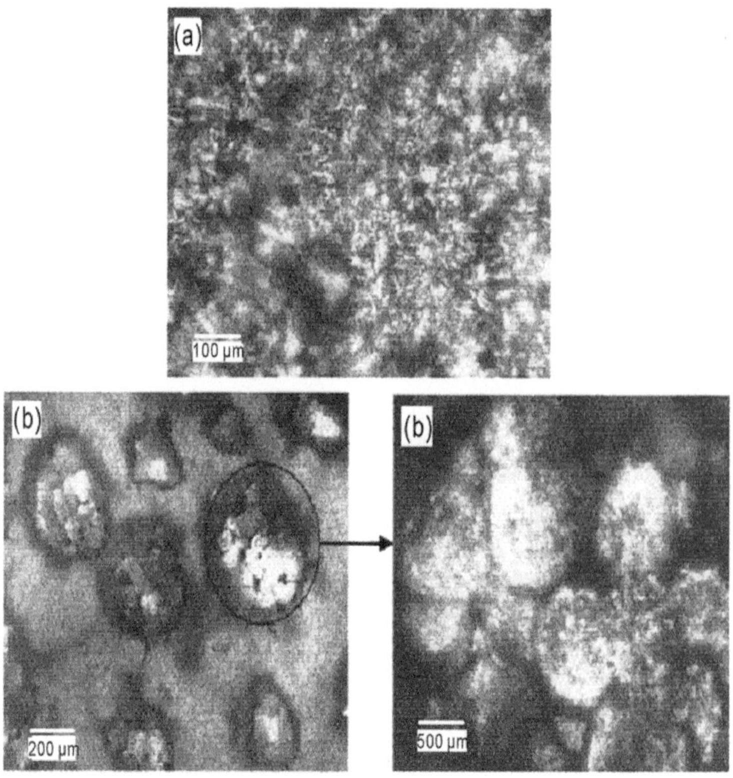

Figure 4: Optical micrographs showing crystallization of (a) pure GAH and (b) 1% Pglass in GAH. (c) represents an enlargement of a section of (b).

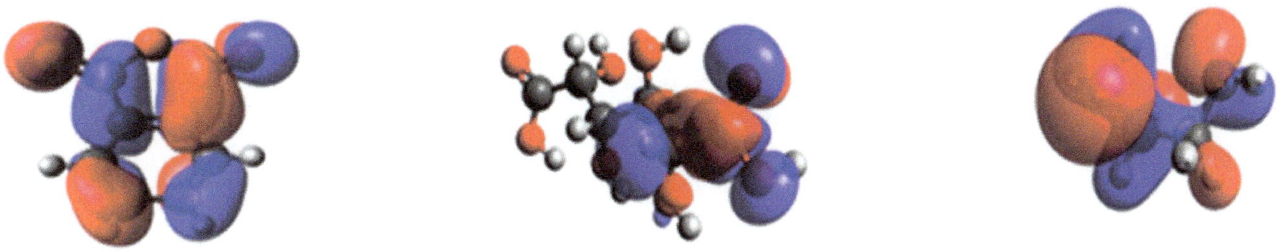

Figure 5: the LUMO and HOMO of phythalicanhydride ,Glyconic acid and Ethylene diamine.

Theoretical Calculations

The stability of the reaction of Ethylene diamine with phthalic anhydride and glyconic acid was qualitatively investigated using molecular orbital calculations. Geometry optimization for ethylene diamine, phthalic anhydride and glyconic acid were carried out using Density Functional Theory with the Backle3LYP hybrid functional, by using a basis set of 6-31G (d) [15] . Geometry optimization was followed by frequency calculations for each structure. The frequency calculations indicated no imaginary frequencies hence emphasizing that the structures are global minimum structures. Computations have been performed using the Gaussian 09 Revision D.01 program package. The optimized geometry of the ethylene diamine along with HOMO and phthalic anhydride and glyconic acid along with LUMO are depicted in the Figure 5. HOMO of ethylene diamine is the electrondonor orbital and is concentrated on the NH2 group. The LUMO of phthalic anhydride reveals electron-acceptor orbital which resides on the carbonyl group and in the case of glyconic acid, the LUMO resides on the carboxylate group. The energies of the frontier orbitals of the donor and the acceptors along with the energy corresponds to the energy gap

$$\Delta E\left(\mathrm{HOMO_{EDA}} - \mathrm{LUMO_{PA}}\right) = -3.24\ \mathrm{eV}$$

$$\Delta E\left(\mathrm{HOMO_{EDA}} - \mathrm{LUMO_{GA}}\right) = -4.88\ \mathrm{eV}$$

The shape of the frontier orbitals shown in the figure and the negative value of the energy gap is indicative of the feasibility of the chemical reaction.

CONCLUSION

The crystallization behavior of semi crystalized polymers GAL and GAH changed due to presence of Pglass which acts as nucleation agent. The optical microscopy and differential scanning calorimetry (DSC) conforms the increase of Avrami index due to the change in growth geometry from 2D to 3D spherulites. From nonisothermal crystallization, it was concluded that the solid Pglass phase induced the formation of narrow distribution of large crystal in Pglass-GAL hybrid instead of narrow distribution of small crystallization in Pglass-GAH hybrid.

REFERENCES

1. Allen, G. and Bevinton, J.C. (Eds.) (1989) Comprehensive Polymer Science. Pergamon Press, Oxford, Vol. 5, 97.
2. Avalos, F., Lopez-Manchado, M.A. and Arroyo, M. (1998) Crystallization Kinetics of Polypropylene III. Ternary Composites Based on Polypropylene/ Low Density Polyethylene Blend Matrices and Short Glass Fibres. Polymer, 39, 6173-6178.http://dx.doi.org/10.1016/S0032-3861(98)00098-6
3. Gupta, A.K., Gupta, V.B., Peters, R.H., Harland, W.G., Berry, J.P. (1982) The Effect of Addition of High-Density Polyethylene on the Crystallization and Mechanical Properties of Polypropylene and Glass-Fiber-Reinforced Polypropylene. Journal of Applied Polymer Science, 27, 4669-4686. http://dx.doi.org/10.1002/app.1982.070271214
4. Tai, H.-J., Chiu, W.-Y., Chen, L.-W. and Chu, L.-H. (1991) Study on the Crystallization Kinetics of PP/GF Composites. Journal of Applied Polymer Science, 42, 3111-3122.http://dx.doi.org/10.1002/app.1991.070421205
5. Guschl, P.C. and Otaigbe, J.U. (2002) An Experimental Study of Morphology and Rheology of Ternary PglassPS-LDPE Hybrids. Polymer Engineering & Science, 43, 1180.http://dx.doi.org/10.1002/pen.10100
6. Khalil, A.A. (2011) Synthesis, Exchange Reactions, and Biological Activity of Poly(8-Methacryloxyquinoline). Journal of Applied Polymer Science, 121, 1160-1165. http://dx.doi.org/10.1002/app.33624
7. Gentle, F.T. and Suter, U.W. (1989) Comprehensive Polymer Science. Pergamon Press, Oxford.
8. Rothon, R.N. (1995) Particulate-Filled Polymer Composites. Longman, Harlow, UK.
9. Denault, J. and Vu-Khanh, T. (1992) Crystallization Behavior and Effect of Thermal Treatment on Fracture Performance of Glass Flake Reinforced Polypropylene. Polymer Composites, 13, 372-379. http://dx.doi.org/10.1002/pc.750130506
10. Frank, J.W. (1965) The Analytical Use of Ethylenediaminetetracetic Acid. Van Nostrand, New York.
11. Badr, S.K. (2005) Synthesis and Characterization of Polyamide(1,6-Gliconic Acid) with Some Metal Ion Complexes. Oriental Journal of Chemistry, 21, 55-60.
12. Guschl, P.C. and Otaigbe, J.U. (2003) Crystallization Kinetics of Low-Density Polyethylene and Polypropylene MeltBlended with a Low-Tg Tin-Based

Phosphate Glass. Journal of Applied Polymer Science, 90, 3445-3456. http://dx.doi.org/10.1002/app.13084

13. El-Mossalamy, E.H. and Khalil, A.A. (1996) Synthesis and Investigation of N-Methylacryloyloxytetrabromophthalimide Polymer Complexes with Some Transition Metal Chlorides. European Polymer Journal, 32, 397-400. http://dx.doi.org/10.1016/0014-3057(95)00197-2

14. Adalja, S.B., Otaigbe, J.U. and Thalacker, J. (2001) Glass-Polymer Melt Hybrids. I: Viscoelastic Properties of Novel Affordable Organic-Inorganic Polymer hybrids. Polymer Engineering & Science, 41, 1055-1067. http://dx.doi.org/10.1002/pen.10806

15. Frisch, M.J., Trucks, G.W., Schlegel, H.B., Scuseria, G.E., Robb, M.A., Cheeseman, J.R., et al. (2003) Gaussian 03, Revision B.04. Gaussian, Inc., Pittsburgh.

CHAPTER 14

MICROPHASE SEPARATION INDUCED IN THE MELT OF PLURONIC COPOLYMERS BY BLENDING WITH A HYDROGEN BONDING UREA–URETHANE END-CAPPED SUPRAMOLECULAR POLYMER

Daniel Hermida-Merino [ab], Gemma E. Newby [c], Ian W. Hamley [d], Wayne Hayes [d] and Andrew Slark [e]

[a]BM26/DUBBLE, CRG, ESRF-The European Synchrotron, CS40220, 38043 Grenoble Cedex 9, France

[b]Dutch Polymer Institute DPI, P.O. Box 902, Eindhoven 5600 AX, The Netherlands

[c]ID09b, ESRF-The European Synchrotron, CS40220, 38043 Grenoble Cedex 9, France

[d]Department of Chemistry, University of Reading, Whiteknights, Reading, RG6 6AD, UK.

[e]Henkel UK Limited, Wexham Road, Slough, SL2 5DS, UK

ABSTRACT

Blending with a hydrogen-bonding supramolecular polymer is shown to be a successful novel strategy to induce microphase-separation in the melt of a Pluronic polyether block copolymer. The supramolecular polymer is a polybutadiene derivative with urea–urethane end caps. Microphase separation is analysed using small-angle X-ray scattering and its influence on the macroscopic rheological properties is analysed. FTIR spectroscopy provides a detailed picture of the inter-molecular interactions between the polymer chains that induces conformational changes leading to microphase separation.

INTRODUCTION

Block copolymers are a fascinating class of macromolecules which have the ability to self-assemble into morphologies with ordered microphase-separated structures in solution,[1–4] and in the melt[2,5] and solid[1,2] state. The ordering of these structures can be controlled with remarkable precision and as such these materials have received considerable interest in recent years[6–8] as a result of their remarkable mechanical and structural properties.[1,9–11]

Blending of polymers provides a further versatile tool to tune morphology and properties. Triblock copolymers have been blended with homopolymers that selectively hydrogen bond with one of the blocks.[12,13] Both of the individual polymers were disordered but upon mixing, a disorder-to-order transition was observed, creating ordered microphase – separated melt structures including cubic and lamellar morphologies. Order to order transitions have also been reported, for example from an FCC to a lamellar phase which is due to an increase in the effective Flory–Huggins interaction parameter between the components.[14]

Here we show that a strategy employing non-covalent interactions via blending with a hydrogen-bonding supramolecular polymer can be used to produce microphase separation in the melt of a well-known Pluronic type polyether block copolymer. This is interesting because the interaction is mainly driven by the urea–urethane functionalities in the polymer end groups, in contrast to existing studies on Pluronics interacting with homogeneous H-bonding polymer chains such as polyacrylic acid.[12] In other words, there are only a few H-bonding interactions per supramolecular polymer chain. We used as a model amphiphilic copolymer Pluronic P123[15] which was blended in the melt with the urea–urethane end-capped supramolecular polymer, PB-UR-OH[16] (see Scheme 1) to

Citation: Daniel Hermida-Merino, Gemma E. Newby, Ian W. Hamley, Wayne Hayes and Andrew Slark. Microphase separation induced in the melt of Pluronic copolymers by blending with a hydrogen bonding urea–urethane end-capped supramolecular polymer, Soft Matter, 2015, 11, 5799-5803, DOI: 10.1039/C5SM01461A.

investigate whether it was possible to create microphase-separated morphologies with novel structural and mechanical properties. Supramolecular end-capped bisurea and urea–urethane polymers possess the capability of forming hydrogen bonds by the association of the end-groups and bisurea moieties of the supramolecular segmented supramolecular polymers.[16–23] Pluronics are a series of amphiphilic polyether block copolymers that continue to be exploited in many applications.[3,7,24]

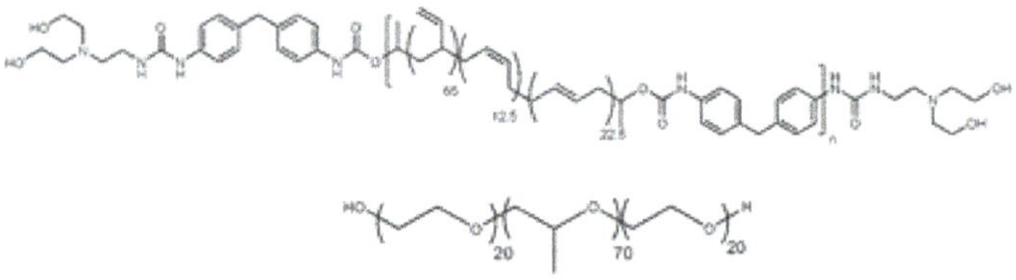

Scheme 1: Molecular structures of PB-UR-OH supramolecular polymer and P123.

To gain an insight into the structure and properties of the polymer blends, rheology, small – angle X-ray scattering (SAXS) and Fourier transform infrared (FTIR) spectroscopy were performed. The rheology and SAXS results can be coupled to provide a link between the macroscopic properties and the microstructure.[4] FTIR provides detailed information at the molecular scale on the nature and strength of the association through hydrogen bonding interactions responsible for the microphase separation generated by the polymer blending. The polymer blends were investigated at three distinct temperatures (25, 45 and 75 °C) to probe the thermal stability of microphase-separated structures.

We first report on the effect of blending on the macroscopic mechanical properties. The frequency dependence of the dynamic shear moduli was measured for blends with different compositions at three temperatures. Fig. 1 shows representative data for a 70/30 (P123/PB_UR_OH) blend. The modulus was found to be $G' > 10^3$ Pa at low frequency and largely frequency independent at 25 °C and 45 °C and also $G' > G''$ at low frequency. These features are consistent with microphase separation. Melting behavior was observed at 75 °C since the slopes of the power-law frequency (ω) dependence of $G' \sim \omega^p$ ($p = 1.82 \pm 0.04$) and $G' \sim \omega^q$ ($q = 0.98 \pm 0.01$) are close to those expected for a disordered block copolymer for which $p = 2$ and $q = 1$.[1]

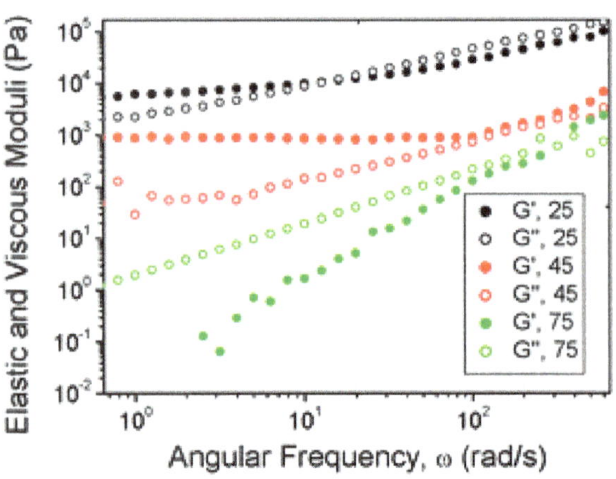

Figure 1 Frequency dependence of the dynamic shear moduli of a 70/30 P123/PB-UR-OH polymer blend at three temperatures. Open symbols represent the viscous modulus and the solid symbols the elastic modulus.

The crossover point of the elastic and viscous moduli appear at a characteristic angular frequency from which the relaxation time of the system may be determined by simply taking the reciprocal of the frequency. The values are presented in ESI,† Table S1 and plotted in Fig. 2 and Fig. S1 (ESI†). Interestingly, the crossover point follows a non-monotonic dependence with respect to the PB-UR-OH content. Both the modulus (G′) and the relaxation time of the system at the crossover point show a sharp increase above 70% P123.

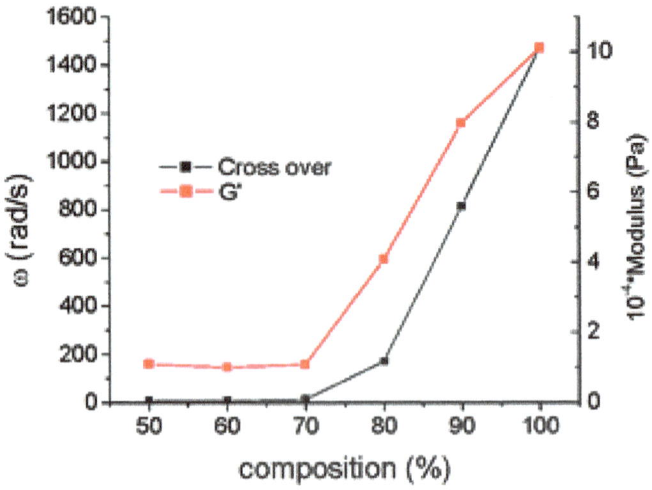

Figure 2 Angular frequency and modulus at the crossover point of the elastic and viscous moduli for blends with varying composition (shown as % P123) at 25 °C.

This is correlated to the extent of microphase separation as shown through SAXS, to be discussed shortly. The values in ESI,† Table S1 show that generally the crossover points appear at higher frequencies at 45 and 75 °C than for the systems at 25 °C suggesting that as the temperature is increased the relaxation time shortens.

SAXS is a powerful tool to examine microphase separation in block copolymers in situ[2] and was used to probe the ordering within the P123/PB-UR-OH blends. Representative SAXS profiles for a series of blends at 25 °C as well as the parent polymers are shown in Fig. 3. These results show no observed peak for P123 consistent with a lack of microphase separation, however, addition of PB-UR-OH leads to a peak at $q^* = 0.08$ Å^{-1}, indicating microphase separation. The corresponding domain spacing is d = 79 Å. The SAXS peak for the blends is significantly shifted from that for PB-UR-OH, discussed in detail in our previous paper[16] (ca. q = 0.089 Å^{-1}, i.e. d = 72 Å). The non-monotonic intensity increase of the SAXS reflection with increasing PB-UR-OH content as well as a slight shift to lower q seems to exclude the possibility of a simple dilution effect of the supramolecular morphology upon addition of the P123 (see Fig. 3), confirming the mechanical properties trend observed previously (see Fig. 2). Moreover, for a given blend composition, the intensity of the SAXS peak decreases upon increasing temperature (Fig. 3 and Fig. S2, ESI†), indicating a reduction in microphase separation.

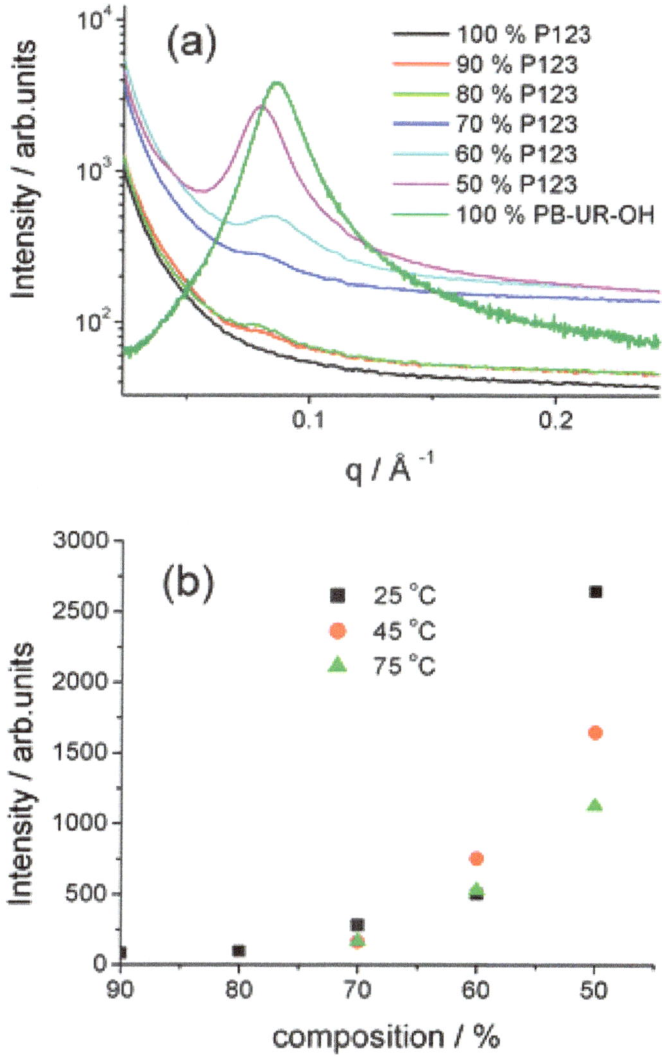

Figure 3 (a) SAXS profiles for the polymers and blends 25 °C (see ESI† for data at 45 °C and 75 °C) (b) SAXS intensity against composition (P123 content) at three distinctive temperatures (25 °C: black, 45 °C: red, 75 °C: green).

It is seen that when a peak is present the position does not change significantly with temperature, the spacing being between 74–79 Å for the entire range of blends and temperatures (see ESI,† Fig. S2 and Table S2). It is noted that some of the systems only have a very weak and broad peak not classed as a lattice spacing. However, the presence of even a weak correlation peak does indicate that phase mixing in the blend is present to a small extent.

To understand the morphology of the blends, the SAXS results have been analysed by fitting the 100 wt% P123 and 50/50 wt% P123 PB-UR-OH SAXS profiles (see ESI,† Fig. S3). Both form and structure factors are required to define the overall structure (see ESI† for further details about the fitting parameters). A standard Gaussian polymer form factor with polydispersity in radius of gyration R_g of 3 to 5 nm was employed in line with the usual polydispersity of ±3% of the P123 polymer chains. In addition, a mass fractal structure factor was used to model the aggregate morphology. Mass fractals are employed to define clustered objects. A fractal dimension is usually observed between 1 and 3, giving an indication of the extent of branching within the system. The fractal dimension obtained from our fits was D = 2.8, signifying a highly branched globular structure. The model used to fit the 50/50 wt% P123 PB-UR-OH SAXS profile includes both the contribution from pure P123 and a second contribution (Lorentzian distribution) to model the PB-UR-OH scattering within the blend (see ESI,† Fig. S4). A decrease in polydispersity of the form factor of the Gaussian polymer (describing P123) due to the monodisperse nature of the added PB-UR-OH polymer is noted. The mass fractal from the structure factor increases slightly, indicating a more branched structure similar to that obtained in analogous blend systems. A Lorentzian structure factor was required to fit the correlation peak in the scattering profile.

FTIR spectroscopy has been employed in order to probe non-covalent interactions associated with the structural organization upon blending. FTIR spectroscopic analysis has been employed extensively to examine the binding of hard and/or soft segments in segmented polyurethanes and polyureas but also the blending with polyethers or polysiloxanes.[25] The analysis of the FTIR absorption bands permits to distinguish among the different interactions related to the functional groups (urea, urethane, hydroxyl and ether groups,[25–28]) featured by the polymers that are involved in the hydrogen bonds. The FTIR spectra of P123/PB-UR-OH polymer blends contain details of concomitant and competitive intra- and inter-polymeric interactions, depending on the overlapping of different vibrational modes (see Fig. 4). The addition of the PB-UR-OH to P123 will promote the interaction of the polar segments of both polymers. In particular, the ethylene oxide (EO) backbone groups of P123 will form hydrogen bonds to the end groups of the PB-UR-OH whilst interrupting the self-assembly of the hard segments of the PB-UR-OH. The interactions between hard segments in the blend will be assessed by a thorough analysis of the absorption bands (see ESI,† Table S3) related to the urea–EO hydrogen bonds to confirm the phase mixing of the blend component.

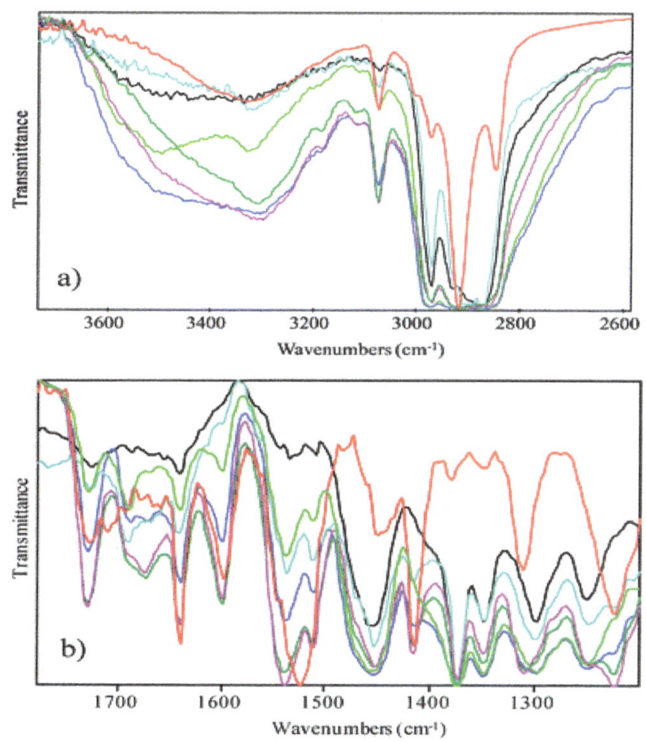

Figure 4 FTIR spectra in two regions at 25 °C: black P123, green 90%, blue 80%, cyan 70%, olive 60%, magenta 50% and red PB-UR-OH.

The absorption band centered at 3400–3000 cm^{-1} of the P123 FTIR spectrum is very broad and diminishes rapidly with increasing temperature, suggesting weak hydrogen bonds involving the hydroxyl groups. As PB-UR-OH was added to P123 the spectra of the polymer blends firstly shows the appearance of the free hydrogen bonding NH stretching absorption band (free N–H 3445–3495 cm^{-1})[25] which subsequently decreases upon further PB-UR-OH addition (from 30% of PB-UR-OH). In addition, a concomitant sharpening of the signal centered at 3300 cm^{-1} characteristic of N–H···C =O hydrogen bonds[29,30] together with a small shift to lower frequencies suggests the interaction of hard segments of the PB-UR-OH polymer with the ether linkage of the PEO groups at higher concentrations of PB-UR-OH (30%). Temperature increase leads to the disruption of the hydrogen bonds in the low and mid concentration regime (reappearance of the free NH signal at 3490 cm^{-1}). The spectral region associated to the carbonyl stretching (amide I) reveals both free urethane and urea stretching band (C=O (urethane) free 1730–1740; C=O (urea) free 1690–1700) up to 30% of the PB-UR-OH whilst the disordered and ordered hydrogen bonded urea is the predominant absorption band at higher PB-UR-OH compositions (40 and 50%, respectively). With increasing temperature, the spectra show similar enhancements of both the free urethane and urea stretching bands. Importantly, the higher content of the free-hydrogen bonded carbonyl urea stretching band when compared to the free-hydrogen bonded N–H stretching band in PB-UR-OH supports our analysis regarding the competition between the interaction of the NH with the carbonyl and the ether oxygen of P123.[29,30] Furthermore, the higher content of the disordered hydrogen bonded urea stretching band in the spectra of the P123–PB-UR-OH blends relative to the PB-UR-OH adsorption band confirms that the strong interaction of the oxygen ether serves to destabilize the urea packing.[25,29,30]

Interestingly, the Amide II absorption band (N–H···C =O 1540–1560 cm^{-1}) shows a clear shift from 1520 to 1540 cm^{-1} related to the reduction in hydrogen bonding as temperature is increased. Variations of the methyl umbrella deformation absorption band (1374 cm^{-1}) are indicative of intermolecular interaction. The general sharpening of the absorption band suggest the inhibition of the rotational motion of the PO chains. In addition, the disappearance of the trans conformation of the out-of-phase bending vibration of methylene group at 1284 cm^{-1}upon PB-UR-OH addition as well as the coalescence with the absorbance band of the gauche conformation at 1300 cm^{-1}, confirms the higher mobility of the PEO groups of the P123 interacting with the hard segment of the PB-UR-OH.[26,27] However, the coexistence of absorption bands that correspond to different block interactions in the complex spectra of the polymer blends indicates that there is not strong phase separation between the phases and a diffuse interface between the obtained microphase separation exists.

CONCLUSIONS

In summary, we have shown that blending with a supramolecular polymer capable of hydrogen bonding interactions is a novel method to induce microphase separation in Pluronic-type polyether block copolymers. This also provides a simple means to tune the mechanical properties of the blends. The microphase separated structures are observed up to higher temperatures (50 °C). The resulting microphase separated structures have small domain spacings, around 8 nm, although no evidence for defined (e.g. lamellar or hexagonal cylindrical) ordering was noted. Microphase separated block copolymer structures with small domains spacings have recently attracted attention for thin film applications in electronics, specifically in the creation of block copolymer mask structures for high density lithographic patterning of semiconductors. Blending of inexpensive polymers such as Pluronics with supramolecular polymers to induce microphase separation is a valuable new addition to the range of methods to induce microphase separated structures with small domain spacings.[31] FTIR spectroscopy provides a detailed molecular-level interpretation of the intermolecular interactions responsible for the ordering within the blends. In particular, the FTIR spectra show hydrogen bonding interactions between the EO groups of the P123 and the hard segment of the PB-UR-OH that decrease with temperature.

ACKNOWLEDGEMENTS

The development of the supramolecular polymer at Reading was supported by Henkel UK, who sponsored the PhD studentship of DH-M. The PhD research of GEN was supported by the STFC Centre for Molecular Structure and Design. We are grateful to the EPSRC (UK) for Platform grant funding. IWH is the recipient of a Royal Society-Wolfson Research Merit Award

REFERENCES

1. I. W. Hamley, The Physics of Block Copolymers, Oxford University Press, Oxford, 1998.
2. I. W. Hamley and V. Castelletto, Prog. Polym. Sci., 2004, 29, 909–948 .
3. I. W. Hamley, Block Copolymers in Solution, Wiley, Chichester, 2005.
4. G. E. Newby, I. W. Hamley, S. M. King, C. M. Martin and N. J. Terrill, J. Colloid Interface Sci., 2009, 329(1), 54–61 .
5. J. M. Dealy and R. G. Larson, Structure and Rheology of Molten Polymers: From Structure to Flow Behavior and Back Again, Hanser Gardner Publications, 2006 .
6. I. W. Hamley, Nanotechnology, 2003, 14, R39–R54 .
7. P. Alexandridis and T. Hatton, Colloids Surf., A, 1995, 96, 1–46 .
8. W. K. Lee, S. H. Cha, K. H. Kim, B. W. Kim and J. C. Lee, J. Solid State Chem., 2009, 182, 3243–3248 .
9. F. Artzner, S. Geiger, A. Olivier, C. Allais, S. Finet and F. Agnely, Langmuir, 2007, 23(9), 5085–5092 .
10. R. K. Prud'homme, G. Wu and D. K. Schneider, Langmuir, 1996, 12(20), 4651–4659 .
11. K. Nakashima and P. Bahadur, Adv. Colloid Interface Sci., 2006, 123–126, 75–96 .
12. V. R. Tirumala, A. Romang, S. Agarwal, E. K. Lin and J. J. Watkins, Adv. Mater., 2008, 20, 1603–1608 .
13. A. Dehsorkhi, V. Castelletto, I. W. Hamley and P. J. F. Harris, Soft Matter, 2011, 7, 10116–10121 .
14. C. I. Huang, H. Y. Hsueh, Y. K. Lan and Y. C. Lin, Macromol. Theory Simul., 2007, 16, 77–85 .

15. C. Chaibundit, N. M. P. S. Ricardo, F. M. L. L. Costa, S. G. Yeates and C. Booth, Langmuir, 2007, 23(18), 9229–9236 .
16. D. Hermida Merino, A. T. Slark, H. M. Colquhoun, W. Hayes and I. W. Hamley, Polym. Chem., 2010, 1, 1263–1271 .
17. P. Woodward, A. Clarke, B. W. Greenland, D. H. Merino, L. Yates, A. T. Slark, J. F. Miravet and W. Hayes, Soft Matter, 2009, 5, 2000–2010 .
18. P. Woodward, D. H. Merino, I. W. Hamley, A. T. Slark and W. Hayes, Aust. J. Chem., 2009, 62, 790–793 .
19. P. J. Woodward, D. H. Merino, B. W. Greenland, I. W. Hamley, Z. Light, A. T. Slark and W. Hayes, Macromolecules, 2010, 43, 2512–2517 .
20. O. Colombani, C. Barioz, L. Bouteiller, C. Chanéac, L. Fompérie, F. Lortie and H. Montès, Macromolecules, 2005, 38, 1752–1759 .
21. R. M. Versteegen, R. P. Sijbesma and E. W. Meijer, Macromolecules, 2005, 38, 3176–3184 .
22. A. Gooch, A. M. McGhee, L. C. Renton, J. P. Plante, C. I. Lindsay and A. J. Wilson, Supramol. Chem., 2009, 21, 12–17 .
23. A. Gooch, C. Nedolisa, K. A. Houton, C. I. Lindsay, A. Saiani and A. J. Wilson, Macromolecules, 2012, 45, 4723–4729 .
24. Amphiphilic Block Copolymers, Self-Assembly and Applications, ed. P. Alexandridis and B. Lindman, Elsevier, Amsterdam, 2000.
25. E. Yilgör, E. Burgaz, E. Yurtsever and I. Yilgör, Polymer, 2000, 41, 849–857 .
26. Y. L. Su, J. Wang and H. Z. Liu, J. Phys. Chem. B, 2002, 106, 11823–11828
27. C. Guo, H. Z. Liu and J. Y. Chen, Colloid Polym. Sci., 1999, 277, 376–381 .
28. M. C. Delpech and G. S. Miranda, Cent. Eur. J. Eng., 2012, 2, 231–238 .
29. M. M. Coleman, D. J. Skrovanek, J. Hu and P. C. Painter, Macromolecules, 1988, 21, 59–65
30. A. Marcos-Fernandez, A. E. Lozano, L. Gonzalez and A. Rodriguez, Macromolecules, 1997, 30, 3584–3592
31. I. W. Hamley, Prog. Polym. Sci., 2009, 34, 1161–1210

CHAPTER 14

SMART POLYMER INVERSE-OPAL PHOTONIC CRYSTAL FILMS BY MELT-SHEAR ORGANIZATION FOR HYBRID CORE–SHELL ARCHITECTURES

C. G. Schäfer [a], T. Winter [a], S. Heidt [a], C. Dietz [b], T. Ding [c], J. J. Baumberg [c] and M. Gallei [a]

[a]Ernst-Berl Institut für Technische und Makromolekulare Chemie, Technische Universität Darmstadt, Alarich-Weiss-Str. 4, D-64287 Darmstadt, Germany.

[b]Center of Smart Interfaces and Department of Materials Sciences, Technische Universität Darmstadt, Alarich-Weiss-Str. 10, 64287 Darmstadt, Germany

[c]NanoPhotonics Centre, Cavendish Laboratory, University of Cambridge, Cambridge CB3 0HE, UK

ABSTRACT

A feasible strategy to achieve large-area mechano-, thermo- and solvatochromic hybrid opal (OPC) and inverse opal photonic crystal (IOPC) films based on polymer hydrogels is described. Silica core particles featuring surface-anchored stimuli-responsive polymers are prepared and advantageously used for the melt-shear organization technique. By this approach hybrid OPC films with adjustable periodicities for photonic applications can be prepared. The large-area OPC films can be furthermore converted into IOPC structures simply by etching the silica particles while maintaining the excellent order of the entire opal film. This herein developed new process seems to be universal and is successfully applied to two thermo-responsive polymers, poly(N-isopropylacrylamide) (PNIPAM) and poly(diethylene glycol methylether methacrylate) (PDEGMEMA) as particle shell materials. Besides the remarkable mechanical robustness of the hybrid OPC and IOPC films, optical properties upon changes of temperature, mechanical stress and different solvents as external triggers are successfully confirmed. The herein described novel strategy for the preparation of inorganic/organic OPC and IOPC polymer films is feasible for a wide range of applications in fields of sensing and photonic band gap materials.

INTRODUCTION

Photonic band gap (PBG) materials have attracted enormous attention due to their wide range of potential applications within the fields of optical (bio)sensing platforms, catalysis, display technologies, actuation systems and security materials.[1–7] Such PBG materials feature the possibility to modulate the propagation of electromagnetic waves due to their periodic structure forming so-called stop bands. Moreover, there is intensive progress in the field of responsive colloidal crystals, i.e., where the optical properties can be switched by using external triggers such as a change of solvent, temperature, ionic strength, light, electric or magnetic fields or mechanical stress.[8–10] The first example of a thermo-responsive opal photonic crystal (OPC) structure was reported by Weissman et al. using poly(N-isopropylacrylamide)-based (PNIPAM) particles almost two decades ago.[11] There are many stimuli-responsive polymer systems which have been used for switching the optical properties of (composite) OPC films. Excellent and recent reviews within that field are given by e.g. Ge et al.,[8] Lee et al.[12] and Lopez et al.[4]Especially organic OPC or inorganic inverse opal photonic crystal (IOPC) structures have attracted enormous attention in the field of PBG materials and for many other applications.[13–20]One of the major issues however is in finding a technique to assemble such opals on a macroscopic length scale.

Citation: C. G. Schäfer, T. Winter, S. Heidt, C. Dietz, T. Ding, J. J. Baumberg and M. Gallei. Smart polymer inverse-opal photonic crystal films by melt-shear organization for hybrid core–shell architectures, J. Mater. Chem. C, 2015, 3, 2204-2214, DOI: 10.1039/C4TC02788D

In general, synthetic opals, which can also be used as templates for the preparation of IOPC structures, typically consist of uniform particles capable of forming periodicities with length scales of 150 to 350 nm. Much effort has been spent in order to improve the macroscopic arrangement of colloidal spheres into an ordered three-dimensional OPC structure. Many techniques exist for this purpose, e.g., based on sedimentation and colloidal crystallization and, in particular, template-directed approaches.[21] Typically, large-area opals can be prepared by drying[22–24] or spin coating[25,26] of the particle dispersions. Nevertheless, high quality OPCs with almost perfect three-dimensional particle order remain a significant challenge. A more recent technique for particle ordering focuses on the melt-shear organization of core–shell particles in the absence of any solvent. This route involves the compression of monodisperse hard core–soft shell particles between the plates of a moderately hot press. In recent years, this technique has been further optimized by using a combination of extrusion, rolling and edge-induced rotational shearing steps, providing access to mono-domain bulk-ordered films on the multi-meter length scale.[27–29] Large-area self-supporting OPC films can be obtained.[30–34] Compared to other well-established methods for the preparation of polymeric opal films, this method is very fast and yields ordered free-standing OPC films in one single step. Until now, this technique was limited to pure organic particle systems featuring a rather low refractive index contrast between the polymer core and the organic matrix material.

Here we report a novel and convenient protocol for the preparation of stimuli-responsive polymer IOPC films on macroscopic length scales. Hybrid core-interlayer-shell (CIS) particles featuring silica cores and surface-anchored responsive polymer shells are prepared. These inorganic/organic particles can be arranged into highly organized hybrid OPC films by using the melt-shear organization technique followed by removing the silica core particles to obtain polymer IOPC structures. This universal methodology for the preparation of IOPC films is exemplarily applied to two different thermo-responsive polymers as matrix-forming materials, poly(N-isopropylacrylamide) (PNIPAM) and poly(diethylene glycol methylether methacrylate) (PDEGMEMA). The obtained films are characterized to reveal their regular particle order using atomic force microscopy (AFM) and scanning electron microscopy (SEM) and in terms of their optical response upon mechanical stress, temperature change and solvent response by using UV-Vis reflection spectroscopy. Expanding this powerful melt-shear organization technique to inorganic particle core materials featuring smart soft-shell polymers is shown to enhance the field of polymer IOPC film preparation in terms of up-scaling strategies and convenient accessibility to novel photonic materials as sensing platforms.

EXPERIMENTAL SECTION

Materials and Methods

Ethyl acrylate (EA), butanediol diacrylate (BDDA) and Irgacure 184 were obtained from BASF SE, methyl methacrylate (MMA) and allyl methacrylate (ALMA) from Evonik Röhm GmbH and Dowfax 2A1 from Dow Chemicals. Diethylene glycol methylether methacrylate (DEGMEMA) and N-isopropylacrylamide (NIPAM) were purchased from Sigma Aldrich. All other chemicals were purchased from VWR and Sigma Aldrich and used as received if not otherwise mentioned. Prior to use in emulsion polymerization protocols, the stabilizer hydroquinone monomethylether was removed from EA and MMA monomers. For this purpose, EA and MMA were extracted with 1 M sodium hydroxide solution, washed with water until the solution was neutral followed by drying over sodium sulfate. Silica@PMMA@P(NIPAM-co-MMA) core-interlayer-shell (CIS) particles were synthesized as described elsewhere.[38]

Transmission electron microscopy (TEM) was performed on a Zeiss EM10 at an operating voltage of 60 kV. Scanning electron microscopy (SEM) was performed on a Philips XL30 FEG at an operating voltage of 20 kV. Atomic force microscopy (AFM) measurements were performed in the PeakForce tapping mode[35] using a Dimension Icon atomic force microscope (Bruker AXS, Santa Barbara, CA) and Scanasyst Fluid cantilevers (spring constant $\approx$ 0.7 N m^{-1}). The maximum force exerted to the sample was set to 5 nN with an oscillation amplitude of 300 nm resulting in tip indentation depths into the sample surface of approximately 2 nm – sufficient to extract the local elasticity of the sample applying the DMT model[36] to the force-versus-separation data. For dynamic light scattering (DLS) measurements of the particle dispersions a Nanophox photon cross-correlation spectrometer (Sympatec) was used. The experiments were carried out at an angle of 90° at 25 °C. Temperature-dependent DLS experiments were carried out with a set-up based on a He–Ne laser (λ = 632.8 nm) as the light source. Polarization of the primary beam was defined by a Glan–Thompson prism. Scattered beam polarization was analysed in vertical–vertical geometry. Scattered intensity was detected with an optical fibre coupled to two avalanche photodiodes that were used in cross-correlation. The intensity autocorrelation functions, calculated with ALV 5000 software for both photodiodes, were measured in steps of 20° of the scattering angle θ. The angular range was from 50° to 130°. Measurement time for each angle was 5 min. The measurements were performed in the temperature range from 5 °C to 20 °C using cylindrical cuvettes (Hellma) for the samples in a temperature controlled index matching bath. The obtained correlation functions were analysed using the cumulant method. The obtained relaxation times vary with scattering vector $q = 4\pi n/\lambda \sin(\theta/2)$ as $\tau \approx 1/q^2$. Here n is the refractive index. The correlation time τ is related to the translational diffusion coefficient D as $D = 1/(\tau q^2)$. The expected linear variation of $1/\tau$ with q^2 was well represented by our data. By using the Stokes–Einstein relation, a hydrodynamic diameter $d_h = k_B T/3\pi\eta D$ could be calculated, with k_B, T, and η as the Boltzmann constant, the absolute temperature, and the solvent viscosity, respectively. Thermal properties of the silica@polymer CIS particles were characterized using a differential scanning calorimeter (DSC) (Mettler Toledo DSC-1) in the temperature range from −50 °C to 125 °C for silica@PEA@P(DEGMEMA-co-EA) CIS particles and from 20 °C to 180 °C for silica@PMMA@P(NIPAM-co-MMA) CIS particles with a heating rate of 10 K min^{-1}. Reflection spectra were recorded using a Vis-NIR fibre spectrophotometer (USB 2000, Ocean Optics). For the reflection measurements a deuterium/tungsten halogen lamp (DT mini 2, Ocean Optics) was used. Reflection measurements in water and water–diethylene glycol (DEG) mixtures were carried out at normal light incidence (θ = 90°). Solvent- and temperature-induced colour shifts of inverse opal photonic crystal (IOPC) films in water and water–DEG mixtures were tracked after fixing the film with sticky tape on the bottom of a double-wall beaker. Film cooling and heating were carried out using an external cryostat (Thermo Haake). For the temperature-dependent measurements of the swelling process in the temperature range from 15.0 °C to 0.0 °C, spectra were recorded at intervals of 2.5 °C. In order to ensure a constant temperature of the solvent and the opal film, UV-Vis spectra were recorded after 5 min of waiting time. Reflection measurements as a function of strain ε were carried out at normal light incidence (θ = 90°). For this purpose, the strain-dependent optical properties of a 2 × 1 cm strip were measured in situby homogeneous uniaxial stretching in a water bath by using a custom-made sample holder.

Synthesis and Functionalization of Silica core Particles

Silica particle dispersions in ethanol with a final silica solid content of 2.5 wt% were prepared following the procedure described by van Blaaderen et al.[37] 1.3 L of this ethanolic dispersion was mixed with 1.6 mL of 3-methacryloxypropyltr imethoxysilane (MEMO), and then diluted with 5 mL ethanol. The mixture was slowly heated to 60 °C and stirred for 1 h and ammonia was removed under reduced pressure. Then water and ammonia were removed by azeotropic distillation of the ethanol–water mixture at 60 °C, while the volume was kept constant by continuous addition of ethanol. When the dispersion was free of water and ammonia, the volume was reduced to 350 mL at 60 °C. In order to transfer the particles into an aqueous dispersion medium, a solution of 50 mg sodium dodecyl sulfate (SDS) in 100 g of water was added and ethanol was removed by azeotropic distillation, while the volume was kept constant by continuous addition of water. When the dispersion was free of ethanol, the volume was reduced to 130–160 mL. The final MEMO-functionalized silica (silica$_{MEMO}$) particle dispersions in water revealed a solid content of 22.1 wt% (silica particles with an average diameter of 230 nm), 20.6 wt% (silica particles with an average diameter of 199 nm) and 17.8 wt% (silica particles with an average diameter of 172 nm), respectively.

Synthesis of Silica@PEA@P(DEGMEMA-*co*-EA) CIS Particles

Under argon, a 250 mL flask equipped with a stirrer and a reflux condenser was filled at 75 °C with the silica$_{MEMO}$ particle dispersion (110.5 g of silica particles with an average diameter of 230 nm, 118.6 g of silica particles with an average diameter of 199 nm, or 137.2 g of silica particles with an average diameter of 172 nm), then 68 mg Dowfax 2A1 were added and the particle dispersion was diluted with water to a total mass of 150.0 g. After heating to 75 °C, the polymerization was initiated by adding a solution of 50 mg sodium peroxodisulfate (SPS) in 5 g water. After 10 min, a monomer emulsion containing 0.2 g ALMA, 1.02 g EA, 0.034 g Dowfax 2A1, 0.04 g KOH and 14.65 g water was added continuously over a period of 1 h (silica@PEA). After 15 min of reaction time, a solution of 10 mg SPS in 5 g water was added. After an additional 10 min, a monomer emulsion of 8.25 g DEGMEMA, 2.75 g EA, 0.035 g SDS and 20.5 g water was added continuously over a period of 1.8 h (silica@PEA@P(DEGMEMA-co-EA)). After an additional 60 min, the product was cooled to room temperature.

Preparation of Silica@Polymer hybrid OPC and Polymer IOPC Films

For the preparation of silica@PEA@P(DEGMEMA-co-EA) opal photonic crystal (OPC) films, the latex was dried at 40 °C. The precursor powder was mixed with 0.05 wt% carbon black (Special Black 4, Degussa), 1 wt% benzophenone, 1 wt% Irgacure 184 and 3 wt% BDDA in a microextruder (micro1, DSM Research) at 80 °C. A 1 g portion of the mixture was covered with the PET film and heated to 80 °C between the plates of a laboratory press (300E, Dr Collin). Melt flow was induced by applying a pressure of 150 bar for 3 min, resulting in OPC films of about 10 cm in diameter and 150 μm thickness. For subsequent cross-linking, the films were irradiated with a mercury lamp (UV Cube 2000, Dr Hoenle) with an output power of 1000 W. The PET-covered films were treated at a distance of 10 cm for 2 min from both sides. The preparation of silica@PMMA@P(NIPAM-co-MMA) OPCs has been reported elsewhere.[38]

For the preparation of polymer IOPC films, the resulting silica@polymer hybrid OPC films were etched with hydrofluoric acid (HF, ~2 wt% in water) for 3 days in order to completely remove the silica cores. Afterwards the IOPC films were washed several times and stored in deionized water.

RESULTS AND DISCUSSION

In general, soft opal photonic crystal (OPC) films can be obtained from core-interlayer-shell (CIS) particles consisting of a hard core and a surface-anchored soft poly(meth)acrylate shell featuring a low glass transition temperature ($T_g \leq 0$ °C) by using the powerful melt-shear organization technique.[39] The resulting opal films are mechanically stable due to the covalent linkage between the cores and the matrix material. But as a second requirement for the melt-shear process, the particle cores must be comparably hard in order to prevent core deformation during melt-shear organization to form highly ordered OPC films. Compared to reports described previously, inhomogeneous opal films were obtained, if poly(meth) acrylates featuring glass transition temperature above room temperature ($T_g \geq 20$ °C) were used. Here we report that this powerful melt-shear organization technique is not limited to pure organic hard core–soft shell polymer architectures. Basically, the same procedures can be applied on silica@ polymer hybrid CIS particles. In order to prepare homogeneous opal films featuring a low or highT_g polymer matrix, silica core particles were used as the starting point and copolymers of poly(N-isopropylacrylamide) (PNIPAM) or poly(diethylene glycol methylether methacrylate) (PDEGMEMA) were proven to be a good choice as a shell-forming material for the preparation of hybrid silica@polymer OPC films. In contrast to previously described particle systems, the use of silica as a core material for the synthesis of CIS particles is particularly advantageous: the synthesis of silica particles is well established since 1968 by Stöber,[40] and silica particles are easy to handle because they do not need the presence of stabilizing agents, e.g., surfactants or surface-anchored polymers to prevent aggregation. Moreover, the rigid silica particles can be functionalized with a high T_g but meltable shell with the possibility for subsequent melt-shear organization in order to form OPC films of silica cores embedded in a hard poly(meth) acrylate matrix. In addition, silica can be easily removed by etching with hydrofluoric acid (HF) after the OPC film formation leading to inverse opal photonic crystal (IOPC) films featuring a well-defined pore structure. The following chapter is divided into different sections introducing the universal methodologies for the generation of polymer IOPC films by using the melt-shear organization technique. First, the general synthesis strategy for the formation of uniform hybrid silica@polymer CIS particles is discussed followed by introducing the melt-shear organization of the novel particles into hybrid OPC and polymer IOPC films. Beyond these general preparation protocols and film characterization, the stimuli-responsiveness of the newly prepared IOPC films with respect to mechanical stress-, solvent- and temperature-induced structural colour changes is presented.

Bottom-up Fabrication and Characterization of Hybrid Silica@Polymer CIS Particles

For the preparation of the silica@polymer CIS particles described here a combination of the well-established Stöber process and convenient semicontinuous and stepwise emulsion polymerization was used. The silica@polymer CIS particles were synthesized in four main steps as illustrated in Fig. 1: first silica particles were synthesized (Stöber process) followed by a functionalization step. Afterwards a cross-linked interlayer is introduced followed by coating with the meltable outer shell (emulsion polymerization). For the synthesis of hard–soft silica@polymer CIS particles P(DEGMEMA-co-EA) as a soft shell material was used, which was grafted via an ALMA-containing PEA interlayer. To prove the universality of our method described here for soft and hard OPC formation and to compare with the best characterized thermo-responsive polymer in the literature, a copolymer of N-isopropylacrylamide and methyl methacrylate (P(NIPAM-co-MMA) as a hard shell polymer featuring an ALMA-containing PMMA interlayer was investigated as well.

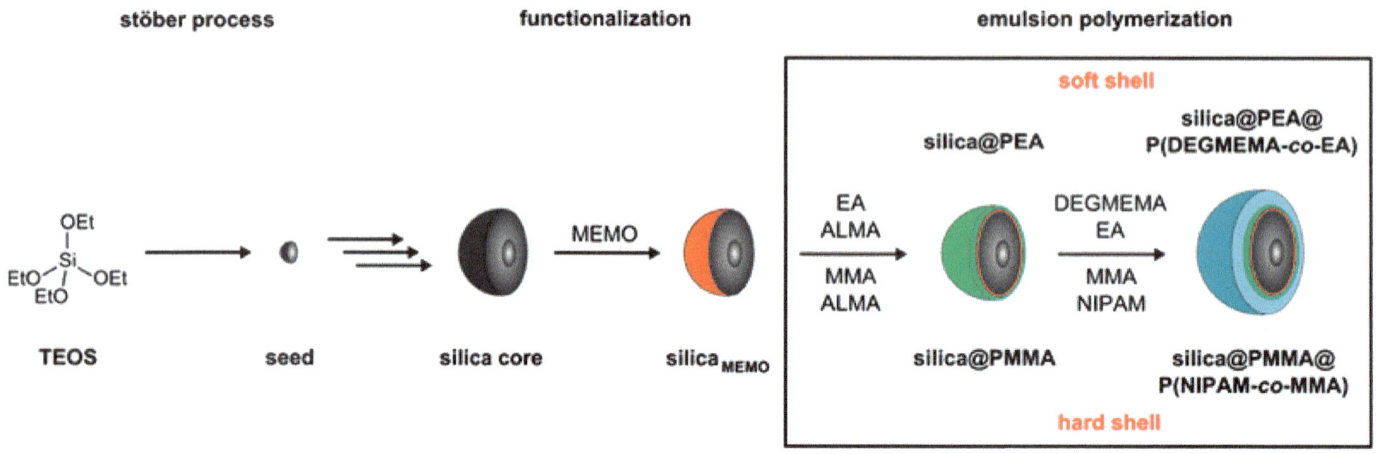

Figure 1 Stepwise preparation of hard–hard and hard-soft silica@polymer hybrid CIS particles by using the Stöber process and emulsion polymerization protocols.

The silica cores were prepared from tetraethoxysilane (TEOS) in ethanol as described by van Blaaderen et al.[37] Corresponding dynamic light scattering (DLS) measurements (Fig. S1†) and transmission electron microscopy (TEM) measurements (Fig. S2) are given in ESI†. The silica core particles were functionalized with 3-methacryloxypropyltrimethoxysilane (MEMO) to compatibilize the silica surface for the polymer shell growth. Afterwards ethanol was completely substituted by water through distillation without isolating the particles. The resulting silica dispersion in water could be introduced in an emulsion polymerization process to form a cross-linked interlayer by the continuous addition of EA and ALMA for hard–soft CIS particles or MMA and ALMA for hard–hard CIS particles. Finally either EA and DEGMEMA or MMA and NIPAM were added to obtain a film formed of soft or hard shells, respectively. The successful generation of the CIS architecture was followed by DLS measurements (Fig. S3–S5†) evidencing an increase of the particle diameter after each synthesis step. The PDEGMEMA-containing CIS particles were additionally investigated by using TEM measurements (Fig. S6 and S7†). Fig. 2 exemplarily shows respective TEM images of silica@PEA@P(DEGMEMA-co-EA) particles starting from 230 nm silica cores after key steps of the applied synthetic protocol.

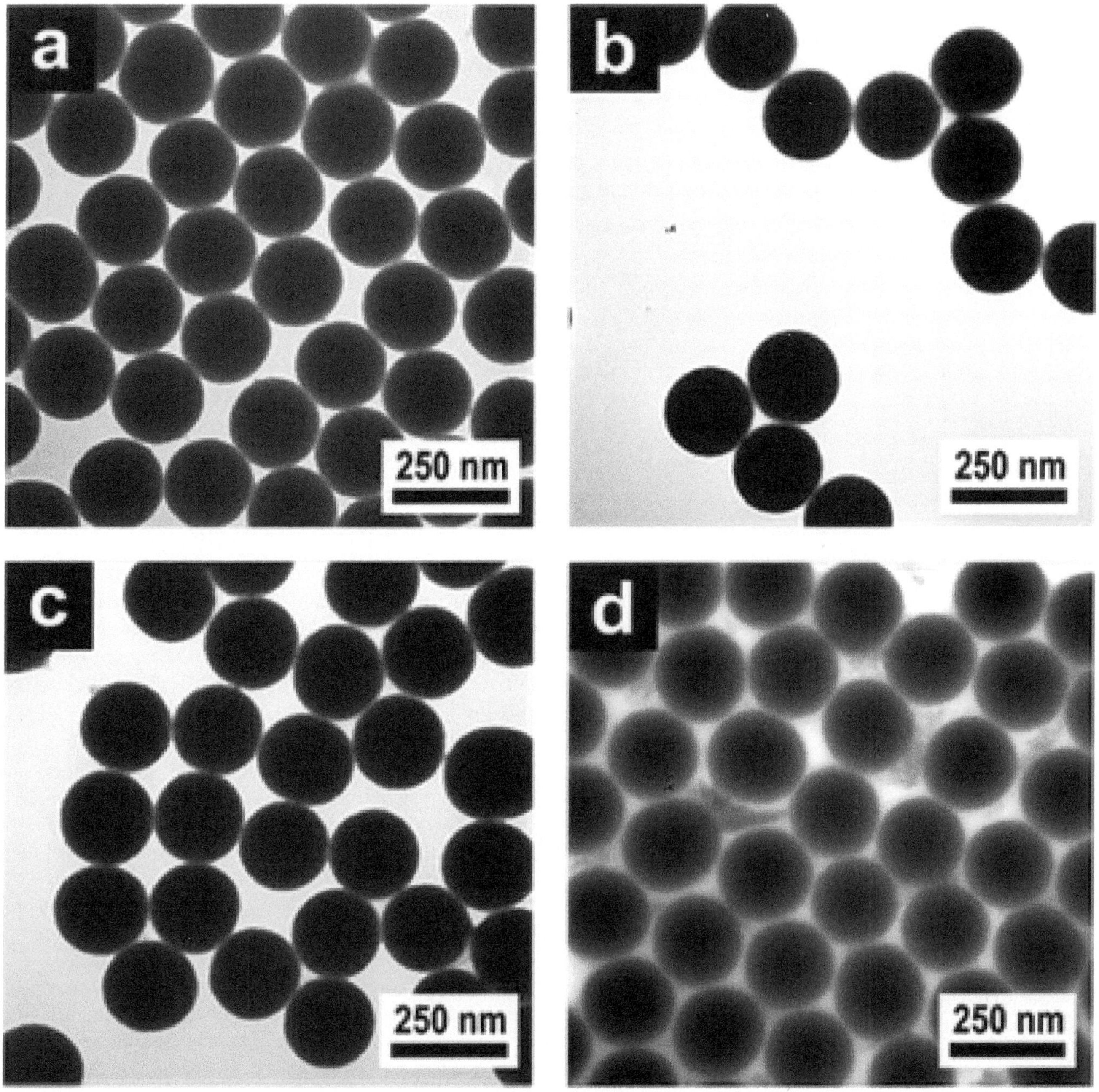

Figure 2 TEM images of (a) 230 nm silica core particles, (b) functionalized silica$_{MEMO}$particles, (c) silica@PEA core-interlayer and (d) silica@PEA@P(DEGMEMA-co-EA) CIS particles.

From the TEM images it could be confirmed that monodisperse and spherical silica cores (Fig. 2a), separated functionalized silica$_{MEMO}$ particles (Fig. 2b), silica@PMMA core-interlayer particles (Fig. 2c) and finally silica@PEA@P(DEGMEMA-co-EA) CIS particles (Fig. 2d) were formed. As can be concluded from the TEM image of the final silica@PEA@P(DEGMEMA-co-EA) particles, particle coalescence occurred when the CIS particles were dried on the supporting grid due to the soft shells and the particles tended to form ordered hexagonal arrangements already at room temperature.

Similar products of silica@PEA@P(DEGMEMA-co-EA) particles were synthesized starting from 199 nm and 172 nm silica cores, respectively. The average diameters obtained by TEM (d_{TEM}) and hydrodynamic diameters (d_{DLS}) obtained by DLS measurements of all particle batches prepared are summarized in Table S8† and were in excellent agreement with expectations. The successful synthesis of silica@PMMA@P(NIPAM-co-MMA) has been recently reported elsewhere,[38] but is also mentioned here for the sake of completeness and for proving the generality of the applied synthetic protocol. The average diameters d_{TEM} and d_{DLS} of silica@PMMA@P(NIPAM-co-MMA) that were prepared from 160 nm silica cores are also given in Table S8.†

In order to investigate the thermo-responsiveness of these hybrid CIS particles, the reversible swelling and shrinking behaviour of the shell polymer in water was studied. The thermosensitivity of PNIPAM-containing hybrid core–shell particles has already been reported in several publications and was found to be repeatedly addressable.[41–43] To demonstrate the temperature-induced volume phase transition of surface-anchored shell polymers of the novel silica@PEA@P(DEGMEMA-co-EA) particles, the particle dispersion was repeatedly brought above or below the LCST of P(DEGMEMA-co-EA) and changes of the hydrodynamic particle diameter (d_h) were followed by using temperature-dependent DLS measurements. In detail, the particles were first cooled from 20.0 °C to 5.0 °C and again heated up to 20.0 °C. Fig. 3 shows d_h of the particles as a function of temperature for the cooling and heating run.

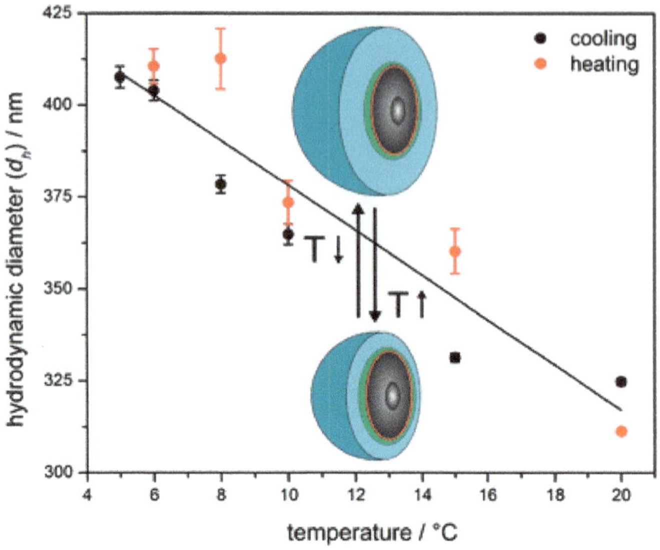

Figure 3 Temperature-dependent DLS investigations of silica@PEA@P(DEGMEMA-co-EA) particles with 230 nm silica cores (the solid line is a guide for the eye).

As expected, there is a strong change in size induced by the change in temperature. The PDEGMEMA-containing hybrid particles revealed a large increase of d_h from initially 325 nm at 20 °C to 408 nm after cooling to 5 °C. Similar to PS@PEA@P(DEGMEMA-co-EA) CIS particles,[39] the transition is also fully reversible for these new hybrid materials. From Fig. 3 it can be concluded that the swelling of the P(DEGMEMA-co-EA) shell can clearly be influenced by temperature and the silica cores do not significantly alter the swelling behaviour of the polymer shell.

The silica@PEA@P(DEGMEMA-co-EA) as well as silica@PMMA@P(NIPAM-co-MMA) have been further characterised by differential scanning calorimetry (DSC) to estimate the glass transition of the final shell polymers (Fig. S9 and 10†). As expected, the measured T_g value of P(NIPAM-co-MMA) (T_g = +112 °C) is significantly higher than the T_g value found for P(DEGMEMA-co-EA) (T_g = −20 °C). As the processing of the hard–soft CIS particles usually takes place at temperatures of about 20–140 °C, P(NIPAM-co-MMA) seemed to be a good choice to form hard OPC films. However, at the higher processing temperatures changes of the mechanical and rheological properties of the shell polymer and the viscoelastic melt have to be critically considered. The melt-shear organization as well as different cross-linking approaches of these hybrid silica@polymer CIS particles are discussed in the ensuing section.

Preparation and Characterization of Hybrid OPC and IOPC Films

Typically, the fabrication of polymer IOPC films requires a multi-step templating strategy, which comprises the (self) assembly of colloidal silica spheres into colloidal crystals, the infiltration of the structure with a solution of radical initiator, monomer and cross-linker, followed by thermo- or photo-polymerization and removal of the colloidal crystal template by etching with HF.[44–48] However, problems have previously occurred by this method since the infiltration and polymerization stage of the process lead to cracking of the structures and the prepared IOPCs are neither compact nor monocrystalline.

In our approach, compact, stable and monocrystalline colloidal crystals that are embedded in a polymer matrix are obtained by the method of melt-shear.[33,34,49–52] Inspired by this work, we now conceived the technique for the novel thermo-responsive silica@PEA@P(DEGMEMA-co-EA) and silica@PMMA@P(NIPAM-co-MMA) hybrid CIS particles to prepare hybrid OPC and polymer IOPC films by applying melt-shear organization, cross-linking (xlink) and subsequent etching of the silica cores, thus avoiding the infiltration stage and the cracking of the films. After etching the silica cores, the resulting IOPC films consist only of crystalline arranged pores embedded in a polymer matrix. A scheme of the preparation process is depicted inFig. 4.

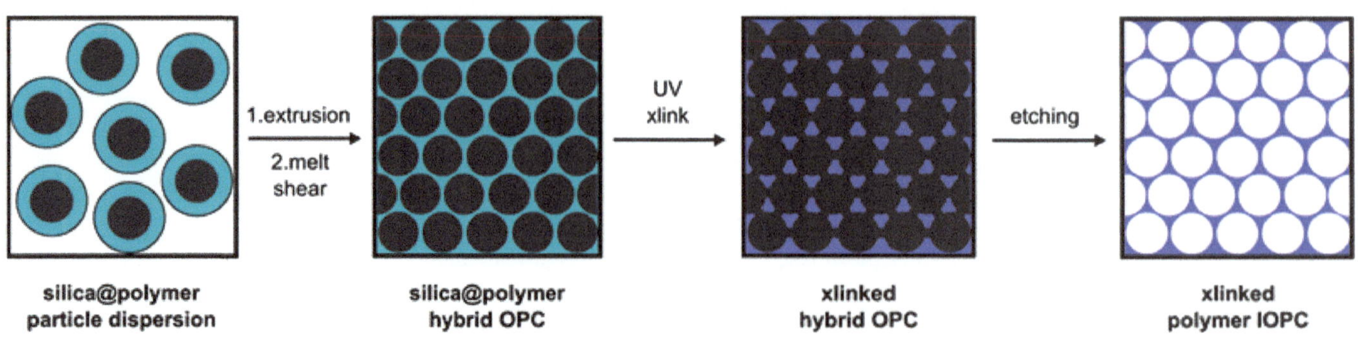

Figure 4 Processing of the CIS particles into close-packed, cross-linked (xlinked) silica@polymer hybrid opal photonic crystal (OPC) and polymer inverse opal photonic crystal (IOPC) films.

Prior to melt-shear organization, the silica@polymer CIS particles were separated from their dispersion by drying at 40 °C. After complete evaporation of water, the silica@PEA@P(DEGMEMA-co-EA) particles had the consistency of a soft and sticky solid, while the silica@PMMA@P(NIPAM-co-MMA) particles featured the consistency of a solid powder at room temperature. Afterwards, both products could be mixed with different cross-linking agents by using extrusion at elevated temperatures, whereby the shells of the CIS particles merged into viscous liquids forming a continuous molten matrix. In order to enhance the reflection colours of corresponding OPCs and IOPCs, 0.05 wt% of carbon black powder were added during extrusion, which has been found to dramatically enhance the perceived reflection colour due to spectrally resonant scattering inside the OPC structure without affecting the lattice quality.[53]

The soft silica@PEA@P(DEGMEMA-co-EA) precursor could be processed at a relatively low temperature of 80 °C, which made it possible to use the temperature-sensitive bifunctional monomer butanediol diacrylate (BDDA) as a cross-linker and commercially available photoinitiators benzophenone and Irgacure 184 (weight ratio: 95 : 3 : 1 : 1), which had already been shown to be very effective for subsequent photocross-linking of polymer OPCs.[39,49,50] In the e of silica@PMMA@P(NIPAM-co-MMA) particles even higher processing temperatures were necessary due to the higher T_g value of the shell polymer (T_g = +112 °C). For that reason the precursor material had to be processed in the extruder at 180 °C. Since BDDA was not suitable for this processing temperature, the bifunctional monomer N,N′-(1,2-dihydroxy-ethylene)bisacrylamide (DHEBA) in combination with benzophenone and Irgacure 184 (weight ratio: 88 : 10 : 1 : 1) were chosen as cross-linking agents to prevent premature thermal cross-linking that would hinder the melt-shear organization. DHEBA monomer starts to melt at 156 °C, so all components formed homogeneous mixtures at the processing temperature of 180 °C. The extruded strands were then heated to the respective processing temperature between the plates of a laboratory press, in which the meltable shells merge into a continuous molten matrix. Then melt-shear was induced by applying pressure perpendicular to the plates and the silica cores arranged themselves in the parabolic flow field into stacked hexagonal layers resulting in a precise close-packed arrangement of the silica cores embedded in the polymer matrix. Prior to the formation of IOPCs by etching with HF, the formation of a mechanically stable hybrid OPC film was achieved by subsequent cross-linking reactions of the matrix polymer. During UV irradiation the free polymer chains in the matrix material were linked with embedded BDDA or DHEBA monomer, respectively, generating a dense network. Afterwards, the silica cores were removed with HF aqueous solution (~2 wt%), and the corresponding IOPC hydrogel was thoroughly washed and stored in deionized water. To ensure that silica cores were completely removed after etching and polymer IOPCs were on hand, cross-sections of the IOPCs were prepared for SEM investigations. Fig. 5 shows respective AFM and SEM images of corresponding OPC and IOPC films.

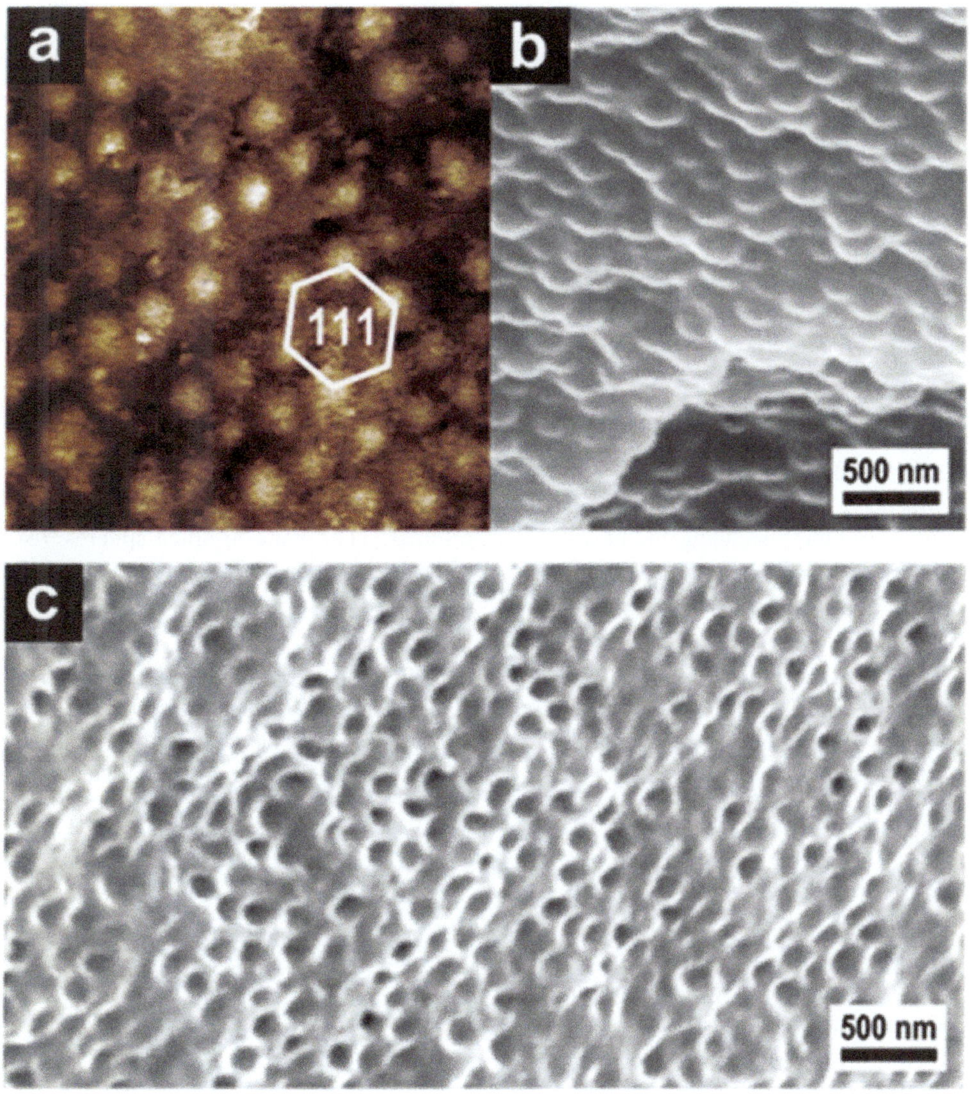

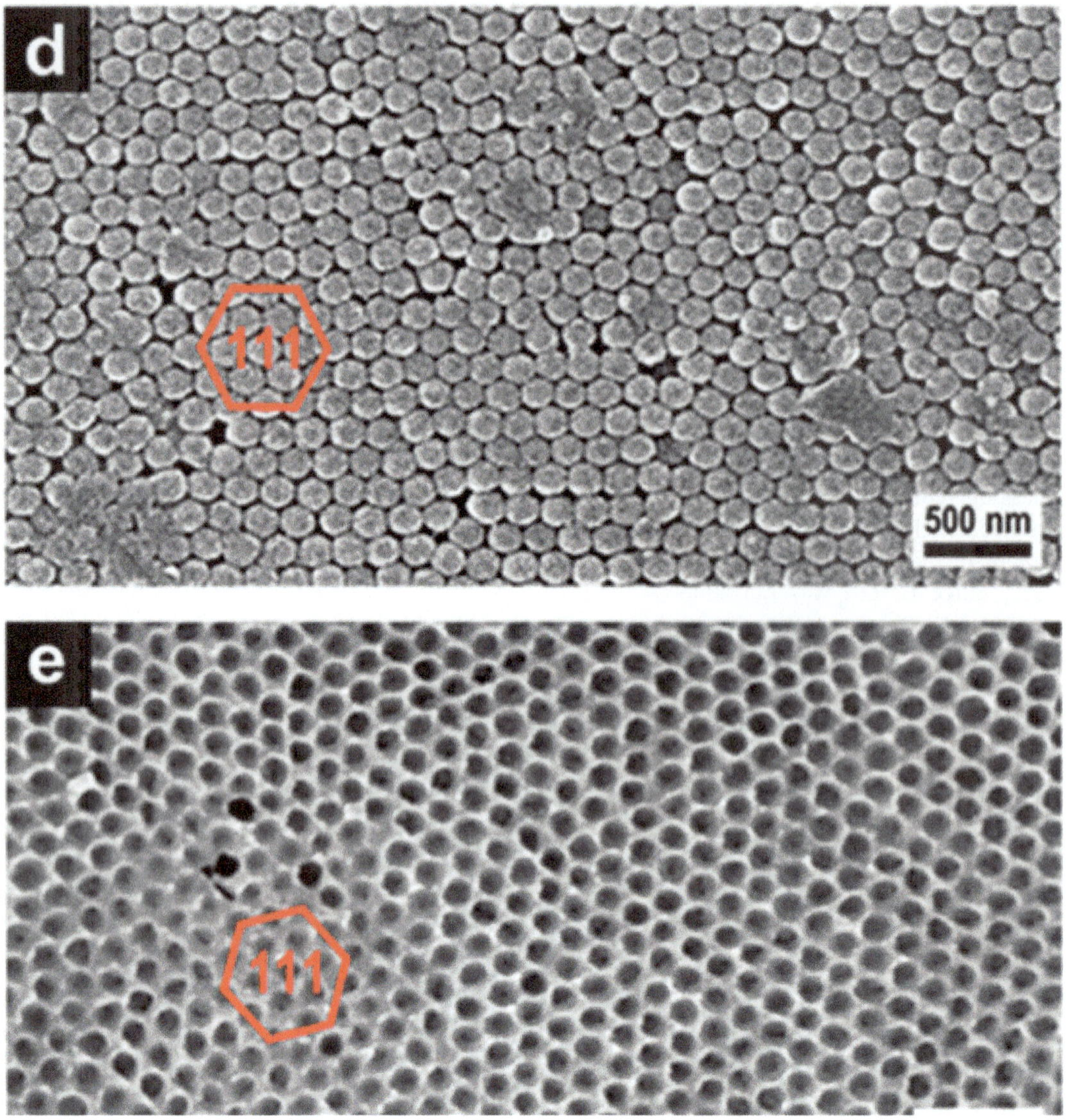

Figure 5 (a) PeakForce tapping AFM image revealing the local elasticity of soft silica@PEA@P(DEGMEMA-co-EA) OPC. A hexagonally arranged (111) plane of silica particles (light) embedded in the polymer matrix (dark) is clearly visible (color scale: 145–307 MPa). (b) SEM image of a cross-section of silica@PEA@P(DEGMEMA-co-EA) OPC showing silica particles embedded in the polymer matrix. (c) SEM image of a cross-section of P(DEGMEMA-co-EA) IOPC after etching with HF showing air pores (dark) embedded in the polymer matrix (light). (d) SEM image of the oxygen plasma-etched surface of silica@PMMA@P(NIPAM-co-MMA) OPC showing the hexagonally arranged (111) plane of silica spheres. (e) SEM image of the surface of P(NIPAM-co-MMA) IOPC after etching with HF showing the hexagonally arranged (111) plane of air pores (dark) embedded in the polymer matrix (light).

From the AFM PeakForce tapping image showing a map of the local elasticity of the surface of the soft OPC film (Fig. 5a) it can be confirmed that the film contains closely packed hexagonally arranged layers of silica spheres that are aligned parallel to the sample surface. The top layer of the hard silica cores is exposed in the AFM image, because the spheres are stiffer than the soft surrounding polymer matrix. The elastic modulus of the silica spheres is approximately 300 MPa, i.e. softer than expected due to a thin polymer film covering the spheres, thus reducing the effective modulus. The spherical silica cores embedded in the polymer matrix can also be seen in the SEM image of a cross-section of the film (Fig. 5b). After etching with HF the porous character of the soft IOPC films could be proven by SEM investigations of film cross-sections, which show that the matrix material is completely pervaded with pores. A comparison of Fig. 5b and c confirms that the initial structure is replicated and a microscopic pore structure appeared after inversion. It should be noted that after etching the silica and subsequent drying for SEM imaging, the surface and the majority of the pores were closed because of the collapse of the soft matrix during drying. Therefore, the resulting IOPC film was directly used for SEM imaging, and no more extra treatments were performed for SEM sample preparation.

For the hard OPCs the precise arrangement of the surface parallel (111) planes could be seen in SEM investigations of the corresponding film surface after etching the matrix material with oxygen plasma (Fig. 5d). Silica cores arranged in hexagonal layers of the (111) planes are evidenced and it is clear that the OPC film has a close-packed structure with long-range regularity. In the SEM image of the IOPC film surface after etching with HF (Fig. 5e), an almost perfectly ordered hexagonal arrangement of air pores embedded in the polymer matrix is visible and the inverted IOPC structure reproduces the lattice of the initial OPC almost completely. It should be noted that the surface of the inverse opal film is open, as shown by the SEM image (Fig. 5e). Additionally, the pores are not connected, but rather are isolated from each other by the matrix material. This is the reason why the next layer cannot be observed under the first layer in the SEM image of the IOPC structure (Fig. 5e).

As can be concluded from Fig. 5, the melt-shear organization of silica@polymer hybrid CIS particles led to regularly ordered OPC films, while the subsequent cross-linking reaction provided inherently stable films which could therefore be completely converted into IOPC structures. Moreover, all of these films featured Bragg reflection colours evidencing the homogeneous domain size.

For that reason, the success of HF etching could also be proven by examining the changes of the Bragg-peak position and peak intensity in UV-Vis reflection measurements. The perceived Bragg peak wavelength λ_{111} of the surface parallel (111) plane (stop band) of the films depends on several factors, including the periodicity a_{111} and the average refractive indexn_{eff} of the OPCs and IOPCs and can be well described by combining Bragg's law with Snell's law (eqn (1)),

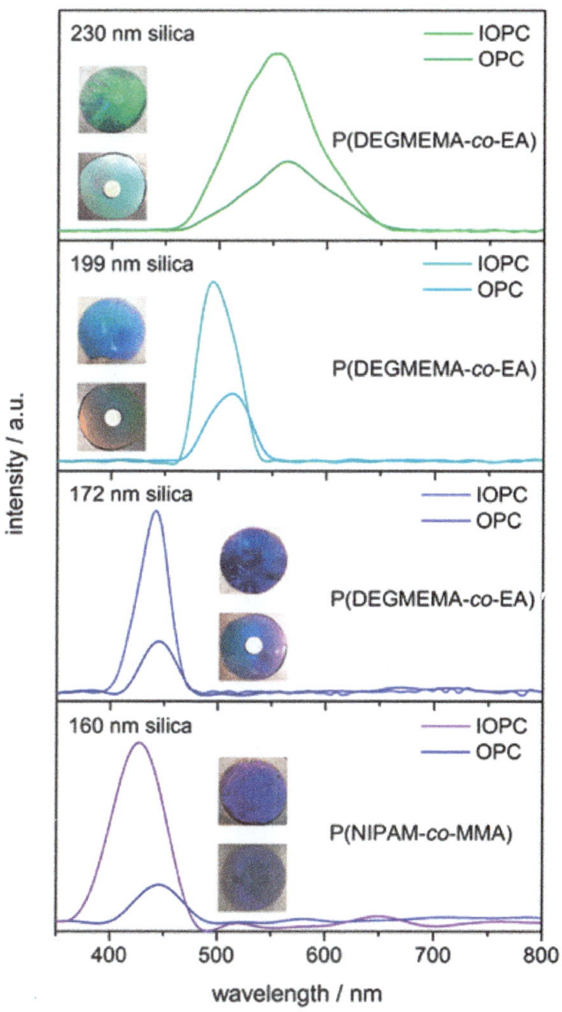

Figure 6 UV-Vis reflection spectra and photographs of OPC films before and IOPC films after etching with HF, prepared from silica@polymer CIS particles with 230 nm, 199 nm, 172 nm and 160 nm silica cores (from top to bottom), respectively. The upper photographs correspond to the swollen IOPCs, while the other photographs correspond to the bulk OPC films.

where θ denotes the angle of incident light:

$$\lambda_{111} = 2a_{111}(n_{eff}^2 - \sin^2 \theta)^{1/2}$$

The average refractive index n_{eff} can be calculated by considering the refractive indices (n_i) and the volume fractions (ϕ_i) for both the silica cores respectively the pores and the polymer shells as well as the water content of the IOPC hydrogel under certain conditions according to eqn (2):

$$n_{eff} = \sum n_i \phi_i$$

In addition, it is well known that the intensity of the reflection peaks is proportional to both the magnitude of the refractive index contrast Δn and the number of ordered lattice planes that result from the periodic close-packed arrangement. According to eqn (1) the Bragg peak corresponds to the light diffraction from the surface parallel (111) plane, which can be measured perpendicular (θ = 90°) to the sample surface. Because a_{111} is directly proportional to the respective particle core size, the colour of the OPCs and corresponding IOPCs can easily be shifted to lower wavelength colours simply by decreasing the size of the silica cores used for the preparation of CIS particles. To demonstrate effects of the core size and silica etching on the colour and reflection intensity of the films, we prepared OPC and IOPC samples from CIS particles, which differed in the size of the silica cores. Fig. 6 shows reflection spectra as well as photographs of OPCs and IOPCs prepared from CIS particles with 230 nm, 199 nm, 172 nm, and 160 nm silica cores before and after etching with HF.

Surprisingly, despite the low refractive index contrast between the silica cores (n ~ 1.45) and the polymer shell (n ~ 1.47), the resultant hybrid OPC samples revealed distinct iridescent colours. Obviously, the films feature well-ordered layers, which was also proven in the UV-Vis reflection spectra in Fig. 6. The hard OPC film showed similar strong reflection colours and, as evidenced by the reflection spectra, was as crystalline as the soft OPC films. The colours of the films were furthermore improved after subsequent etching of the silica cores, resulting in brilliant IOPC films that appear green (λ_{111} = 554 nm), turquoise (λ_{111} = 495 nm), blue (λ_{111} = 442 nm) and purple (λ_{111} = 427 nm), respectively. Two general spectral characteristics have to be noted as the OPC films are inverted: compared to the initial OPC films, the IOPCs reflect very strongly and the intensity of the Bragg peaks increases greatly after inversion (up to threefold), but nevertheless the Bragg peak position changes only slightly. Both effects can be explained as follows: the intensity of the peak increased due to an increase in Δn, since the silica cores (n ~ 1.45) inside the OPCs are replaced by water featuring a much lower refractive index (n ~ 1.33). At first glance, one may wonder about the fact that λ_{111} shifts only slightly to lower wavelength after inversion, although n_{eff} of the compact OPC and the porous IOPC films was quite different. This phenomenon is based on the compensation of two different effects. According to eqn (1), λ_{111} depends on n_{eff}, which is lower in the IOPCs, but it also depends on a_{111}, which is slightly higher in the IOPC films, because they are capable of slight swelling in water already at room temperature forming IOPC hydrogel films.

It can be concluded that well-ordered soft or hard OPC films were accessible from silica@polymer CIS particles by applying the melt-shear organization technique. These films could be completely converted into IOPC hydrogels after etching the silica cores. After inversion, the colour brilliance of the films is strongly enhanced, however, the position of the Bragg peaks changes only slightly due to the swelling of the polymer matrix in water.

Mechano-, thermo- and Solvatochromic Response of Inverse Opal Photonic Crystal Films

The Bragg peak positions of the stop bands of IOPCs can be varied in a predictable manner by altering certain variables in eqn (1). Since the P(DEGMEMA-co-EA) IOPCs feature a highly porous structure that is embedded in an elastic and thermo-responsive hydrogel matrix, a_{111}, n_{eff} and Δn can be influenced by applying various stimuli, such as mechanical deformation, IOPC hydrogel infiltration, swelling and deswelling or thermally induced polymer volume phase transitions. Consequently, λ_{111} can effectively be modulated. Therefore the optical response of the prepared P(DEGMEMA-co-EA) IOPC film was investigated with regard to its mechano-, solvato- and thermochromic response and its usability for optical sensing applications. The possible stimuli-induced structural changes of the IOPC film are illustrated in Fig. 7.

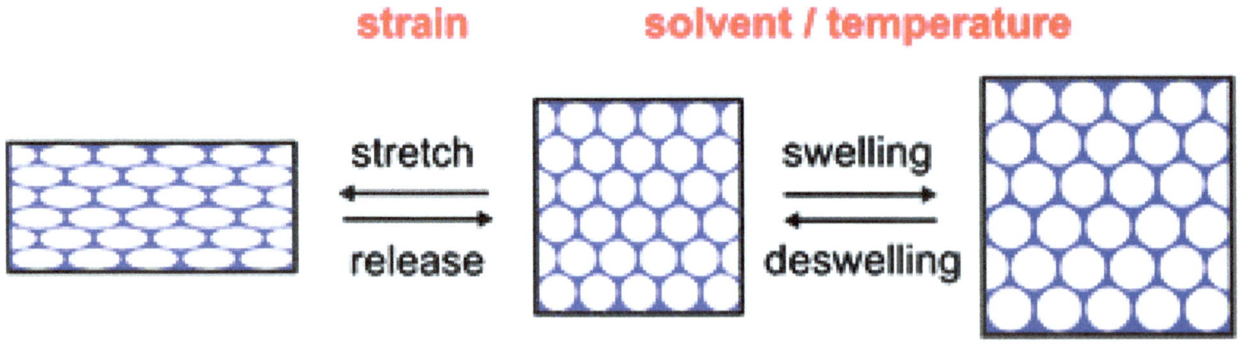

Figure. 7 Reversible strain-, solvent- and temperature-induced structural changes of the polymer IOPC structure.

As depicted in Fig. 7, strain-induced structural colour tuning of the IOPCs mainly depends on the lattice deformation, changing both the symmetry and a_{111} of the lattice. It is worth noting that mechanical stretching of the IOPC embedded in the elastomeric matrix induces an anisotropic change of the lattice constant, which results in a reduced symmetry of the crystal lattice. Since P(DEGMEMA-co-EA) reveals a temperature-sensitive behaviour that undergoes a reversible volume phase transition between a hydrated state and a dehydrated state around its LCST (cf. DLS measurements in Fig. 2), the polymer swells or deswells in water and expands or contracts, so that a_{111} in the IOPC can be reversibly modulated by changing the temperature. Typically, IOPC structures also show tunable diffraction when they are subjected to solvent infiltration, swelling or deswelling, with the diffraction wavelength dependent on the degree of swelling of the hydrogel and degree of pore infiltration. Generally, the swelling or infiltration with solvents is accompanied by the change of n_{eff} as well as Δn between the filled pores and the matrix material. So if the IOPC structure is repeatedly immersed into a solvent, whose refractive index closely matches that of the matrix material ($\Delta n \sim 0$), the diffraction intensity can be switched on and off reversibly. To demonstrate this interesting behaviour experimentally, we chose diethylene glycol (DEG, $n \sim 1.45$) as the solvent to have almost equal refractive indices.

For the determination of the stimuli-induced structural colour shifts we used the green P(DEGMEMA-co-EA) IOPC film, which was swollen in water, corresponding to a photonic hydrogel film. Since elasticity is given due to matrix cross-linking, the mechanochromic behaviour of the entire film was still available in the swollen state. For the determination of stress-, solvent- and temperature-induced colour shifts, the IOPC film was either stretched up to strains of $\varepsilon = 20\%$ or cooled down from T = 15.0 °C to 0.0 °C or the film was immersed in mixed solutions of water and DEG with a decreased water–DEG ratio. The corresponding UV-Vis reflection measurements are shown in Fig. 8.

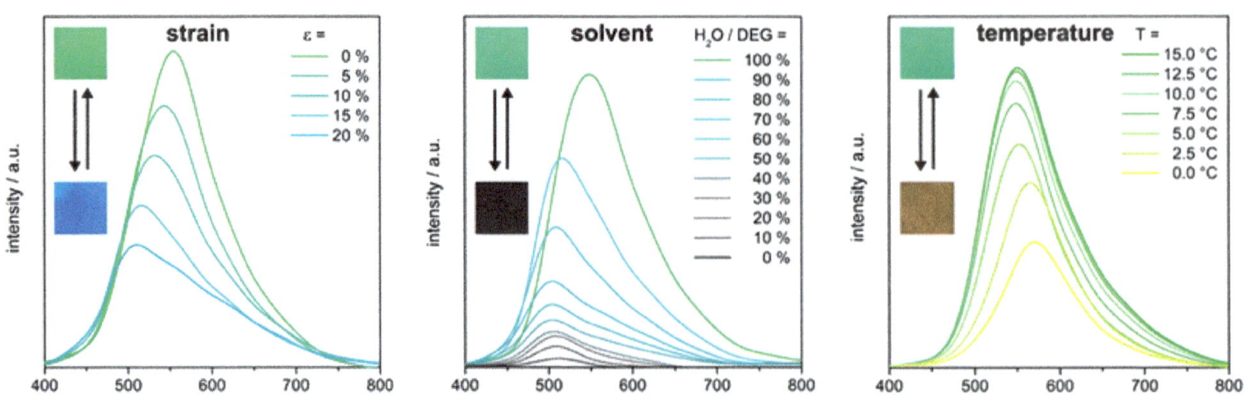

Figure 8 UV-Vis reflection spectra of the IOPC hydrogel film prepared from silica@PEA@P(DEGMEMA-co-EA) CIS particles with 230 nm silica cores showing stress- (left), solvent- (middle) and temperature- (right) induced colour shifts (insets: pho-

tographs of the IOPC hydrogel film before and after stress-, solvent- and temperature-induced colour tuning).

As strain was increased, a blue-shift of the (111) plane stop band was clearly visible in reflection spectra, while the maximum in the reflection intensity decreased (Fig. 8, left). The blue-shift of λ_{111} under increasing strain is based on decreasing a_{111} (cf.eqn (1)). Strains of approximately 20% shifted the stop band by about 45 nm; i.e. the reflected colour changes from its initial green (λ_{111} = 554 nm) to turquoise (λ_{111} = 509 nm). We note that the decrease in reflection intensity in Fig. 8 (left) during stretching is due to a loss of crystalline order from the lattice deformation. But when the film was allowed to relax to its original state, the IOPC structure reassembled giving identical position and intensity of the (111) plane stop band and the green film colour was fully recovered (Fig. S11†).

In addition, the temperature-dependent UV-Vis reflection spectra shown in Fig. 8 (right) evidence a temperature-induced swelling accompanied by a stop band shift of slightly more than 15 nm, i.e. a change from green (λ_{111} = 554 nm) to yellow (λ_{111} = 572 nm) of the Bragg peak under normal light incidence after cooling to 0.0 °C. An accompanying decrease in reflection intensity and reduced visual colour brilliance are seen with the red-shift of λ_{111} (Fig. 8, right). Both effects can be explained as follows: at low temperatures, a significant swelling of the IOPC matrix with water takes place while reaching the LCST of P(DEGMEMA-co-EA). Hence, the volume fraction of water featuring a comparably low refractive index was increased in the matrix material. This led to a lower Δn between the water filled pores and the hydrogel matrix causing a decrease in the reflection intensity. Moreover, the temperature-induced swelling caused a volume expansion of the macroscopic film resulting in the increase of a_{111} of pores embedded in the cross-linked matrix and therefore a red-shift of λ_{111} (cf.eqn (1)). It is important to note that due to the presence of a significant amount of different cross-linking additives, the LCST of the entire IOPC film is expected to further decrease compared to the purified single CIS particle dispersion (cf. DLS measurements Fig. 3). After subsequent heating to 15 °C, the IOPC hydrogel revealed inverted colour shifts and the original IOPC structure was totally restored (Fig. S12†).

In the next step, the optical response of the IOPC film was investigated for its solvatochromic behaviour and to gain insight into wettability and infiltration properties. It is known that P(DEGMEMA-co-EA) is hydrophilic and tends to absorb water already at room temperature. But this effect is restrained by adding DEG in water (Fig. 8, middle). When the IOPC film was immersed into mixtures of water–DEG with a decreased water–DEG ratio, changes of the structural colour within a few seconds could be observed and λ_{111} was blue-shifted from its original green (λ_{111} = 554 nm) to turquoise (λ_{111} = 503 nm) until the water content reached 70% (Fig. 8, middle). Simultaneously, the peak intensity decreased sharply and, at even higher DEG contents, the reflection peak almost disappeared. The blue-shift of λ_{111} can be attributed to the decrease in a_{111} of the IOPC structure due to the collapse of the hydrogel with increasing DEG content. The isotropic change in the volume of the hydrogel caused continuous shifting of λ_{111}, indicating that the porous IOPC gel maintained its nanostructure. Furthermore, the refractive index of the IOPC matrix changed only slightly when the IOPC is soaked in different water–DEG solutions, however, the infiltration of the pores with different solutions caused a tremendous change in refractive index between 1.33 (pure water) and 1.45 (pure DEG) inside the pores. After complete collapse of the hydrogel network the infiltration of solutions with higher DEG content into the IOPC structure, with the average refractive index rising, led to complete suppression of the reflection peak on the time scale of seconds, providing further evidence for the very fast and efficient wetting and infiltration of the IOPC structure. Since the refractive index of pure DEG closely matched that of the P(DEGMEMA-co-EA) matrix (Δn~ 0), the reflection colour completely disappeared and the film appeared grey (due to incorporated carbon black powder) and transparent in pure DEG. This distinct change in colour demonstrates that the response of the IOPC film to water–DEG is perceptible with the naked eye. After subsequent immersion into pure water, the IOPC hydrogel revealed inverted colour shifts and the original IOPC structure with identical position and intensity of the (111) plane stop band and the green film colour was totally restored within at least one minute (Fig. S13†). To summarize, the intense reflectivity of the IOPC film can be switched easily between "on" and "off" states by alternating immersion into pure DEG and pure water. The easily scalable preparation and the excellent stability of the polymer IOPCs together with the capability of relatively fast on–off switching of the diffraction make these films excellent candidates for security materials and optical sensing applications.

CONCLUSIONS

In conclusion, this work demonstrates a novel feasible approach for using the melt-shear organization technique in order to prepare large-area inorganic/organic opal photonic crystal (OPC) and polymer inverse opal photonic crystal (IOPC) films. For applying this convenient technique, surface immobilization of meltable (stimuli-responsive) polymers at silica core particles was necessary and possible by combining the Stöber reaction and seeded emulsion polymerization protocols. By using these techniques, monodisperse and adjustable particles with average diameters in the range of half the wavelength of visible light were obtained. Poly(N-isopropylacrylamide) (PNIPMA) and poly(diethylene glycol methylether methacrylate) (PDEGEMA) were used as meltable and thermo-responsive shell materials for the core–shell-particles providing access to hybrid OPC films with 10 cm in diameter with an excellent particle ordering after melt-shear organization. The high order of the hard/soft particles was investigated and evidenced by using atomic force microscopy, SEM and UV-Vis measurements. Thermo-responsiveness of the pristine particles prior to opal film preparation was evidenced by using temperature-dependent dynamic light scattering measurements of the particle dispersion in water. A reversible increase and decrease of about 80 nm of the average particle diameter were observed in water upon changing the temperature from 20 °C to 5 °C. This remarkable thermo-responsive behaviour was also observed for the stimuli-responsive polymer IOPC films after etching the silica core particles. Moreover, the IOPC films revealed excellent stretch-tunable properties and solvent response accompanied by a photonic band gap switching. This developed melt-shear ordering technique for inorganic/organic core–shell particle systems for the preparation of hybrid opal and inverse opal polymer films is expected to be a universal platform applicable to a broad range of different materials. Thus such stimuli-responsive photonic band gap materials are promising candidates in the field of optical sensing devices and actuation systems.

ACKNOWLEDGEMENTS

The authors want to thank the Landesoffensive zur Entwicklung Wissenschaftlich-ökonomischer Exzellenz (LOEWE Soft Control), UK EPSRC EP/L027151/1, and ERC LINASS 320503 for ongoing financial support of this work. We thank Christina Lederle (Department of Physics, TU Darmstadt) for performing temperature-dependent DLS measurements and Prof. Bernd Stühn for helpful discussion. The authors gratefully acknowledge technical support of Stefano Nieddu for constructing the reflection measurement setup.

REFERENCES

1. Y. Xia, B. Gates, Y. Yin and Y. Lu, Adv. Mater., 2000, **12**, 693–713 .
2. A. P. Hynninen, J. H. Thijssen, E. C. Vermolen, M. Dijkstra and A. van Blaaderen, Nat. Mater., 2007, **6**, 202–205 .
3. L. Gonzalez-Urbina, K. Baert, B. Kolaric, J. Perez-Moreno and K. Clays, Chem. Rev., 2012, **112**, 2268–2285 .
4. J. F. Galisteo-Lopez, M. Ibisate, R. Sapienza, L. S. Froufe-Perez, A. Blanco and C. Lopez, Adv. Mater., 2011, **23**, 30–69 .
5. C. Fenzl, T. Hi h and O. S. Wolfbeis, Angew. Chem., Int. Ed., 2014, **53**, 3318–3335 .
6. J. H. Lee, C. Y. Koh, J. P. Singer, S. J. Jeon, M. Maldovan, O. Stein and E. L. Thomas, Adv. Mater., 2014, **26**, 532–569 .
7. A. Stein, B. E. Wilson and S. G. Rudisill, Chem. Soc. Rev., 2013, **42**, 2763–2803 .
8. J. Ge and Y. Yin, Angew. Chem., Int. Ed. Engl., 2011, **50**, 1492–1522 .
9. Y. Takeoka, J. Mater. Chem. C, 2013, **1**, 6059–6074 .
10. C. I. Aguirre, E. Reguera and A. Stein, Adv. Funct. Mater., 2010, **20**, 2565–2578 .
11. J. M. Weissman, H. B. Sunkara, A. S. Tse and S. A. Asher, Science, 1996, **274**, 959960 .
12. J. H. Lee, C. Y. Koh, J. P. Singer, S. J. Jeon, M. Maldovan, O. Stein and E. L. Thomas, Adv. Mater., 2014, **26**, 532–569 .
13. K. R. Phillips, N. Vogel, Y. Hu, M. Kolle, C. C. Perry and J. Aizenberg, Chem. Mater., 2014, 1622–1628 .
14. K. Rhee do, B. Jung, Y. H. Kim, S. J. Yeo, S. J. Choi, A. Rauf, S. Han, G. R. Yi, D. Lee and P. J. Yoo, ACS Appl. Mater. Interfaces, 2014, **6**, 9950–9954.
15. J. Zhao, Y. Liu and Y. Yu, J. Mater. Chem. C, 2014, **2**, 10262–10267 .
16. J. Kim, S. A. Bencherif, W. A. Li and D. J. Mooney, Macromol. Rapid Commun., 2014, **35**, 1578–1586 .
17. J. Cui, W. Zhu, N. Gao, J. Li, H. Yang, Y. Jiang, P. Seidel, B. J. Ravoo and G. Li, Angew. Chem. Int. Ed., 2014, **53**, 3844–3848 .
18. M. Retsch and U. Jonas, Adv. Funct. Mater., 2013, **23**, 5381–5389 .
19. J. Zhou, H. Li, L. Ye, J. Liu, J. Wang, T. Zhao, L. Jiang and Y. Song, J. Phys. Chem. C, 2010, **114**, 22303–22308 .
20. B. Hatton, L. Mishchenko, S. Davis, K. H. Sandhage and J. Aizenberg, PNAS, 2010, **107**, 10354–10359 .
21. G. von Freymann, V. Kitaev, B. V. Lotsch and G. A. Ozin, Chem. Soc. Rev., 2013, **42**, 2528–2554 .
22. P. Jiang, G. N. Ostojic, R. Narat, D. M. Mittleman and V. L. Colvin, Adv. Mater., 2001, **13**, 389–393 .
23. B. Griesebock, M. Egen and R. Zentel, Chem. Mater., 2002, **14**, 4023–4025 .
24. H.-L. Li and F. Marlow, Chem. Mater., 2006, **18**, 1803–1810 .
25. P. Jiang and M. J. McFarland, J. Am. Chem. Soc., 2004, **126**, 13778–13786 .
26. A. Mihi, M. Ocaña and H. Míguez, Adv. Mater., 2006, **18**, 2244–2249 .
27. C. E. Finlayson, P. Spahn, D. R. Snoswell, G. Yates, A. Kontogeorgos, A. I. Haines, G. P. Hellmann and J. J. Baumberg, Adv. Mater., 2011, **23**, 1540–15444 .
28. A. Kontogeorgos, D. R. E. Snoswell, C. E. Finlayson, J. J. Baumberg, P. Spahn and G. P. Hellmann, Phys. Rev. Lett., 2010, **105**, 233909 .
29. H. S. Wong, M. Mackley, S. Butler, J. Baumberg, D. Snoswell, C. Finlayson and Q. Zhao, J. Rheol., 2014, **58**, 397–409 .
30. T. Ruhl and G. P. Hellmann, Macromol. Chem. Phys., 2001, **202**, 3502–3505 .
31. T. Ruhl, P. Spahn and G. P. Hellmann, Polymer, 2003, **44**, 7625–7634 .
32. O. L. J. Pursiainen, J. J. Baumberg, H. Winkler, B. Viel, P. Spahn and T. Ruhl, Adv. Mater., 2008, **20**, 1484–1487 .
33. C. G. Schäfer, B. Viel, G. P. Hellmann, M. Rehahn and M. Gallei, ACS Appl. Mater. Interfaces, 2013, **5**, 10623–10632.
34. C. G. Schäfer, M. Gallei, J. T. Zahn, J. Engelhardt, G. P. Hellmann and M. Rehahn, Chem. Mater., 2013, **25**, 2309–2318 .
35. B. Pittenger, N. Erina and S. Chanmin, Bruker Application Note #128, 2009.
36. B. V. Derjaguin, V. M. Muller and Y. P. Toporov, J. Colloid Interface Sci., 1975, **53**, 314–326 .
37. C. Graf and A. van Blaaderen, Langmuir, 2002, **18**, 524–534 .
38. C. G. Schäfer, S. Heidt, D. Scheid and M. Gallei, AAP Archives, 2014 , in press.
39. C. G. Schäfer, C. Lederle, K. Zentel, B. Stuhn and M. Gallei, Macromol. Rapid Commun., 2014, **35**, 1852–1860 .

40. W. Stöber, A. Fink and E. Bohn, J. Colloid Interface Sci., 1968, **26**, 62–69 .

41. N. Hatto, T. Cosgrove and M. Snowden, Polymer, 2000, **41**, 7133–7137 .

42. M. Karg, I. Pastoriza-Santos, L. M. Liz-Marzan and T. Hellweg, ChemPhysChem, 2006, **7**, 2298–2301 .

43. C. Liu, J. Guo, W. Yang, J. Hu, C. Wang and S. Fu, J. Mater. Chem., 2009, **19**, 4764 .

44. P. Jiang, K. Hwang, D. Mittleman, J. Bertone and V. Colvin, J. Am. Chem. Soc., 1999, **121**, 11630–11637 .

45. P. Jiang, J. F. Bertone and V. L. Colvin, Science, 2001, **291**, 453–457 .

46. H. Saito, Y. Takeoka and M. Watanabe, Chem. Commun., 2003, 2126 .

47. Y. Takeoka and M. Watanabe, Adv. Mater., 2003, **15**, 199–201 .

48. Y. Takeoka and M. Watanabe, Langmuir, 2003, **19**, 9104–9106 .

49. C. G. Schäfer, D. A. Smolin, G. P. Hellmann and M. Gallei, Langmuir, 2013, **29**, 11275–11283 .

50. C. G. Schäfer, M. Biesalski, G. P. Hellmann, M. Rehahn and M. Gallei, J. Nanophotonics, 2013, **7**, 070599 .

51. C. G. Schäfer, S. Vowinkel, G. P. Hellmann, T. Herdt, C. Contiu, J. J. Schneider and M. Gallei, J. Mater. Chem. C, 2014, **2**, 7960–7975 .

52. D. Scheid, C. Lederle, S. Vowinkel, C. G. Schäfer, B. Stühn and M. Gallei, J. Mater. Chem. C, 2014, **2**, 2583–2590 .

53. O. L. J. Pursiainen, J. J. Baumberg, H. Winkler, B. Viel, P. Spahn and T. Ruhl, Opt. Express, 2007, **15**, 9553–9561

INDEX